Quantity Surveying

a fully metricated text

by **Colin Dent** FRICS

Senior Lecturer in Quantity Surveying
Southampton College of Technology
former R.I.C.S. examiner in Quantities

OXFORD UNIVERSITY PRESS 1970

Oxford University Press Ely House London W.1

GLASGOW NEW YORK TORONTO MELBOURNE WELLINGTON
CAPE TOWN SALISBURY IBADAN NAIROBI LUSAKA ADDIS ABABA
BOMBAY CALCUTTA MADRAS KARACHI LAHORE DACCA
KUALA LUMPUR SINGAPORE HONG KONG TOKYO

MADE AND PRINTED IN GREAT BRITAIN BY
WILLIAM CLOWES AND SONS, LIMITED
LONDON AND BECCLES

Preface

This book is primarily intended for students preparing for the intermediate Chartered Quantity Surveyors' examination, and is based on the syllabus of the Royal Institution of Chartered Surveyors. It has been written to appeal also to candidates reading for degrees and diplomas recognized by the Institution as qualifying for exemptions from the professional examinations, and for whom the discipline of a sound theoretical approach to the subject is thought to be especially desirable.

Nevertheless, it is the author's conviction that students of quantity surveying at all levels, from National Certificate upwards, are in need of more guidance on basic principles than is currently available in the literature on this subject, and it is hoped therefore that this book will prove a useful aid to the understanding of worked examples in the lecture-room on the part of students working for such qualifications as HNC, HND, AIOB, and AIQS, as well as those engaged on RICS courses.

A thorough knowledge of construction technology is an indispensable background to the understanding of quantity surveying, and when in the 1964 syllabus the RICS introduced quantities into the first professional examination, to be studied by the student prior to a thorough grounding in construction, misgivings were heard in many quarters. It is now generally conceded however that, provided an intensive study of construction technology is carried out alongside that of quantities, and the two are carefully phased together, the student gains rather than loses by an early introduction to the discipline of taking-off. In these circumstances it will be found that to be really intelligible to the first-year student, quantity surveying must be taught with a distinct constructional bias; for in order to vindicate the old adage 'the best way to learn it is to take it off', it is now more than ever necessary at each stage of a quantity surveying problem to direct the student's mind to the constructional principles involved, and it is within this framework that the following chapters have been written.

In presenting this book in a metricated form the aim has been to simulate office practice with regard to the presentation of metric calculations, dimensions, and size variables. It will be appreciated, however, that during the transitional period of changing to metric in the construction industry the descriptions in metric bills of quantities may in some cases retain imperial size variables, pending the availability of all component catalogues in metric terms. In this present book all descriptions are in metric, and it may be that in certain cases these may not tally exactly with metric sizes ultimately selected by manufacturers and suppliers. It may also happen that certain conventions regarding the setting down of metric dimensions may vary in different offices. This is also true to some extent with regard to such matters as taking-off sequence, framing of descriptions, etc.

For reasons of this kind it is not intended that this book should be regarded as an attempt to display the ultimate and perfect method of taking-off in metric. It is intended to do no more than lay down the basic principles of a viable and systematic method on which the student can build a sound approach to the analysis of construction projects within the framework of the Standard Method of Measurement.

In preparing the drawings for this book, the nominal thickness of brick walls has been taken as 110 mm per half-brick thickness, in accordance with BS 1192:1969 (*Recommendations for Building Drawing Practice — Metric Units*); but it is appreciated that some designers may well use the parameter of, e.g., 112·5 mm to represent half-brick plus mortar joint, and figure their drawings accordingly. Such slight differences would then be reflected in the taking-off.

The formal analysis of constructional work required in the preparation of quantities in accordance with the Standard Method of Measurement is traditionally taught by means of a series of practical examples for the student to emulate. Unfortunately, however, no two buildings are exactly alike, and working through set examples requires to be supplemented by a background of sound theoretical knowledge, if a true understanding of the subject is to be achieved. This book therefore concentrates mainly on basic principles and their application to general practical problems rather than relying on set examples of specific projects. The present volume deals with elementary structures.

The general treatment of the subject has been based on the assumption that the present use of conventional dimension paper in lecture-rooms and examination halls will continue, and that conversion to slip-sorting or computer systems at a later stage will present no serious problems to the student who has mastered conventional principles. It may be that the complete dominance of one particular new method of processing might perhaps change this position, in course of time. Meanwhile, however, the traditional method of working remains a suitable one for study purposes, owing partly to its convenience as a method of displaying the student's knowledge of the subject.

In view of this it has not been thought necessary to give a detailed analysis of the effect of modern working-up techniques on the preparation of dimensions; although it is desirable that the student should be made aware of the general scope of these techniques at an early stage. With this in mind an introductory section has been included which gives an outline of new techniques, and also briefly reviews the general scope of quantity surveying services; for it is felt that the full-time college-based student in particular benefits from an initial broad picture of the profession, against which to relate his studies. Experience tends to show that this also applies to office-based part-time students, who under the revised syllabus now study this subject in their first year.

Opportunity has also been taken to attempt some definition of the boundaries of knowledge required by the student of this subject at various stages in his studies, and it is hoped that the appendix containing this material will be of some practical help in avoiding embarrassing gaps in the student's curriculum.

The recent history of the quantity surveying profession has seen the development of building economics and cost planning, as well as the introduction of computer techniques, and the preparation of bills of quantities may perhaps be thought to constitute a secondary role in the professional life of the quantity surveyor of the future, in the same way as conveyancing might be regarded as a secondary role for the solicitor.

The analogy is not quite complete however, for the whole structure of cost planning, like that of computer processing, is based upon data obtained from (or derived during the preparation of) bills of quantities. It is by a thorough understanding of the principles involved in BQ preparation work therefore, preferably acquired through the intellectual discipline of taking-off quantities for various types of building and civil engineering work, that the quantity surveyor of the future will be best equipped for whatever role he may be called upon to play.

COLIN DENT

Contents

part one

The quantity surveying profession

Historical background

Quantity surveying originated in England in the nineteenth century. It emerged as a product of the system of competitive tendering by general building contractors, which came into vogue at this time, and resulted in the need for a document known as a bill of quantities.

This document was a schedule or list of all the quantities of materials and labour necessary to erect a building. A building contractor who desired to submit an estimate for a given project would first prepare a bill of quantities by analysing the architect's drawings, and then, by pricing each individual item in the bill, arrive at a total estimated sum of money for the project. This total sum would be his tender price, which, if accepted by the architect on behalf of his client (the future owner of the building) would form the basis of the contract, the 'contract price'.

This system of competitive tendering, by which the accepted estimate becomes a legally binding contract price payable in respect of goods and services rendered, is in wide general use in industry today. The contract is normally awarded to the lowest bidder, and the preparation of each estimate forming the individual tenderer's bid is left entirely to the responsibility of each tendering contractor. In the construction industry however, the practice of basing the tender bid on a detailed bill of quantities leads to a slight departure from this method of tendering, which came about as follows.

The arrival of the quantity surveyor

By about 1850 it was realized in the construction industry in Great Britain that a method of tendering in which each contractor prepared his own detailed analysis of the job in hand, was a wasteful one. In order to arrive at a reasonably accurate estimate, the preparation of such an analysis was imperative on all but the smallest jobs; and yet its production was costly. Moreover, the cost to the contractor was completely abortive unless he was awarded the contract, and so the mounting costs of preparing unsuccessful tenders had to be added to each subsequent tender if the contractor was finally able to show a profit, and stay in business.

In more specific terms; if we assume that each substantial contract went out for tendering, on average, to twelve different contractors, then each bidding contractor could expect on average to obtain one contract in every twelve. To every successful tender therefore had to be added the cost of preparing eleven different bills of quantities, in addition to the successful one. Thus a substantial unnecessary cost was incurred, and this had to be passed on to the client.

The expense of this method of tendering is not its only disadvantage however. There is also a problem arising from the fact that this can hardly be called a fail-safe method of tendering. This follows from the fact that an error of omission in the bill of quantities will tend to produce the lowest tender. The result will be that instead of giving good value for money, the winning bidder will need to concentrate on recovering his losses during the construction period, with every incentive to shoddy workmanship, and dubious claims for extra payment. Such a state of affairs will not benefit the contractor, the architect, or the client.

The obvious solution to these problems is to prepare one accurate bill of quantities and distribute a copy to each tendering contractor. In this way the cost of tendering (and hence of building) is reduced, and each tenderer is bidding on the same basis, thus making for fair competition.

This method was duly adopted, and by 1880 was gaining recognition in the industry. One of the early buildings for which a bill of quantities was prepared was the Houses of Parliament in London, and in due course the preparation of a bill of quantities by an independent surveyor was recognized as being the most efficient and economical way of obtaining tenders for construction work. By the turn of the century the independent quantity surveyor had arrived.

The Standard Method of Measurement

From these beginnings the profession of quantity surveying grew. But much development was necessary before the full potential of quantity surveying services could be realized. Not until the bill of quantities had been standardized and perfected as a tender document could its use as a basis of financial control be fully exploited. Only then would it be possible for the quantity surveyor to turn his attention from measuring quantities to the more advanced tasks of cost planning and operational analysis which were to come later, and be made possible by the use of contracts based on scientifically prepared bills of quantities.

The first step was to make the bill of quantities a contract document. Until this was done, its validity as an accurate analysis of the contract drawings could not be safely assumed by the tenderers, and neither could its use as a basis for payment be properly exploited. In 1909 this was achieved, and a clause was incorporated into the standard form of building contract which made the bill of quantities form part of the contract. By now the independent quantity surveyor was well established as a producer of bills of quantities which contractors could price with confidence.

This confidence was diminished by one factor however, which highlighted the need for a greater measure of uniformity in the preparation of bills at that time. This was the tendency of different surveyors to apply differing methods of measurement and description to similar work. Thus it was not always apparent to a contractor's estimator when pricing an item of (for example) brickwork in a bill of quantities, to what extent the item included (say) rough cutting; or whether this was measured separately elsewhere. This could lead to inaccurate and unfair tendering at times, and it was not until 1922 that a standardized method of measuring and describing constructional work was introduced, and subsequently incorporated into the standard form of contract as being mandatory (unless otherwise stated) for the bills of quantities.

The document codifying this uniformity was called the Standard Method of Measurement of Building Works. It remains so called today, although the latest edition is a much more sophisticated document than the original version. It forms the basis of most quantity surveying procedures and is published jointly by the Royal Institution of Chartered Surveyors and the National Federation of Building Trades Employers. In addition to the United Kingdom edition there are a number of overseas versions now in use.[1]

The quantity survey contract

The building contract incorporating independently prepared bills of quantities based on a standardized method of project analysis has many advantages over a contract based on specification and drawings only, especially in the case of substantial projects. The cost of submitting tenders is reduced (thereby reducing the cost to the client), and the tendering procedure simplified (sets of drawings and specifications are not required by the tenderers, who rely solely on the bill). All tenderers bid on exactly the same basis, so that the contract can be awarded to the one most truly competitive. Moreover, the 'quantity survey' contract (as this type is often known) contains provisions for using the quantities and rates in the 'contract bills' for the purpose of calculating the value of variations in the design, unforeseen extra work, adjustment of provisional sums and quantities, etc., thus making for efficient post-contract management, as will be described later.

[1] At the time of writing, separate standard methods of measurement are available in the RICS library in respect of the following countries: Aden, Australia, Belgium, Canada, Germany, Hong Kong, Iraq, Israel, Jamaica, Malaysia, New Zealand, South Africa.

6 To prepare the bill of quantities and perform post-contract services requires the appointment of a 'contract quantity surveyor' who is appointed by the architect and paid by the client. This quantity surveyor will be available to give cost advice at any time to the client and the other members of the building team.

It is because the fee charged by the quantity surveyor can be clearly seen to be more than offset by the economy and efficiency of this type of contract that the quantity survey contract has grown in popularity. The advantages of bills of quantities to the contractor are only too obvious, but the advantages to the client (who has to pay the quantity surveyor's fee) are also significant, as was shown by an interesting experiment[1] carried out in Singapore in 1966 in respect of a contract for a twelve-storey block of offices valued at around $1 400 000. The tendering contractors were offered the choice of *either* a lump sum contract *or* a quantity survey contract based on fully itemized elemental bills of quantities. In the event, some of the contractors chose the lump sum contract, and others the bill of quantities. The object of giving the choice was to convince the client that bills of quantities would save him money. When the tenders were opened, the lowest one was a BQ tender, and was 5% lower than the first, 9% lower than the second, and 14% lower than the third lump sum tenders. The client was convinced, and the BQ tender was accepted.

The above case history is mentioned in order to illustrate to the student that even if the role of the quantity surveyor today were confined to the preparation of bills of quantities, his job would be one which was vital to the economy of the construction industry in helping to minimize the ever-increasing costs of complex modern building programmes. In fact, however, the quantity surveyor has a great deal more to offer the industry than bill-production alone, as will be apparent from the following chapter.

[1] Reported in *The Chartered Surveyor*, Vol. 99 (No. 4) October, 1966.

The scope of the profession

The services which a present-day quantity surveyor provides to the construction industry arise from his ability to analyse the cost-components of a construction project in a scientific way, and to be able to apply the results of his analysis to a variety of financial and economic problems confronting the developer and the designer. In order to provide these services efficiently and economically, the quantity surveyor has developed special techniques, of which the use of a standard method of measurement in the preparation of bills of quantities is one example. It is not proposed in this chapter to discuss these techniques in detail, but merely to give a broad outline of the variety of services currently offered by the profession, in order to indicate to the student the scope to which his knowledge of basis principles will eventually have to be applied, in addition to the preparation of bills of quantities.

Contract financial management

During the construction and maintenance periods of any major building contract many problems are likely to arise with which the quantity surveyor is equipped to deal, and many of which he is in fact obliged to deal with under the standard form of contract.

The adjustment of accounts will be one of his major tasks. This consists of calculating and itemizing the final sum due to the contractor on completion of the works. The final sum is likely to differ from the contract price for a variety of reasons, most of them provided for under the contract. In addition to the measurement and evaluation of variations in the design as a consequence of changes of mind on the part of the client or the designer, or the substitution of different materials, there will also be the adjustment of provisional quantities and provisional sums, the checking of quotations and accounts in respect of nominated subcontractors and suppliers, and the evaluation of claims for increased costs of labour and material, where these are admissible. These duties will involve the negotiation of prices with the contractor and with subcontractors and suppliers and the evaluation of daywork accounts, the working- and standing-time of plant and equipment, and consultations with all members of the building team. They will also involve the evaluation of any extra-contractual financial claims made by the contractor or the sub-contractors or suppliers. On the rare occasions when these result in arbitration or litigation procedures, the quantity surveyor must be prepared to stand as an expert witness. From this it will be apparent that the quantity surveyor engaged on contract financial management will need to be something of an expert in the interpretation of building contracts, and required to be trained in basic contract law.

Contractors are normally paid for their work by monthly instalments based on the amount of work executed and materials delivered. The preparation of monthly valuations detailing the amounts due to the contractor and also to nominated sub-contractors and suppliers is another of the quantity surveyor's jobs. Towards the latter end of a large contract involving many outstanding claims and variation orders and much work under provisional quantities and sums, the monthly valuation can be a tricky and complex affair, and skill is needed to ensure that the correct amount of money will be left outstanding when the final account is completed and the final payment becomes due.

It is important to ensure that during the construction period the costs of the project to the client do not escalate so as to exceed the original contract price, for if this should happen, the client will need to raise additional funds which may or may not be forthcoming. This dangerous situation could involve bankruptcy. For this and other reasons it is wise to monitor the costs of any large project, so that the exact cost to date and an accurate forecast of the ultimate cost are available to all parties at all times. This involves a system of cost control on the part of the quantity surveyor. By means of a 'running final account', and a system of regular cost-prediction carried out by the surveyor with the collaboration of the architect, it is possible to ensure that the designer does not inadvertently authorize expenditure in excess of the contractural funds available without early warning to the building owner. This service is one which can prove of great value to both the designer and the developer.

Some quantity surveyors specialize in 'post-contract management', as it is often called, and many public bodies employ their own quantity surveyors to deal with this aspect of the work, forming a 'contract management' section within the department. Others consider that the best man to deal with the financial management of a contract is the one who originally prepared the bill of quantities. In this way a surveyor sees the job through from 'inquiry to final account', and many public and private quantity surveying offices are organized in this way.

Pre-contract work

Contract financial management, as described above, starts when the contract has been signed and the successful tenderer commences building operations: it is therefore referred to as post-contract work. Pre-contract work is that which is performed by the quantity surveyor before the signing of the contract, and includes the preparation of the bill of quantities together with other services.

If the quantity surveyor is appointed at an early stage in the project, he is available to give cost advice to the designer regarding the comparative costs of alternative design-solutions. The present-day quantity surveyor is trained as a building economist, and is qualified not only to assess the future building costs of various projected designs, but also to advise on maintenance and user costs, and the overall implications of alternative planning solutions. To do this he requires a knowledge of development economics, and an acquaintance with the principles of investment policy, and with techniques such as cost benefit analysis.

After the architect has selected a suitable design-solution, it will be necessary for a fairly accurate forecast of the total cost to be made, before embarking on the expensive business of preparing a detailed constructional design with full working drawings. The client will be required to approve the sketch design (presented to him in the form of a draft plan with accompanying impressive perspective drawing in full colour) and will assuredly want to know how much a building of such elegance is going to cost him. For this purpose the quantity surveyor prepares what is known as a preliminary estimate. The total amount of this estimate is the figure which the client will subsequently remember with uncanny accuracy: it is imperative that it should be justified by events.

When the design has been accepted and the constructional drawings begin to filter through to the quantity surveyor's office, the preparation of the bill of quantities can commence. Specification notes will also be required by the quantity surveyor, but the actual specification itself is not needed at tender stage, and will be printed later by the architect; for it will be remembered that only the bills of quantities are sent to the bidders (though they may inspect the plans in the designer's office if they wish). The bill of quantities is in theory based on the specification, since the specification is by way of being an amplification of the drawings. Since the bill is written first, however, the specification might in practice be considered as being based on the bill. It may therefore be held that the best man to write the specification is the quantity surveyor. He is in fact trained in specification writing, but it is not part of his official duties under the contract, and is included in the architect's fees.

When the bills are printed they are sent out to the tenderers, but during the tender period the quantity surveyor is not idle, for it is established practice for one copy to be retained by the quantity surveyor himself, who 'tenders' in competition with the builders. This is known as the 'comparable estimate', since the quantity surveyor's priced bill is compared with the list of tender figures on the day when the tenders are opened, as a check against their validity. The quantity surveyor will aim to pitch his tender figure just above that of the lowest tenderer, and below the second lowest. Knowledge of the ensuing post-mortem in the event of this not happening usually sharpens the quantity surveyor's already uncanny sense of market prices, and his comparable estimate will usually be found to be remarkably accurate. He is trained in price analysis as well as market trends. This is just as well, because his next task will be to subject the priced bills submitted by the lowest tenderer to a careful arithmetical and technical examination, prior to their acceptance. The purpose is to search for errors or deliberate 'weighting' of price-rates which might invalidate the tender as being ultimately the most competitive. The discovery of arithmetical errors will induce certain procedures which will ensure both rational pricing of variations and also scrupulous fairness to all tendering contractors, while the detection of malpractices may invalidate the bid; in which case the next lowest bid is examined. It should be borne in mind that tenders for important projects are called in from a selected list of tenderers in the best practice. Thus, only contractors of good reputation are invited to tender in the first place. In these circumstances the lowest tender is almost always the one which is accepted.

Cost planning

This is a technique which is used by the quantity surveyor to monitor the cost of a project at *design* stage. It requires the co-operation of the architect, and its purpose is to ensure that (a) the contract sum does not exceed the preliminary estimate, and (b) the available money is used most effectively to achieve the designers purpose. It works in the following way:

When the sketch plans and preliminary estimate have been accepted by the client, the quantity surveyor prepares the *cost plan.* This is a document analysing the cost of the project in terms of its structural elements. The cost of each element is related to a fixed parameter (usually one square metre of gross floor area) which is both meaningful to the architect, and can be used to compare the costs of alternative design-solutions in respect of each element, relating these costs to the total financial budget. By means of a series of regular *cost checks* during the course of the detailed constructional design, the design can be monitored from the economic standpoint while it is on the drawing-board; the overall cost plan being continually up-dated to accord with each design-decision as

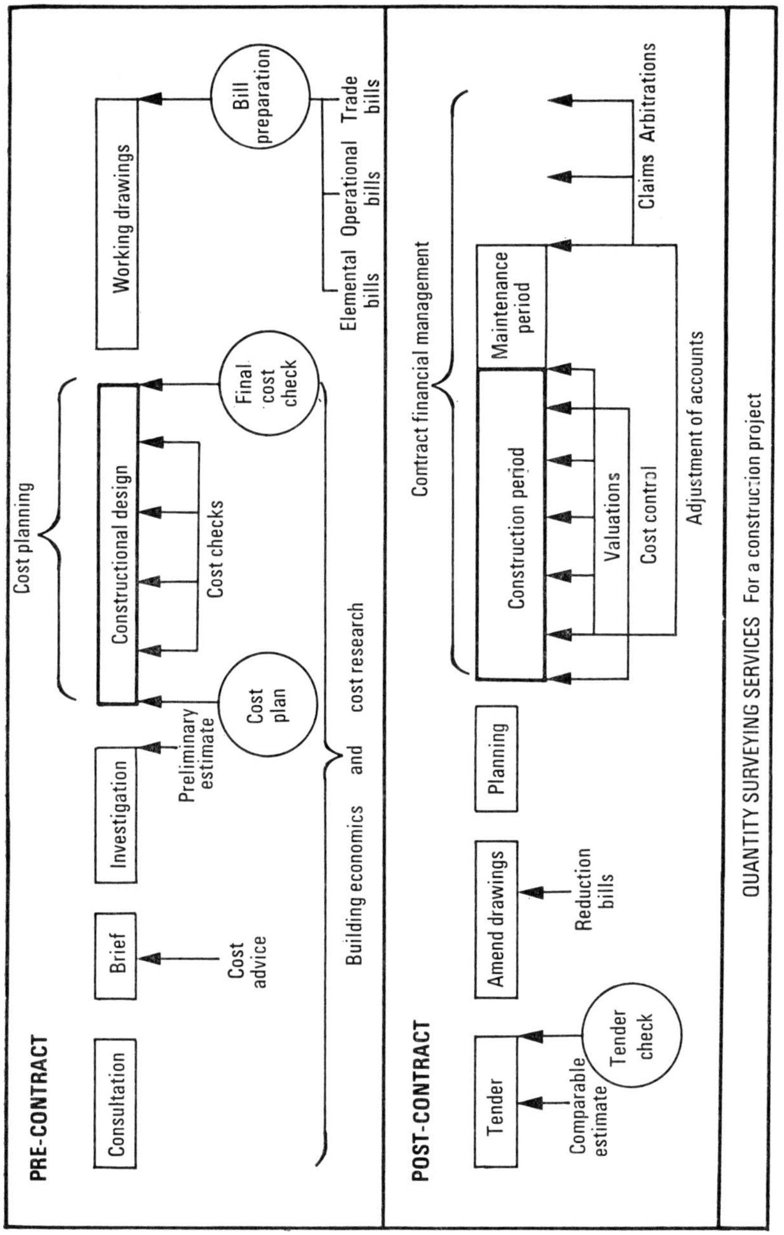
PRE-CONTRACT
Consultation
Brief
Investigation
Constructional design
Working drawings
Cost advice
Preliminary estimate
Cost plan
Cost checks
Final cost check
Bill preparation
Cost planning
Elemental bills
Operational bills
Trade bills
Building economics and cost research
POST-CONTRACT
Tender
Amend drawings
Planning
Construction period
Maintenance period
Comparable estimate
Tender check
Reduction bills
Contract financial management
Valuations
Cost control
Adjustment of accounts
Claims
Arbitrations
QUANTITY SURVEYING SERVICES For a construction project

Chart A

it is taken, and the designer informed immediately of the economic effect of each decision on the overall costs, together with the effect of any alternative decisions. In this way not only is the expensive necessity to redesign (and prepare 'reduction bills') owing to the tender exceeding the estimate avoided, but the design is assured an economic balance, with the client getting the best value for his money.

To do this satisfactorily it is necessary for the quantity surveyor to be able to call upon a library of cost data assembled for the purpose, much of it being derived from the elemental analysis of priced bills of quantities. The RICS Cost Information Service aims at providing such data, but in addition to this, the quantity surveyor will also require his own library of data, and much skill in the interpretation of cost information.

The above description of cost planning is a very abridged one, and is intended to give the student no more than a broad conception of one of the methods now in use; but it should suffice to indicate the importance of this growing branch of the quantity surveyor's functions, for it will be apparent from the above that the ability to aid a designer to plan scientifically the cost of a project at the same time as planning its physical design is one which can be of great value to the architect and his client, and to the construction industry as a whole.

Other services

The services of the quantity surveyor are perhaps seen at their best when dealing with the normal quantity survey contract: it must not be thought that this is the only type of contract on which his services are used however. Maintenance contracts based on schedules of rates, 'serial' contracts, target-cost contracts, cost-plus contracts, all of these call for quantity surveying services of one form or another. Even the ordinary lump sum contract without quantities, of any size, requires monthly valuations and adjustments of accounts, as well as advice on tendering; and preliminary and comparable estimates are desirable in this case as well, to say nothing of cost planning.

System-building, using factory-made components, has lead to contractor-designed buildings being offered in the form of 'package-deal' contracts. The contractor produces his own drawings and bills of quantities, and offers a building to the client for a price inclusive of professional services. In this case the independent quantity surveyor will be required to assess the economic merits of the various schemes put forward, and advise on the selection of tenders, as well as negotiating the value of any extra costs arising at settlement of accounts. It must not be forgotten also that the package-dealer will himself require to employ quantity surveying staff in order to analyse the cost of his proposed design before tendering. Even if the superstructure costs remain constant, the substructure and external services will require reanalysing for each project, since no two building sites are alike.

At this point it may be desirable to have a glance at the duties of the quantity surveyors who, instead of operating as independent professional men (either in private practice or in public service) choose to join the staff of a contracting company.

The contractor's surveyor

The general contractor who undertakes major construction projects will require a quantity surveying staff who are trained to deal with matters involving measurement and costs. In countries where quantity survey contracts are in operation, one of the tasks of the contractor's surveyor will be to represent the contractor in the negotiation of the final account with the contract quantity surveyor. The contract quantity surveyor, although paid by the developer, is acting in an independent capacity under the contract, and must adjust the account with fairness to all parties. Nevertheless, the contractor has the right to bring to his notice any claims for payment which may otherwise be overlooked, and to check his measurements and calculations and raise any queries regarding the bills of quantities. This is the job of the contractor's surveyor, who will also wish to agree the monthly valuations and negotiate the value of extra works. These financial matters are, after all, of prime importance to the contractor, and he will want to feel that a qualified man on his staff is keeping an eye on them. To do this effectively it is obvious that the contractor's QS will require a training similar to that of the contract quantity surveyor. This is why the examination syllabus of the Institute of Quantity Surveyors is very similar to that of the RICS.

A chartered quantity surveyor was until recently prohibited from joining the salaried staff of a building or civil engineering contractor without resigning from the RICS, after which he was of course no longer a chartered quantity surveyor. Should he subsequently leave the employment of a contractor, he could apply for readmission to the RICS. This regulation has now been revoked, to enable contractors to engage the services of chartered quantity surveyors in the same way as chartered accountants.

In addition to the duties outlined above, the contractor's quantity surveyor will often be called upon to perform other tasks for his employer which his training in quantity surveying matters qualifies him to undertake. These duties may include the preparation of estimates, measurement for the purpose of ordering materials or the operation of incentive schemes, assisting in operational planning and cost-control systems; in fact advising and assisting in any sphere concerned with measurements and costs. Since costs are a prime concern in any business undertaking, the chief quantity surveyor of a contracting firm is likely to be a key man in the organization, and a quantity surveying background is usually considered a suitable one from which to aspire to boardroom level.

Although the contractor's quantity surveyor may find himself concerned with preparing estimates, his job should not be confused with that of an estimator. The task of pricing bills of quantities at tender stage usually devolves upon a specialist estimating section of the firm, under the control of the chief estimator. The estimates prepared by the quantity surveyor will normally be only those which, through special circumstances, are not dealt with by the estimating section. There may be exceptions to this rule, however, depending on the organization.

The sub-contractor's surveyor

Sub-contractors often employ surveyors to measure their work and negotiate prices, etc. The work of such surveyors is often confined to a specialized range of trades, or perhaps one trade only, and their main function is often the measurement of work on site. Hence they are usually referred to as 'measuring surveyors' rather than quantity surveyors.

Civil engineering work

This is now included in the examination syllabus of both the RICS and the IQS. Contracts of a mainly civil engineering character are normally carried out in the United Kingdom under a form of contract and method of measurement published by the Institution of Civil Engineers. The functions of the quantity surveyor under a civil engineering contract are similar in principle to those under a building contract.

FUTURE DEVELOPMENTS

Quantity surveying is a comparatively young and growing profession which is constantly seeking to develop new techniques with which to improve its services, and keep abreast of the changing pattern of industry. In these times of rapid change it is not feasible to predict very far ahead. Nevertheless, by giving a brief review of the way in which some of the most recent techniques are developing, it is possible to indicate to the student the likely path of some of the future trends with which he will eventually be concerned.

Building economics

Mention has been made of the training which the quantity surveyor receives as a building economist, to enable him to advise on costs at pre-contract stage. The subject of *building economics* embraces both *construction* economics and *development* economics, and enables the surveyor to apply cost planning techniques at both the planning stage and the design stage of a contract, and to act as a building cost adviser in the widest sense. The first examination paper in Building Economics and Cost Planning was set by the RICS in 1965, and from this it will be evident that the subject is a comparatively new one, and capable of development in the years ahead. Research is being carried out under the auspices of the RICS with a view to quantifying and rationalizing cost data and perfecting new techniques of cost analysis to enable the quantity surveyor to improve further his cost advisory service to the industry.

The role of the future quantity surveyor as a building cost consultant is seen as one calling for a higher degree of truly professional acumen than that required by the demands of routine technical accountancy; and there seems little reason to doubt that the extension of the surveyor's functions in this way is likely to be of increasing benefit to both the industry and the profession in the future. The need for such a service is highlighted by the increasing complexity of modern building processes, the introduction of new materials and forms of construction, and the increasing emphasis on economy and the efficient use of financial resources.

Computer processing

Another step forward that the profession is now making is in the use of the computer for quantity surveying services: we may be sure that in quantity surveying, as in other fields, the future will see the use of these machines exploited to a much greater degree than at present.

It must not be thought that the advent of computers is a danger to the livelihood of the surveyor, and that these machines will eventually replace him. The reverse is in fact the case, for by using the computer as a tool, the surveyor can increase his scope and range of usefulness to the industry. The computer will relieve him of much laborious routine work, and enable him to concentrate on tasks requiring professional judgement as well as technical skill.

At present the use of the computer within the profession is confined mainly to the production of bills of quantities. This involves two main processes, (a) dimension-preparation (known as 'taking-off') and (b) dimension-processing ('working up'). The first of these processes (described in the following sections of this book) must be done by hand,

and it is unlikely that taking-off will be amenable to computer working in the foreseeable future, except in certain special instances. Taking-off may be defined for our present purposes as preparing the data for the computer, which takes over stage (b), the working-up stage. There are a number of systems already in use which enable either a part or the whole of the working-up to be carried out automatically by the computer, and these are further discussed in Chapter 4. In future years the use of the computer for this stage of the work is certain to increase, and is already doing so. One reason for this is that processes which previously took weeks when done by hand in the traditional manner can now be carried out by the computer in a time which is calculated in minutes.

Another reason is that computer working enables certain important and far-reaching applications to be carried out in a way which would not be feasible otherwise. For example; if a computer is used to produce the bill for a specific contract, it is fairly easy to program the machine to assist with the post-contract work as well, and it can be made to price the valuations (automatically giving a break-down of labour and materials costs) and carry out adjustments to the final account. It will also price, extend, and cast the comparable estimate.

In addition to this, it will not only print a trade-order bill for the estimators, but simultaneously reorganize the same bill into structural elements, and provide an elemental print-out for cost planning purposes. Or it can be made to re-sort the items into operational order, and provide an operational print-out for contract planning purposes.

Operational bills

An operational bill is one which arranges the cost-components of a project into actual site operations, and is usually based on a network analysis, which is a device used in the 'critical path' method of planning, now in use on large projects. A bill processed in operational form not only has certain advantages from the tendering point of view, in that it reflects the costs of constructional work in a way directly related to the building process (thus making for a more flexible approach, especially when using new constructional forms) but it also analyses the work in a form which is of direct assistance to the contractor for both contract planning and cost-accounting purposes. It may well be that an increasing number of these bills will be prepared in the future, and with the aid of a computer it may be possible to produce critical path schedules as an automatic by-product of such bills.

There are still some problems to be solved in this field, and their solution will depend on many things, including the greater availability and

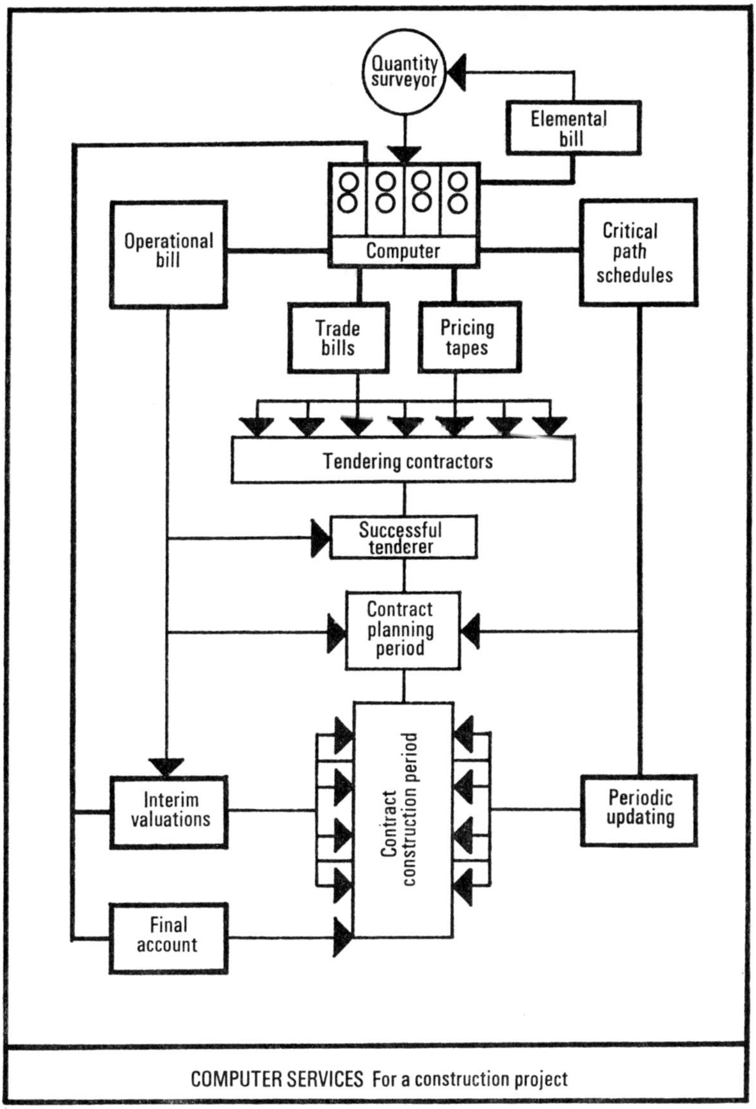
Quantity surveyor
Elemental bill
Operational bill
Computer
Critical path schedules
Trade bills
Pricing tapes
Tendering contractors
Successful tenderer
Contract planning period
Contract construction period
Interim valuations
Periodic updating
Final account
COMPUTER SERVICES For a construction project

Chart B

compatibility of suitable computers, and the close co-operation of all members of the building team.

Integrated data systems

Given the right circumstances there is no fundamental reason why computer systems should not be introduced which even price the bills of the tendering contractors for them, automatically applying suitable rates from the data-tapes of each contractor in a way which accurately reflects the intentions of the estimators. This would require the contractor's cost accounting system to supply feedback information which was continually up-dated by the computer: large contractors are now beginning to acquire computers which are capable of being used in this way.

One stumbling-block in the way of systems which integrate measuring and estimating functions of the kind described above is the present lack of a common coding system, within the industry, which is suitable for sophisticated computer systems, and research effort is currently being put to bear on this problem. Such a coding system could be used to produce the specification as well as priced bills of quantities.

A coding system known as CBC (Co-ordinated Building Communications) is in use, and shows promise. Based on an international library code known as Sfb, the CBC system, which uses a multi-facet code programmed for computer working, originated in Denmark, and is at present undergoing trials in the United Kingdom. Other attempts at rationalization are also being made, and libraries of standard phraseology and coded descriptions for bills of quantities are being brought into use for both manual and computer systems of bill production. Research into the rationalization of the Standard Method of Measurement by the RICS may well ultimately result in a document oriented more towards computer systems than is the present Method of Measurement, and there seems no doubt that future years will see the increasing use of such systems, working towards the more efficient documentation of the construction process, and to better financial and managerial control; a process in which the quantity surveyor of the future has a vital part to play.

Meanwhile, however, it is salutory to remember that any system which can be devised is only as good as the data fed into it. This data largely consists of the quantity surveyor's original dimensions, the preparation of which is examined in detail in this book.

The bill of quantities

The process of taking-off must necessarily be governed by the demands of the end product—the bill of quantities. Before discussing methods of preparing this document it will be desirable to examine the structure of the bill, and to tabulate the various forms it may take in practice.

BASIC COMPONENTS OF THE BILL

The term 'bill' is used here to denote a document prepared in accordance with the United Kingdom Standard Method of Measurement, for use in connexion with the RIBA Standard Form of Contract with Quantities or similar form of quantity survey contract. Such a bill will be found to contain some or all of the following basic components:

(1) Preliminaries.
(2) Preambles.
(3) P.C. and provisional sums.
(4) Headings.
(5) Measured items.
(6) Provisional quantities.

We must deal with each one of these in turn, but before doing so it is important to realize that a bill of quantities is a document containing every single cost element in a construction project. Not only must the quantities for every item of labour and materials be included, but so also must the items of plant and equipment, as well as any temporary work necessary, or expenses incurred by way of insurances, fees, provision of power and water supplies, and such-like intangible outgoings.

The fact is that, since the contractor's estimator will not normally be in possession of any contract document other than the bill of quantities when he makes up his tender figure which will eventually form the Contract Sum, the bill of quantities must contain every conceivable cost-component implied by the project. And if any other documents are supplied, this will not, in principle, absolve the quantity surveyor from the duty of making his bills accurately represent the true nature of the cost of the works to be executed.

Preliminaries

The *Standard Method of Measurement of Building Works*, fifth edition, (metric) July 1968 (hereinafter referred to as the SMM[1]), is divided into sections. Section B is headed 'Preliminaries', and the remaining sections group the work into an order roughly corresponding to the trade of the operatives involved. These are known as the 'work-sections'.

The preliminaries will be found to deal with all general items of cost not specifically related to any one work-section. Thus, for example, 'Water for the Works' under clause B7, would be required both for concrete work (section F) and brickwork (section G) as well as for other work-sections.

As a general rule preliminary items, consisting as they do of such generalities, cannot be measured, and do not therefore require quantities. Also, many of them, like the water referred to above, will be required on all projects, and consequently will tend to be a repetition of those items previously used under a similar form of contract, especially if the architect is the same. It will therefore be rather convenient to copy the preliminaries from a previous set, taking care to redraft inappropriate clauses, rather than to start again from scratch, and such standard sets of preliminaries are in use in most offices.

As an aid to standardization of wording the RICS has recently published a standard set of preliminaries, recommended for use by practitioners.

For these reasons the dimensions as prepared originally by the taker-off do not normally contain ordinary preliminary clauses, the preparation of which may be regarded as a separate operation, although 'special' preliminaries might sometimes require to be drafted at taking-off stage.

The drafting of preliminaries, which form a separate section at the beginning of the bill, is not included in the syllabus of the intermediate examination, and will therefore not be dealt with in this book.

Preambles

These may be broadly described as non-measurable items, each of which applies to a particular work-section. They fall mainly into two categories:

(1) Items specifying quality of materials and workmanship.
(2) Pricing instructions to the estimator.

These items form a separate sub-section at the beginning of each work-section in the bill, before commencement of the measured items and, like

[1] Obtainable from the Royal Institution of Chartered Surveyors, 12 Great George Street, London S.W.1. Clause numbers relating to this document are denoted by the prefix SM in the following pages.

preliminaries, do not normally form part of the dimensions written by the taker-off, though there are exceptions to this rule. Although candidates will not usually be expected to draft preambles to the taking-off in an intermediate examination, however, it will be necessary for the student to be familiar with their content, in order to be able to frame correctly his descriptions relating to the measured items.

The student might be surprised to find that no mention is made in the SMM of preambles. Thus the surveyor is left free to decide for himself to what extent descriptive phrases in his measured item descriptions may be relegated to the preamble sections. The extent to which this is done is largely governed by standard practice, and will be dealt with in this book when discussing the individual items concerned.

Mention should perhaps be made here of *protection items*, which may be thought of as a class of preamble item occurring at the end of most work-sections. Protection items are expressly provided for in the SMM (see SM F68, etc.), and, since their use is more or less standardized, it is common practice to insert these at working-up stage, like the ordinary preambles.

The drafting of preamble clauses, as such (except in the case of special preambles forming part of the taking-off), will not require further discussion at this stage.

P.C. and provisional sums

These groups of items are defined in SM A7, and may affect the taking-off in the following ways:

(1) PROVISIONAL SUMS. Items containing provisional sums are drafted by the taker-off in respect of work which cannot be adequately defined and measured at pre-contract stage. Since any taking-off required of the student in an intermediate examination is likely to be set for the purpose of testing his ability to measure defined work, it will not be necessary to comment further on provisional sums at this stage.

(2) P.C. SUMS. Prime cost sums are those provided in respect of work to be carried out by a nominated sub-contractor (or statutory authority) or materials to be supplied by a nominated supplier. Items containing these are written by the taker-off in the form of

(a) lump sum items, giving a total sum of money to be spent as directed, or

(b) measured items, the descriptions for which contain a unit price which the estimator must allow for in his price build-up.

(a) and (b) above are usually referred to as P.C. sums and P.C. measured items respectively, and although the P.C. sums are often collected together in a separate section of the final bill, both P.C. sums and P.C. measured items are often written by the taker-off as part of the dimensions.

The object of these items is to provide the estimator with a firm price for the work concerned, the amounts being subsequently deducted from the contract sum, and the actual expenditure added back. Thus each initial bid by the tendering contractors is based on the same data, and a fair tender ensured.

The drafting of P.C. sums and P.C. measured items will be dealt with in detail as appropriate when discussing taking-off procedures later.

Headings

The conventional bill of quantities is arranged into the same order, broadly speaking, as the SMM, each work-section being displayed under a separate heading. It is desirable that the student should acquire a copy of a complete trade-order bill of quantities from his tutor at an early stage, and familiarize himself with its layout. He will notice that within each work-section the items are arranged into sub-sections, under sub-headings which may or may not accord with the sub-headings of the SMM.

As will be explained in the next chapter, the work is not taken-off in bill order, however. Under the London Method of taking-off, now becoming universal, the order of taking-off is designed to facilitate accurate measurement, and when the dimensions are completed they require sorting and merging before bill order is established. It follows therefore that, in general, bill headings cannot be inserted at taking-off stage, since bill order is not established until later. With certain exceptions (dealt with later) bill headings are inserted at working-up stage, and do not concern the taker-off.

Headings of another kind very much concern the taker-off however, and these are the taking-off headings which must appear in their correct places in the dimensions. These will be discussed in detail later.

Measured items

These will form the bulk of the taking-off as, indeed, they form the bulk of the bill of quantities. These are the items which will subsequently require squaring, sorting and merging and reducing to form what might be called the bill-proper. In the bill each measured item will comprise: a reference symbol, a prose description and its associated quantity, followed by the unit to which the quantity has been reduced. From the point of view of the student of taking-off however, each measured item

					£	s	d
	CONCRETE WORK						
	REINFORCED CONCRETE AND ASSOCIATED FORMWORK:						
	CONCRETE (1:2:4, 20 mm aggregate)						
A	100 mm Cantilever balcony slab with tamped surface	7	Sq. M.				
B	100 mm Ditto to falls	11	Sq. M.				
C	150 mm Suspended slab with tamped surface	58	Sq. M.				
D	110×75 mm Kerb	11	Lin. M.				
E	135×75 mm Splayed Kerb	9	Lin. M.				
	CONCRETE (1:2:4, 10 mm aggregate)						
F	Staircase	1	Cu. M.				
G	150 mm Suspended landing to receive granolithic paving whilst in unset condition	2	Sq. M.				
	MILD STEEL BAR REINFORCEMENT as described						
H	20 mm Diameter bars in beams	31	Kg.				
I	20 mm Ditto in staircase	60	Kg.				
J	10 mm Ditto in floors	29	Kg.				
	STEEL FABRIC REINFORCEMENT as described						
K	Reference No. 108 in staircases with 75 mm side and 375 mm end laps	13	Sq. M.				
L	Reference No. 107 in floors ditto	56	Sq. M.				
M	Reference No. 106 in floors ditto	25	Sq. M.				
N	Reference No. 104 in strip 1·25 metres wide with 75 mm laps	4	Lin. M.				
		To Collection		£			
	24						

Chart C. A typical page from a conventional bill of quantities. (Courtesy of Southampton City Architect's Dept.)

to be written on the dimension paper will comprise some or all of the following components:

(a) waste calculations,
(b) dimensions,
(c) description,
(d) side-notes.

These components are discussed in Chapter 5.

Provisional quantities
The Standard Method of Measurement (Clause A1) provides for the inclusion in the bill of quantities of work which, for some reason or other, cannot be measured accurately at pre-contract stage. Unlike provisional *sums* (for work which cannot be measured), provisional *quantities* are in respect of work which can be defined as to description, but not as to extent. The result is a normal measured item (or items) as above defined, but marked 'provisional' with regard to the actual quantity, which is adjusted later when the work is executed.

As with provisional sums, the student is unlikely to encounter the need to measure provisional quantities in an intermediate examination.

ALTERNATIVE FORMS OF BILL PRESENTATION

The bill of quantities may be regarded primarily as a device for obtaining a competitive tender. Its secondary uses are for the purposes of (a) assisting in the preparation of valuations for stage payments to the contractor (interim valuations) and (b) forming a basis on which variations to the original contract may be measured and valued. In recent years, however, it has been increasingly realized that a bill of quantities may have other legitimate uses, and that its scope may be profitably widened in the interests of the contract, and the construction industry as a whole. The following types of bill may therefore be encountered in current practice.

Trade bills
This term denotes the traditional bill of quantities, divided into work-sections, or 'trades'. Unless otherwise qualified, the term bill of quantities usually refers to a trade bill, the order of items in which follows the general sequence of clauses in the SMM (although elemental bills appear to be growing in popularity at present). The system of taking-off described in this book presupposes the preparation of a trade bill. The estimators employed by the tendering contractors price the bill by means of a unit rate for each measured item, which is then multiplied by the quantity for the item concerned, and the result extended into the cash column.

Estimators traditionally divide the work into trades, and this facilitates the letting out of whole trades to sub-contractors specializing in an individual trade. This system of analysing the cost of building work into unit rates and grouping the rates under trade headings is the expression of a craft-based industry, and is not necessarily the most effective way of expressing the cost of building work. Nevertheless, the current Standard Method of Measurement is based on this method of estimating.

The order of items within each trade of the bill is governed by tradition. Work is broken down into sub-sections, and within each sub-section the order (1) cube (2) super (3) linear (4) number is the rule, referring to the unit of measurement. Within each of these categories the sequence of (a) labour only (b) extra over (c) labour and material is followed. Groups of items falling within each of these last categories are then placed in the order; cheapest first, dearest last. Certain small numbered items are traditionally 'written short', i.e., taken out of the above sequence and made to follow the items to which they relate (e.g., mitred angles to skirtings) for convenience of pricing; such items being indented in the bill to assist recognition.

Thus it may be said that there is a rigid order for all the measured items in a trade bill, and the advantage of this is that a trained estimator or surveyor can very quickly locate any one of hundreds of items in a bill. Similarly, an estimator knows which item to expect next, and can easily find his way about when dealing with a very complex document analysing a scheme in respect of which he has not even seen the drawings.

Elemental bills

In this type of bill, the work is not divided into SMM work-sections, but into structural elements of the following nature[1]:

Work below lowest floor level.	Ceiling finishings.
Frame.	Decoration.
Upper floors.	Joinery fittings.
Roof.	Sanitary fittings and wastes.
Staircases.	Cold water services.
Walls.	Hot water services.
Windows.	Heating.
External doors.	Gas.
Internal walls and partitions.	Electrical services.
Internal doors.	Drainage.
Ironmongery.	External works.
Wall finishings.	Preliminaries.
Floor finishings.	Contingencies.

[1] For current standard list of elements see *Standard Form of Cost Analysis* published by the RICS.

The main purpose of this type of sub-division is to assist cost planning techniques by enabling the cost of each structural component to be readily displayed, thus admitting a comparison between the cost of various structural designs at the planning stage of future projects to be made.

Some offices prepare many of their bills in elemental form, when the quantities are usually taken-off by elements, instead of in the general order of taking-off used under the group system for trade bills. Within each structural element, however, the taking-off is generally in normal group-system order. The practice of analysing a normal trade bill into elements after submission of tenders, instead of preparing an elemental bill in the first instance, is quite common. Computer systems can produce an elemental print-out from a suitably coded trade bill, and, since a trade bill is usually considered more suitable for actual tendering purposes than an elemental bill, this method has much to commend it.

Sectionalized bills

These are trade bills, with each trade sectionalized into elements. This is said to make for a better pricing document than the true elemental bill, since it retains the work-section format, at the same time facilitating an elemental analysis for cost planning purposes. Taking-off procedure is the same as for elemental bills.

Annotated bills

Some firms produce bills of quantities containing specification notes. Other notes sometimes refer to the physical location of items, and such like. A combined bill and specification ('billification') may result. The taking-off procedure is not materially affected.

Operational bills

This type of bill is designed to split the work into actual site operations, as opposed to trades, or structural design elements. It is used for operational planning and control by network analysis, etc. The taking-off technique required for operational bills will not be dealt with in this book.

Activity bills

This term is sometimes given to a type of operational bill in which the work is measured in accordance with the standard method of measurement, but in which items are sorted into site activities in accordance with a previously agreed network analysis. The taking-off procedure is not materially affected, but the items are coded in accordance with their general location on the network. The result is sometimes called a *loca-*

tional bill. Its use for site planning and cost-feedback purposes is more restricted than that of a true operational bill.

Master bills

This type of bill, which may be in ordinary work-section form, is used in connexion with obtaining tenders for serial contracts. It is based on notional quantities of work, and forms a basis upon which further bills are prepared at a later stage. The taking-off procedure is not materially affected.

Computer bills

A bill produced on the computer may be in any of the above forms. In general the practice is for the computer to produce a trade bill for tendering and afterwards to produce elemental and/or operational print-outs as required, either blank, or fully priced out. Most computer systems require some slight modification to the taking-off procedure, as previously indicated, but most systems in use at present are designed to require a minimal departure from normal taking-off procedure.

Chapter Four

Review of office systems

Before embarking upon a detailed study of taking-off procedures it is desirable that the student should be aware of the principal operations involved in the preparation of a bill of quantities, so that the process of taking-off may be placed in its proper perspective. As indicated in Chapter 2, the main processes involved are (1) taking-off and (2) working-up. At one time an examination paper in practical working-up was included in the syllabus of the intermediate professional examination. Modern developments, however, are tending to reduce the technical skills involved in this field, and increase the professional skills involved in other fields (taking-off, cost-planning, building economics, etc.) and RICS intermediate candidates are no longer examined in practical abstracting. Instead, a general knowledge of working-up systems is required, and the purpose of this chapter is briefly to review the main systems now available to the practitioner.

It will be realized that in respect of any one system, practice may vary slightly from office to office, depending on the organization. Again, some offices may use a combination of more than one system, or may adopt a variation of a well-known system, incorporating new ideas. Thus there are offices which combine the process of abstracting and billing into one operation, and others which make use of punched cards, or adopt a variation of the 'trade-by-trade' system, or combine the processes of billing and specification-writing. The following notes, which do not claim to be exhaustive, deal only with the principal systems, of which an examination candidate might reasonably be expected to have a broad knowledge.

The Northern Method (or trade-by-trade system)

The most direct method of preparing a trade-order bill of quantities is to take off the quantities in the order in which they will subsequently appear in the bill. Or at least in trade order, allowing for some rearrangement within each trade before final printing.

This is known as the Northern Method, and has the following principal advantages:

(1) Sorting of items into bill order is minimal, or unnecessary.
(2) Each trade can be prepared for printing (total quantities calculated, descriptions edited, etc.) while the next trade is being taken-off.

This method is generally considered to have the following disadvantages however, especially in the case of large and complex schemes:

(1) The grouping of work into trades, while convenient for pricing purposes, is not necessarily convenient for taking-off purposes.
(2) Each taker-off may require to examine the complete set of drawings, to ensure that no item is missed in the trade he is measuring. Hence the work tends to be difficult to split up among a team of takers-off.
(3) The same section of work on the drawing needs to be examined several times, once for each trade involved.
(4) Since the sequence of measuring bears little relation to either (a) order of site operations or (b) layout of work on the drawings, the possibility of overlooking or duplicating items may be difficult to exclude. Hence the system may be regarded as error-prone.
(5) The tracing of items from the bill back to the original dimensions may prove difficult, or well-nigh impossible.
(6) Adjustment of quantities due to design changes during bill preparation stage tends to present difficulties.

(7) Adjustment of variations at post-contract stage tends to present difficulties.

It is possible to overcome the above disadvantages in practice, and the Northern Method still has its protagonists: they are now relatively few in number however.

The London Method (or group system)
Here the work is grouped into a special order for taking-off purposes, and re-grouped into trade order before the bill is printed. The order in which the taking-off is carried out is designed to assist speedy and accurate taking-off, and bears little relation to the order in which items will finally appear in the bill. Thus a window, for example, is measured as a complete unit, for joinery, glass, ironmongery, paint, concrete lintel, cavity tray; together with appropriate deductions of brickwork, facework, plaster, etc., necessary to form the opening. This not only makes for a coherent taking-off sequence, ensuring that nothing is missed or measured twice, but allows brickwork, plaster, etc., to be measured without regard for window or door openings, etc., thus promoting speed and simplicity of working. Other elements in the building are grouped together in a similar way; roofs for example being dealt with as units, and measured for coverings, timberwork, plumbing and paintwork in a sequence best suited for taking-off purposes. When the taking-off is completed the dimensions form a coherent pattern, and such a system can incorporate an efficient reference-back from bill to dimensions, and allows easy adjustment of quantities at any time by reason of its relationship to design elements in the building.

In fact the disadvantages of the Northern Method are no longer present. However, when taking-off is complete the items require sorting to trade order. Moreover, like items require merging together and adjusting in respect of 'deduction' items. This is achieved by a process known as abstracting. Until taking-off is complete the bill cannot be commenced, for an item measured at the end of the taking-off may affect a quantity which will appear towards the front of the bill although in practice it is often possible to finalize some trades in advance of others.

The process of manual abstracting is rather laborious. Since it essentially comprises the operations of recording, classifying, sorting, calculating and summarizing, however, it is particularly amenable to electronic processing methods. Even when working-up is executed manually, the London Method is generally considered to be the most efficient, especially for major schemes. The sequence of operations (using traditional manual methods) is as follows:

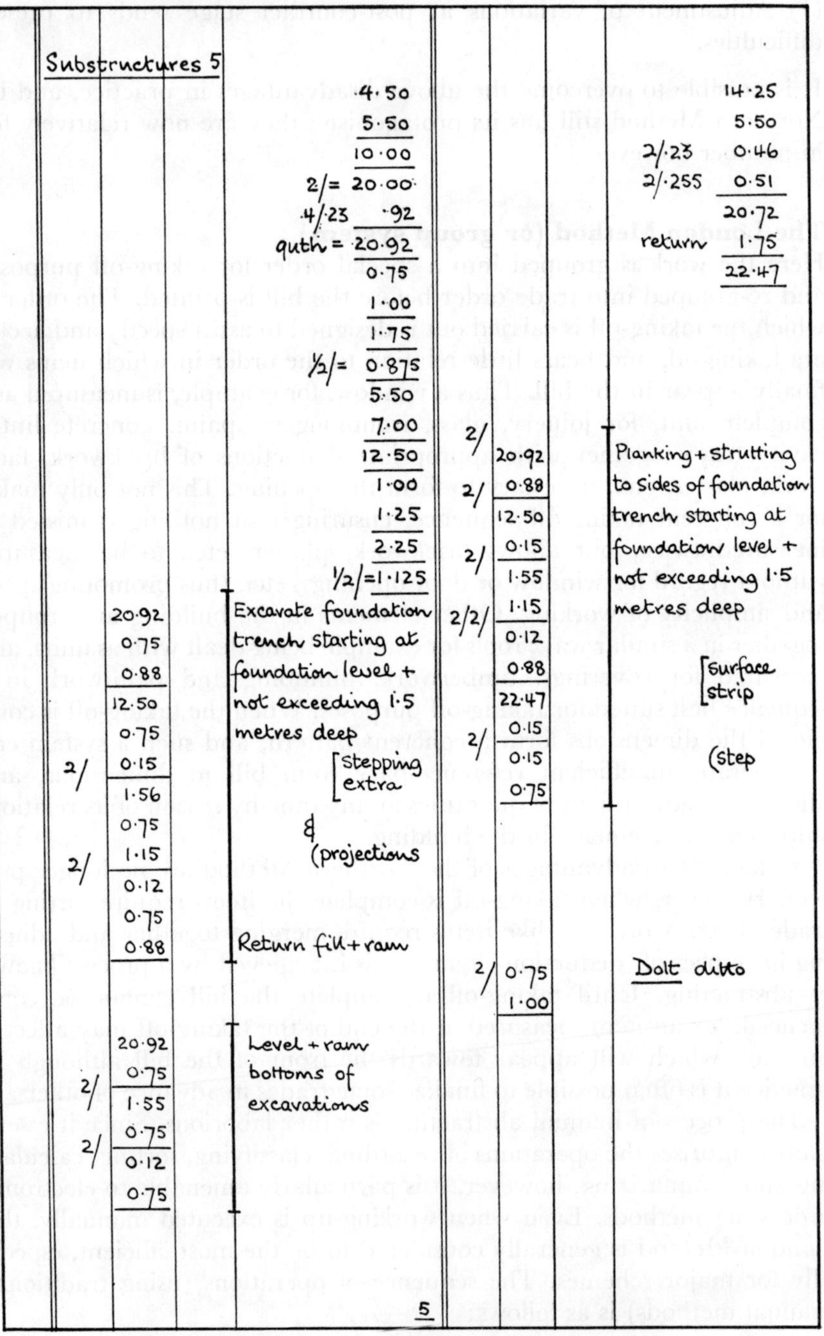
Substructures 5
4·50
5·50
10·00
2/= 20·00
.4/.23 ·92
guth = 20·92
0·75
1·00
1·75
½/= 0·875
5·50
7·00
12·50
1·00
1·25
2·25
½/=1·125
20·92
0·75
0·88
Excavate foundation trench starting at foundation level + not exceeding 1·5 metres deep
12·50
0·75
2/ 0·15
1·56
[stepping extra
0·75
2/ 1·15
&
(projections
0·12
0·75
0·88
Return fill + ram
20·92
2/ 0·75
1·55
2/ 0·75
0·12
0·75
Level + ram bottoms of excavations
14·25
5·50
2/·23 0·46
2/·255 0·51
20·72
return 1·75
22·47
2/ 20·92
2/ 0·88
Planking + strutting to sides of foundation trench starting at foundation level + not exceeding 1·5 metres deep
12·50
2/ 0·15
1·55
2/2/ 1·15
0·12
0·88
22·47
[surface strip
2/ 0·15
0·15
0·75
(step
2/ 0·75
1·00
Dolt ditto
5

Chart D. A typical page of traditional dimensions (group system of taking-off)

(1) TAKING-OFF. Standard dimension paper is used. Measurements and their associated descriptions are recorded by the taker-off in 'taking-off' order.

(2) SQUARING AND CHECKING. As each section of taking-off is completed the sheets are handed to an operator who 'squares' the dimensions on a desk calculator. Squaring consists of calculating and recording the areas and volumes represented by the dimensions. Results are written on the taking-off sheets and independently checked and ticked.

Preliminary calculations made by the taker-off ('waste' calculations) are also checked and ticked at this stage.

(3) ABSTRACTING AND CHECKING. The descriptions are transferred to abstract paper, where they are assembled in bill order. Associated squaring results are inserted under their descriptions, repeat items forming columns for subsequent casting. Addition and deduction columns enable final results to represent net quantities.

Each squaring result on the abstract is indexed with its taking-off page number, to enable reference back from abstract to dimensions.

Certain sections of the work may bypass the abstracting stage, being 'billed direct'. Joinery fittings, drainage manholes, and alteration works, for example, may require little sorting and merging prior to billing; taking-off grouping being similar in these cases to bill grouping. Collections can therefore be made on the dimension sheets, and a note made in the appropriate place on the abstract that these sections are to be carried direct to the draft bill.

The completed abstract is independently checked by ticking each individual entry from the dimension sheets.

(4) CASTING AND REDUCING. When the abstract has been completed and checked, the columns of squaring results are cast, and deduction items deducted from the totals. The resultant net totals are then reduced to metres, kilogrammes etc. (as appropriate), in accordance with the unit of measurement required by the Standard Method of Measurement.

The operation of casting and reducing is then independently checked and ticked.

(5) WRITING THE DRAFT BILL. Each description on the abstract is now transferred to standard bill paper, the format of which is suitable for pricing by the tendering contractors. The total reduced quantity in respect of each item is transferred to the quantity column of the bill, and suitable trade and section headings are inserted, as customary. During this process the taking-off descriptions are monitored and expanded as necessary, ready for printing.

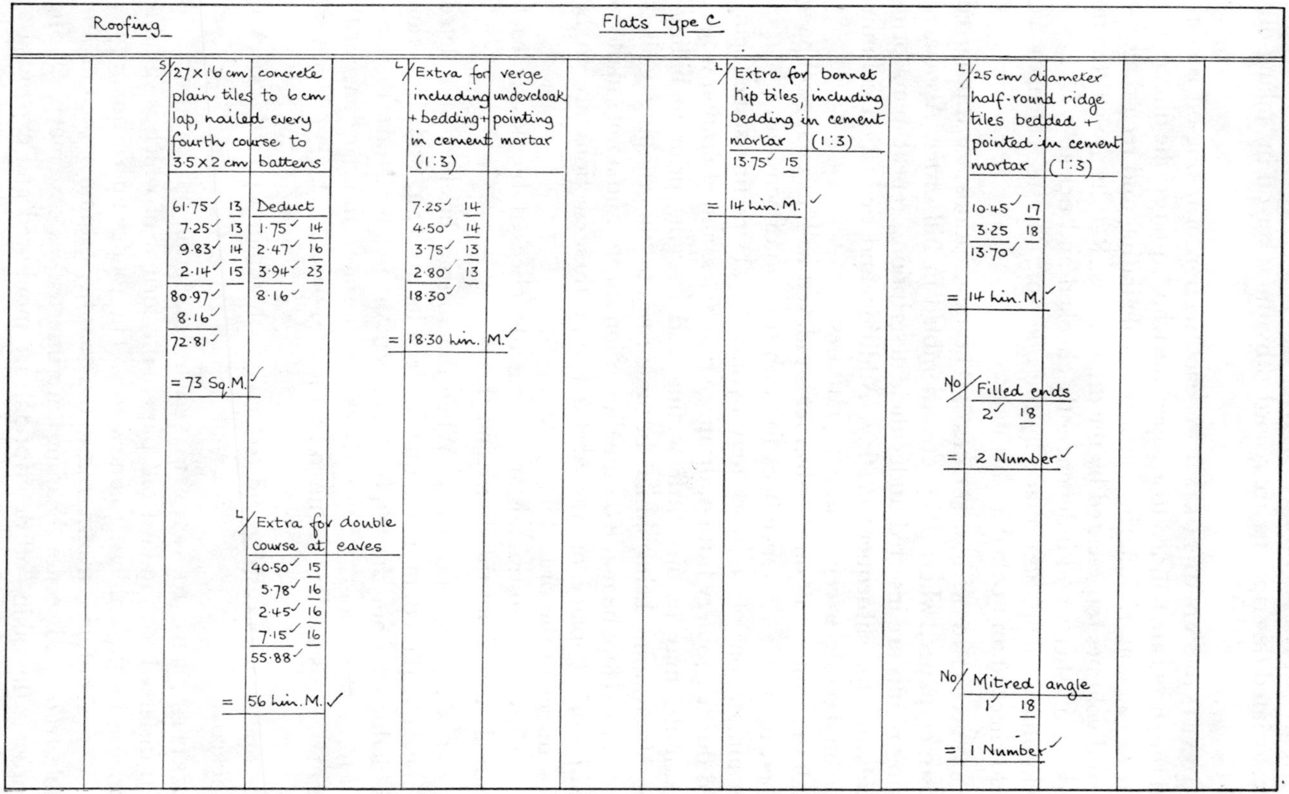
Roofing
Flats Type C
S/ 27 x 16 cm concrete plain tiles to 6 cm lap, nailed every fourth course to 3.5 x 2 cm battens
61.75 13
7.25 13
9.83 14
2.14 15
80.97
8.16
72.81
Deduct
1.75 14
2.47 16
3.94 23
8.16
= 73 Sq. M.
L/ Extra for double course at eaves
40.50 15
5.78 16
2.45 16
7.15 16
55.88
= 56 Lin. M.
L/ Extra for verge including undercloak + bedding + pointing in cement mortar (1:3)
7.25 14
4.50 14
3.75 13
2.80 13
18.30
= 18.30 Lin. M.
L/ Extra for bonnet hip tiles, including bedding in cement mortar (1:3)
13.75 15
= 14 Lin. M.
L/ 25 cm diameter half-round ridge tiles bedded + pointed in cement mortar (1:3)
10.45 17
3.25 18
13.70
= 14 Lin. M.
No/ Filled ends
2 18
= 2 Number
No/ Mitred angle
1 18
= 1 Number

Chart E. *A typical page of traditional abstract*

The transfer of items from abstract to draft bill is independently checked.

(6) DRAFTING PRELIMINARIES AND PREAMBLES. These are inserted at draft billing stage, preliminary clauses appearing at the front of the bill, and preambles at the beginning of each trade.

(7) EDITING THE BILL. Before the draft bill is sent to the printers it is read over by a senior surveyor, and the quantities subjected to a series of feasibility checks.

(8) PRINTING. Reproduction is carried out by typing the draft bill on to stencil or offset lithography masters. At this stage item reference numbers (or letters) are inserted.

(9) READING OVER. In offices where reproduction is carried out on the premises, it will be necessary to check the masters by reading over against the draft bill, before final reproduction is carried out on the machine.

(10) BINDING. The printed copies of the bill, after collation of pages, are bound in manilla or cardboard covers. Copies for the tendering contractors are bound loose-leaf, to facilitate division into sections for pricing by sub-contractors.

The Slip-sorting Method (cut and shuffle system)

This is a variation of the above system which eliminates the operation of abstracting and also enables the major part of draft billing to be avoided. This not only accelerates the process of working-up, but reduces the number of technical staff required. The sequence of operations is as follows:

(1) TAKING-OFF. The dimension paper used is similar in principle to standard sheets, but each sheet is divided into four identical sections suitable for separation by cutting. Taking-off follows the London Method, but certain procedures are observed:

(a) one description only (with associated dimensions) is written in each section; four times per sheet, used one side only.

(b) Each section is coded with a trade reference to facilitate sorting.

(c) Repeat items are cross-referenced to their original sections, to facilitate subsequent collation.

(d) If draft billing is to be avoided, descriptions must be written in full by the taker-off, except in the case of repeat items. Normal use of the word 'ditto' must be avoided, since subsequent sortation of the sections will render it invalid.

(2) SQUARING AND CHECKING. This is carried out in the usual way, modern practice being to use a printing-calculator, i.e., a desk calculator which records results on a continuous slip of paper, totals being written

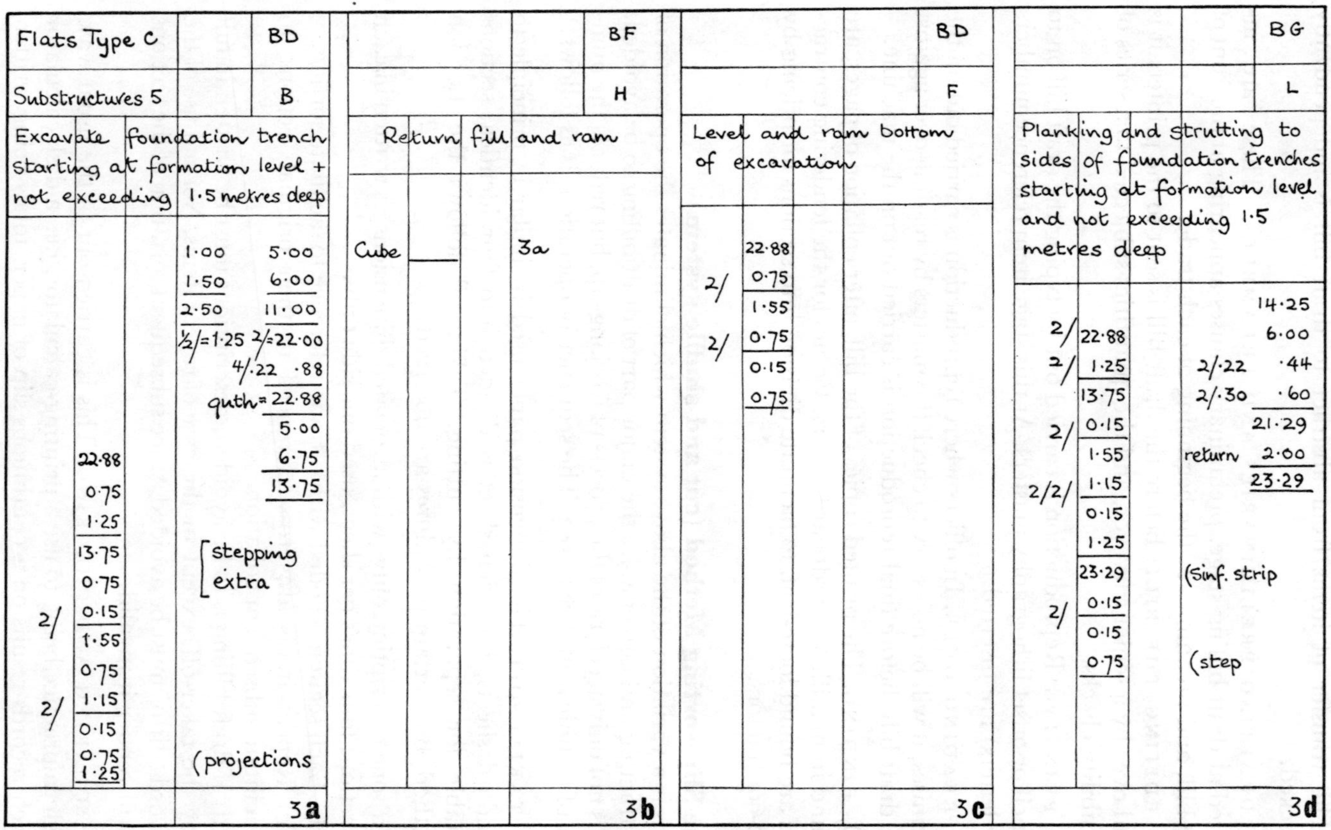
Flats Type C
BD
Substructures 5
B
Excavate foundation trench starting at formation level + not exceeding 1·5 metres deep
1·00 5·00
1·50 6·00
2·50 11·00
½/=1·25 2/=22·00
4/.22 ·88
guth= 22·88
5·00
6·75
13·75
[stepping extra
22·88
0·75
1·25
13·75
0·75
2/ 0·15
1·55
0·75
2/ 1·15
0·15
0·75
1·25
(projections
3a
BF
H
Return fill and ram
Cube
3a
3b
BD
F
Level and ram bottom of excavation
22·88
2/ 0·75
1·55
2/ 0·75
0·15
0·75
3c
BG
L
Planking and strutting to sides of foundation trenches starting at formation level and not exceeding 1·5 metres deep
14·25
6·00
2/·22 ·44
2/·30 ·60
21·29
return 2·00
23·29
2/ 22·88
2/ 1·25
13·75
2/ 0·15
1·55
2/2/ 1·15
0·15
1·25
23·29
(Sinf. strip
2/ 0·15
0·15
0·75
(step
3d

Chart F. A page of dimensions on slip-sorting stationery

on the dimension sheets. Checking is usually carried out by means of a second machine and operator.

By squaring the dimensions before copying, the original sheets show squaring results, which may be useful when making subsequent adjustments.

(3) COPYING. Before each batch of squared dimension sheets is cut, the sheets are copied; the usual method being by means of a dyeline machine, using translucent dimension paper.

(4) CUTTING. The original dimension sheets having been returned to the taker-off for record purposes, the copied sheets are cut by guillotine to produce 'slips', each slip containing one item.

(5) TRADE SORTATION. Slips are sorted into trade order using a rack of pigeon-holes, one for each trade.

(6) ITEM SORTATION. When each trade is complete, its slips are removed from the pigeon-hole and sorted to bill item order. Each slip containing a repeat item (a slave slip) is clipped to the back of the slip containing the original item (the master slip), thus merging like items together.

(7) REDUCING. The squaring totals on the slips are now reduced to the unit of measurement required in the bill, a printed 'box' being provided for this purpose on each slip. In the case of slave slips the totals are transferred to the attached master slip, for addition or deduction as necessary.

The operation is checked, as for squaring.

(8) INSERTION OF HEADINGS. If draft billing is to be avoided, the trade and section headings required to appear in the bill must be written on slips, which are inserted in the relevant positions; the groups of slips for each trade (with the relevant headings slips inserted) being bound together with a string tag.

(9) RENUMBERING OF SLIPS. The slips are now renumbered serially, slave slips having been discarded, and heading slips inserted. At this stage a careful check on the total number of slips is made, to ensure that none is missing.

(10) DRAFTING PRELIMINARIES AND PREAMBLES. These general items are drafted in the normal way, for delivery to the printer with the slips.

(11) EDITING. This will require more care than with normal abstracting, since the descriptions on the slips will require to be examined for consistency and edited as necessary in the interests of continuity (e.g., dittos inserted as appropriate).

(12) PRINTING AND BINDING. When final editing is complete the bill may be printed direct from the slips, and bound as usual.

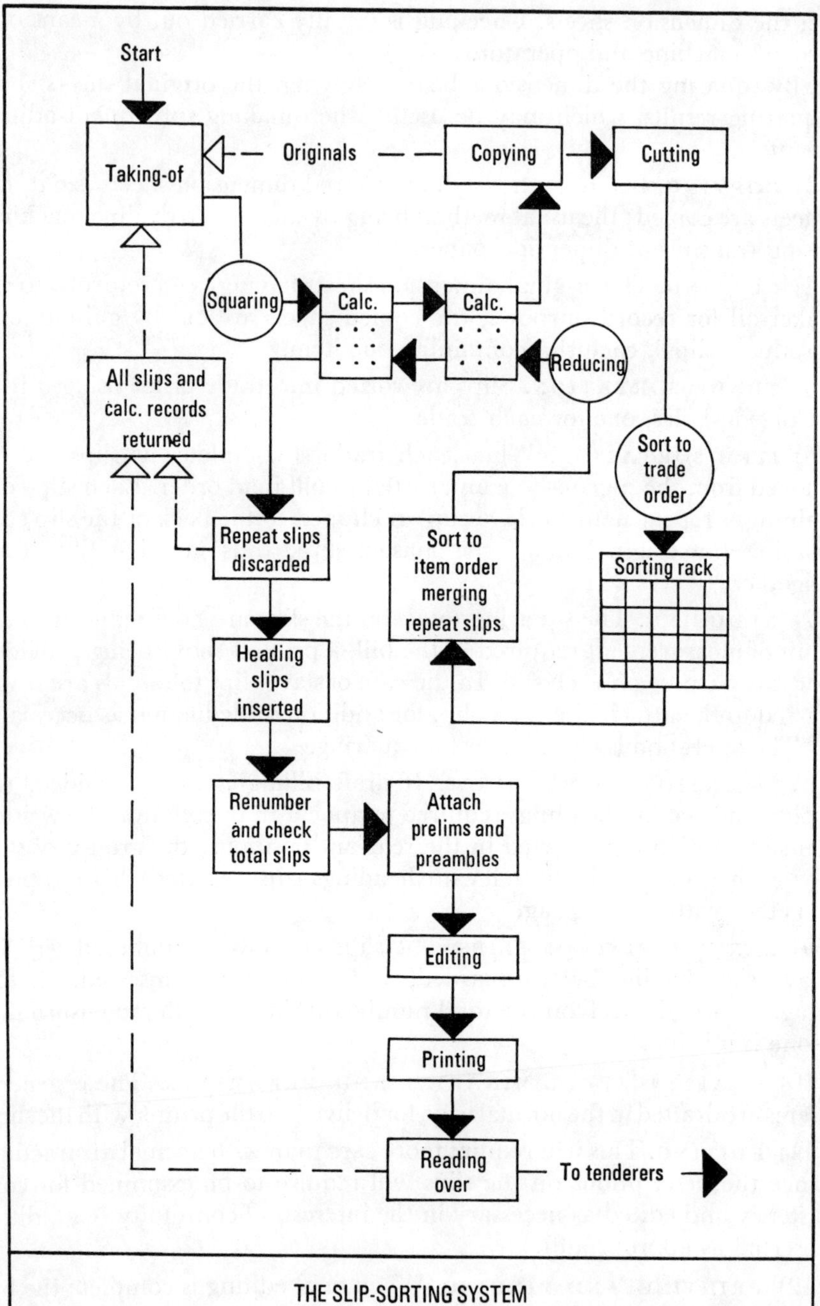

THE SLIP-SORTING SYSTEM

Chart G

Abstract processing by computer

The process of sorting and merging of items into bill order (i.e., abstracting) may be carried out by means of a digital computer, which also squares, casts, and reduces the items at the same time producing a printed abstract from which the bill may then be drafted. The computer 'processes' the dimensions into the form of an abstract—hence the term *abstract processing* used to describe systems of this kind.

Such systems take a variety of forms, depending upon the type of computer equipment available. In general, however, the basic sequence of operations remains the same.

Input of data to the computer may take one of two forms:

(1) CARD INPUT. Data is fed into the computer on individual punched cards, prepared on a keyboard machine (*card punch*) similar to a typewriter.

(2) TAPE INPUT. Data is fed into the computer on a continuous reel of punched paper tape, prepared on a special electric typewriter (*tape punch*) which simultaneously produces a separate typed copy of the data.

The general sequence of operations for abstract processing by computer (tape input assumed in this instance) is as follows:

(1) TAKING-OFF. The London Method is adopted in principle, but special dimension paper is used, having a coding column in lieu of a squaring column (squaring being carried out automatically by the machine).

(2) CODING. This is usually carried out by the taker-off during the measuring process. A code reference is allocated to each item on the dimension sheets in accordance with a prearranged coding system, suitable for sorting the items and reducing the quantities (alternatively the actual reducing may be carried out manually).

The coding system required for abstract processing (as opposed to bill processing) is a relatively simple one, and may require as few as four digits (maximum) per item, depending on the system. Since the computer is not required to print the actual descriptions themselves (only to arrange the quantities in bill order under code numbers whch can be matched against the taker-off's descriptions for the purpose of draft billing) the individual code numbers selected by the taker-off for a particular description need not be the same for each job. Hence the term *ad hoc* coding is applied to this system.

(3) DATA PREPARATION. The data which the computer requires in order to produce an abstract of items in bill order must now be prepared in suitable form, from the taking-off sheets. To achieve this the dimensions

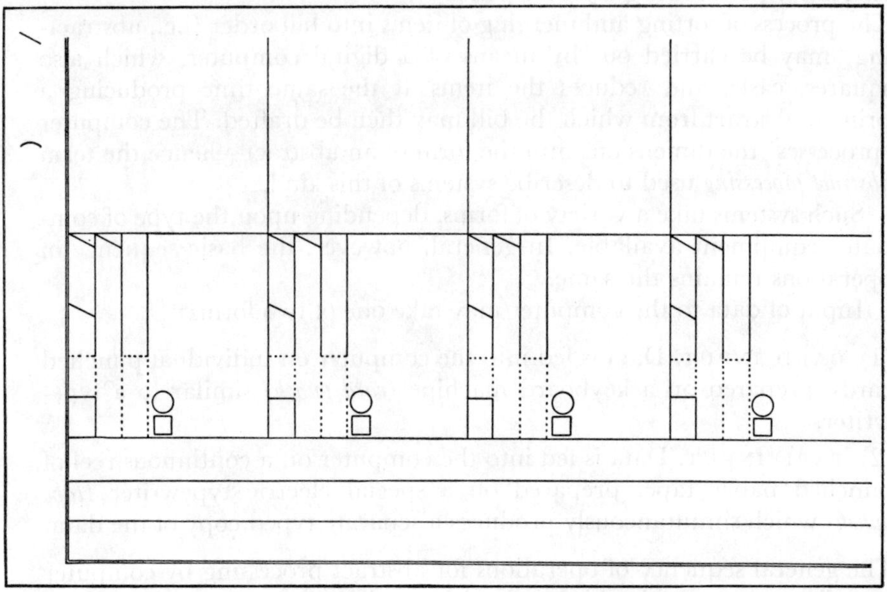

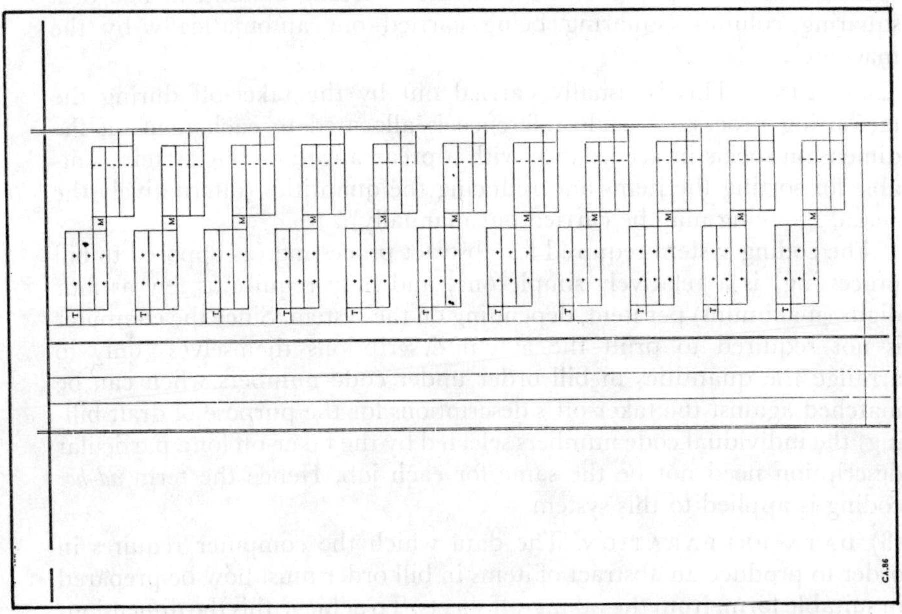

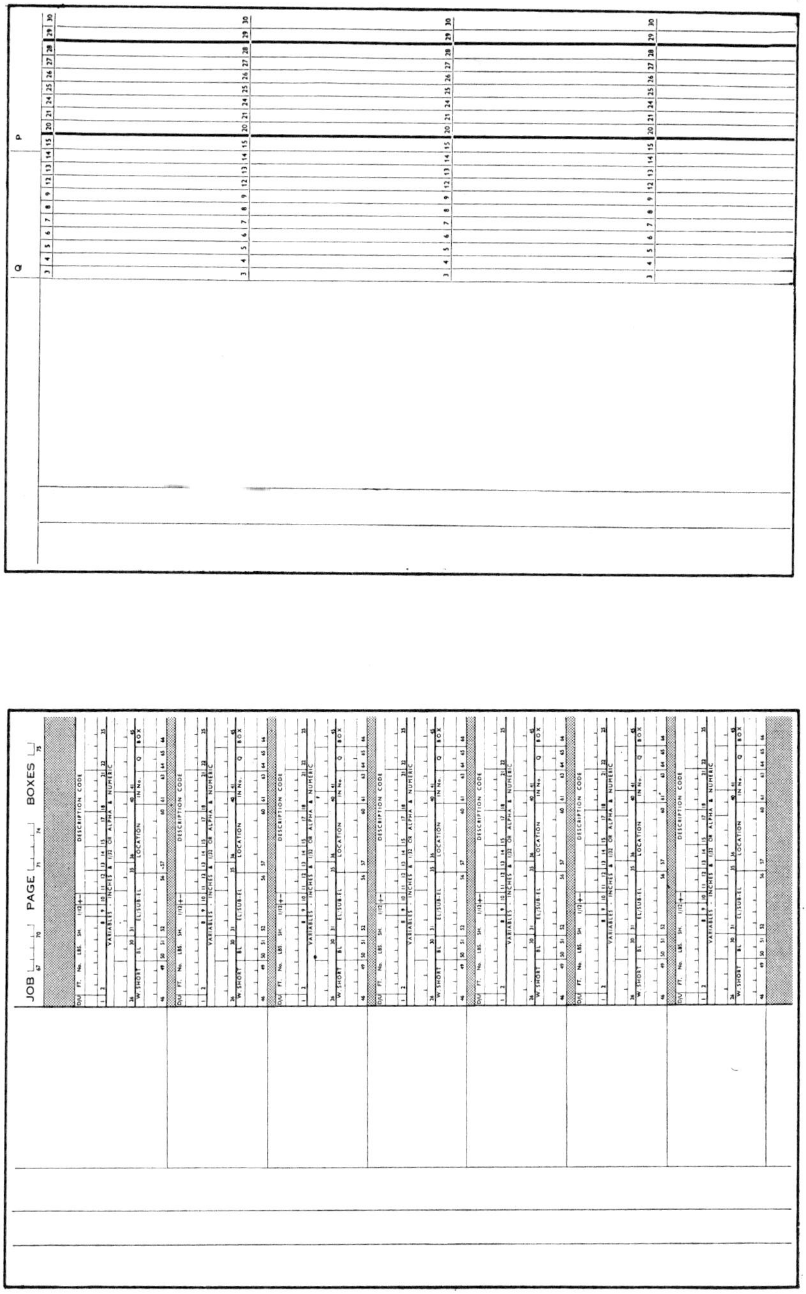

Chart H. Four different examples of taking-off stationery printed for use with computer processing systems

and their associated codes are transferred to paper tape by an operator using a tape punch; the resultant tape is known as the data tape.

(4) DATA CHECKING. The data on the tape must now be checked before input to the computer. This may be carried out by reading over the print-out (the typescript produced by the tape punch) against the original dimensions.

The checking of waste calculations, if not done previously, must also be carried out at this stage. Any errors discovered are corrected by editing the tape, or preparing an 'error tape' instructing the computer to make the necessary adjustments.

(5) COMPUTER PROCESSING. The checked data type may now be taken to the computer room for processing. The machine is first set up for quantity surveying by input of a 'program tape' (standard for all jobs) after which the data tape for the current job is run in on the tape-reader.

The computer now automatically squares the dimensions, sorts and merges the items, casts and reduces the quantities, and outputs the results. Processing speeds vary, but as a rough indication a medium size job may take up to half an hour.

Output from the computer may take one of the following forms:

(a) Print-out of the abstract direct from a line printer attached to the computer.

(b) Output of the abstract on a paper tape (the *output* tape) which is used by the surveyor to actuate the tape punch in his office, which, (working in reverse, so to speak) will then automatically type out the abstract from the output tape.

(6) DRAFT BILLING. The bill is drafted from the computer print-out, full descriptions from the original dimensions being inserted, in lieu of code numbers (code numbers may be retained as well, and used as bill item numbers).

To facilitate matching of descriptions with their codes, some offices draft the bill during taking-off stage (*pre-billing*), leaving only the insertion of quantities to be carried out after processing. Often a standard library of descriptions is built up, from which the taker-off can select his codes during measurement; draft bill descriptions being subsequently selected from the library in accordance with the code number on the abstract.

The above computer system may be extended to provide for a separate print-out of individual squaring results against code and/or taking-off folio numbers (a *squaring print-out*), or a display of indexed squaring casts on the abstract, if desired. The code may be enlarged to provide alternative sortations (trade order, element order, etc.) or to enable the computer automatically to price, extend, and cast the abstract in order to provide a comparable estimate of costs.

Programming for abstract processing is relatively simple. The surveyor may use his own computer, or hire time on a locally available computer. Alternatively he may make use of one of the service bureaux offering computer services to the quantity surveyor.

Bill processing by computer

This is a system whereby the computer performs the entire working-up process, including automatically printing the bill, with full descriptions, ready for tendering.

Initial programming of the computer is based on the use of a library of coded phraseology or bill descriptions (a number of such libraries are now in use, and some are being made generally available to the profession) which can be manipulated by the computer with the aid of auxilliary storage equipment. This equipment may take the form of magnetic tape machines, magnetic card devices, or random access disc stores. Most modern computers of any size now provide such facilities.

Bill processing systems vary fairly widely at present, and it is not proposed to do more than give the briefest of outlines of such a system here,[1] the main processes of which usually take the following form:

(1) TAKING-OFF. The London Method is employed, the taker-off using paper with a coding column (no squaring column is required).

(2) CODING. This may be done by the taker-off, or as a separate operation by a 'coder'. Coding is done from a standard library, special provision being made for 'rogue items' (items not in the library). The standard library will be based on the Standard Method of Measurement, and may even use SMM clause numbers in the code. The coding of descriptions is usually checked as a separate operation.

(3) DATA PREPARATION. Dimensions and their associated codes are punched on to tape (or cards as the case may be) and 'verified', i.e., the data tape is checked by means of dual punching procedures, using verifier or comparator equipment to detect errors. Waste calculations must be checked at this (or an earlier) stage.

In the case of some systems each actual description may be punched on to the tape, headings only being supplied by the computer. In others, abbreviated descriptions may be punched, for the computer to expand into proper form during processing. In others again, only codes are input, the computer supplying the descriptions.

[1] For a detailed treatment of computer working see the author's *Quantity Surveying by Computer*, London, O.U.P.

	ACT/OP/EV/ELEM**RAA		
EXCAVATE OVER SITE TO REDUCE LEVELS	F(348)D	705.8546	FC D 7
EXCAVATE OVER SITE AVERAGE 6" DEEP TO REMOVE TURF AND VEGETABLE SOIL	F(360)A	81.0000	FS D 7
EXCAVATE OVER SITE TO REDUCE LEVELS AVERAGE 1" DEEP	F(347)B	2414.9622	FS D 7 / :1
DITTO AVERAGE 2" DEEP	F(347)C	559.6258	FS D 7 / :2
DITTO AVERAGE 3" DEEP	F(347)D	4366.0905	FS D 7 / :3
DITTO AVERAGE 4" DEEP	F(347)E	2248.5870	FS D 7 / :4
DITTO AVERAGE 5" DEEP	F(348)A	2339.3371	FS D 7 / :5
DITTO AVERAGE 6" DEEP	F(348)B	2450.2539	FS D 7 / :6
DITTO AVERAGE 7" DEEP	F(348)C	554.5842	FS D 7 / :7
EXCAVATE FOUNDATION TRENCH NOT EXCEEDING 5'0" DEEP	F(360)E	162.9375	FC D 10A
EXTRA OVER EXCAVATION BELOW WATER TABLE LEVEL	F(381)B	27.0000	FC D 13 / :1
DITTO IN RUNNING SAND	F(381)A	270.0000	FC D 13 / :2
DITTO IN RUNNING SAND BELOW WATER TABLE LEVEL	F(381)C	135.0000	FC D 13 / :3
EXTRA OVER TRENCH EXCAVATION FOR BREAKING UP BRICKWORK	F(381)D	27.0000	FC D 13 / :A
DITTO CONCRETE	F(381)E	27.0000	FC D 13 / :B
RF&R SELECTED EXCAVATED MATERIAL AROUND FOUNDATIONS	F(350)B	757.9152	FC D 16C / :1
RF&R SELECTED EXCAVATED MATERIAL AROUND FOUNDATIONS	F(360)F	162.9375	FC D 16C / :1
RF&R SELECTED EXCAVATED MATERIAL AROUND FOUNDATIONS	F(361)A	92.1946-	FC D 16C / :1
EXCAVATE FROM SPOIL HEAP AND SPREAD AND LEVEL VEGETABLE SOIL AROUND FOUNDATIONS 6" THICK	F(360)C	65.9988	FS D 16C / :2
LOAD UP AND CART AWAY EXCAVATED MATERIAL TO TIP	F(348)E	294.4557	FC D 16F / :1
LOAD UP AND CART AWAY EXCAVATED MATERIAL TO TIP	F(349)A	4788.5656	FC D 16F / :1
LOAD UP AND CART AWAY EXCAVATED MATERIAL TO TIP	F(350)A	757.9152-	FC D 16F / 1

52

Chart J. A page of computer abstract print-out (imperial units), with full descriptions for final editing prior to bill print-out. (Courtesy of English Electric Computers)

				£	S	D
	GRADE C CONCRETE					
00044	75 MM BLINDINGS	1620 SQ M				
	GRADE DV VIBRATED REINFORCED CONCRETE					
00045	PILE CAPS OVER 300 MM THICK	34 CU M				
00046	COLUMNS OVER 0.10 SQUARE METRE	5 CU M				
00047	BEAMS OVER 0.05 BUT NOT EXCEEDING 0.10 SQUARE METRE	64 CU M				
00048	GROUND BEAMS OVER 0.10 SQUARE METRE	383 CU M				
00049	100 MM BEDS : IN BAYS AVERAGE 10 SQUARE METRES	105 SQ M				
00050	220 MM DITTO	13 SQ M				
00051	100 MM BEDS SLOPING NOT EXCEEDING 15 DEGREES	14 SQ M				
00052	150 MM WALLS	63 SQ M				
00053	150 X 70 MM (AVERAGE) KERBS	25 M				
00054	150 X 200 MM DITTO	17 M				
00055	450 X 150 MM ADDITIONAL THICKNESSES TO BEDS	13 M				
00056	600 X 300 X 50 FILLINGS	38 NO				
00057	900 X 400 X 150 MACHINE AND SUNDRY BASES : PREPARED FOR MONOLITHIC FINISHES	10 NO				
	BAR REINFORCEMENT					
	IN IN-SITU CONCRETE					
	MILD STEEL					
00058	10 MM DIAMETER BARS IN BEDS - (PROVISIONAL)	9652 KG				
00059	15 MM DIAMETER BARS IN WALLS - (DITTO)	9856 KG				
00060	5 MM DIAMETER BARS IN LINKS STIRRUPS AND BINDERS	860 KG				
00061	15 MM DIAMETER DOWEL BARS 600 MM LONG : TIN CAP 300 MM LONG PACKED WITH COTTON WASTE	38 NO				
65042/2	0004		£			

Chart K. A page from a bill of quantities printed automatically from a computer output tape. (Courtesy Messrs Monk & Dunstone)

(4) COMPUTER PROCESSING. On insertion of the data tape, the computer (having been supplied with its program tape and coded library files) now proceeds to perform the entire working-up process—the printed bill of quantities duly emerging from the output printer, working at the nominal rate of something like 1000 lines of print per minute.

Ancilliary print-outs of squaring results and/or abstracts may be produced at the same time, from auxiliary output devices, if required. Often an abstract (with full descriptions) is produced first, and this can then be taken away and edited prior to the final bill print-out.

Alternatively, print-outs may be obtained from off-line printers in the surveyor's office, actuated by output tape produced by the computer.

Two examples of such print-outs are shown on the previous page.

Processing systems always provide sophisticated checking routines, so that the machine will check its own working and also that of the operator, as well as searching for errors on the data tape. Some computers are capable of processing more than one job at a time, a master-program automatically preventing the jobs from getting mixed up with one another.

(5) PRINTING. By loading the output printer with suitable offset lithography master paper, the computer can be made to produce the bill in suitable form for direct reproduction on an office machine. The master can be backed with copy paper, so as to produce a carbon copy. Alternatively the computer output can be copied on a Xerographic copier, or by other means.

(6) EDITING. A final edit is carried out before reproduction. It is possible to incorporate certain feasibility checks associated with editing in the computer program, enabling the machine to do its own editing to some extent. A final human edit is likely to remain necessary however, even with the most sophisticated of systems.

Bill processing systems are inherently capable of providing alternative sortations, and of pricing, extending, and casting the bill (though programming and coding facilities must, of course, be provided for these services). When such a system is used to produce the original bill, it may well be used to process data in respect of periodic valuations of the work and for adjustment of final accounts.

part 2

The Preparation of Dimensions

Dimension Layout

A set of dimensions is a means of communication. It is for this reason that special care must be taken by the student to ensure that his taking-off is not only neatly written and well spread out on the page, but that it complies in every respect with the conventions which will render it instantly intelligible to those to whom it is addressed. As previously indicated, these conventions may vary in practice, depending upon the particular system of bill preparation adopted.

It is customary to use conventional dimension paper for study and examination purposes, since this is the most convenient method of displaying the student's knowledge of the subject. What follows, therefore, is based on the use of standard dimension paper, and presupposes traditional methods of working-up (using the group method of taking-off to produce a conventional bill of quantities).

Use of dimension paper

A sheet of dimensions is shown on page 30. On the left-hand side can be seen a narrow binding-margin; the remainder of the page being divided into two identical halves. Each half contains three narrow columns and a wider one: the *timesing* column, *dimension* column, and *description* column, respectively. The taker-off commences at the top of the left-hand half of of the page, working downwards, and continues at the top of the right-hand half. It is customary to use both sides of the paper. The columns are used as follows:

Timesing column. The left-hand narrow column is used for multiplying ('timesing') the dimensions when necessary, each multiplying factor being followed by an oblique stroke. A factor followed by a dot instead of a stroke indicates addition instead of multiplication. Later examples will make this clear.

Dimension column. The centre narrow column is for recording the dimensions in metres and centimetres. Each *linear* dimension is underlined. A pair of dimensions, one above the other thus:

$$\begin{array}{c} 2{\cdot}65 \\ \underline{4{\cdot}30} \end{array}$$

indicates an area (a *superficial* item) resulting from multiplication of the two dimensions. A group of three dimensions thus:

3·90
1·66
4·75

indicates a volume (a *cubic* item) resulting from multiplying the three dimensions together.

A number (without a radix mark) underlined indicates an enumerated item.

Squaring column. The resulting areas and volumes are subsequently calculated and inserted in the right-hand narrow column, opposite their respective dimensions. This operation, known as 'squaring' the dimensions, is not normally carried out by the taker-off.

Description column. The wide column is for writing descriptions of the work measured.

Waste column. This is an imaginary column occupying the right-hand side of the description column, in which the taker-off inserts any preliminary ('waste') calculations which may be necessary in order to arrive at his dimensions.

In recording each item the taker-off will usually find it convenient to work in the sequence (1) waste calculation, (2) dimensions, (3) description. The following notes concerning the display of work on the taking-off sheets deal with each of these in turn, and offer guidance regarding setting-out conventions generally, in accordance with widely established practice.

Waste calculations

The term *waste* is a misnomer. Preliminary calculations (sometimes referred to as 'side casts') need to be set down carefully and accurately so that they can be checked. They should be written either above or below the description *not level with it*, or confusion may result. Waste calculations will be necessary in respect of each dimension, except in the following circumstances:

(1) when a scaled or figured dimension can be transferred direct from the drawing to the dimension column,
(2) when a dimension has been derived from a previous waste calculation,
(3) when a dimension has been copied from a previous dimension.

In all other instances the preliminary calculation, however simple, should be displayed on waste, thus allowing an independent visual check to be made. It is not customary, however, to display the intermediate arithmetical stages of a waste calculation which are implied by the figures involved. Thus the multiplication of (for example) 17·95 metres by 13 would be shown on waste as

$$13/17{\cdot}95 = 233{\cdot}35$$

the working being carried out on scrap paper, subsequently destroyed. The same procedure applies in the case of formulae. The particular formulae should be clearly shown on waste, together with the answer and sufficient information to enable the working to be checked. The working itself, however, is not shown.

As far as possible the taker-off should confine his waste calculations to the description column, keeping to the imaginary waste column on the right-hand side where possible. On no account should the calculations be allowed to infiltrate round the sides of the descriptions (a common fault with beginners).

Dimensions

After rounding-off to the nearest 10 mm the results obtained from waste calculations are transferred to the dimension column; the first dimension being positioned so that the first line of the description opposite will come well clear of the lowest figure on waste.

Certain conventions exist which employ diagrams in the dimension column to denote areas of circles, quadrants, and the like; but this merely shifts the responsibility for correct choice of formulae from the taker-off to somebody else, which may be undesirable, especially with computer working. Dimensions in diagrammatic form should be avoided therefore, and the formula translated into dimensional form, using the timesing column as necessary. Thus two quadrants of 5·00 metres radius each, followed by a semi-circle 20·00 metres diameter, for example, would be set down as

$$\begin{array}{r} 2/\tfrac{1}{4}/3\tfrac{1}{7}/5{\cdot}00 \\ \underline{5{\cdot}00} \\ \tfrac{1}{2}/3\tfrac{1}{7}/10{\cdot}00 \\ \underline{10{\cdot}00} \end{array}$$

while the volume of three cylindrical columns 1·50 metres radius and 20·00 metres high would be denoted by

$$\begin{array}{r} 3/3\tfrac{1}{7}/1{\cdot}50 \\ 1{\cdot}50 \\ \underline{20{\cdot}00} \end{array}$$

Further examples of the translation of mathematical formulae into dimensional form are given in an appendix to this book.

A common error made by students in the setting down of dimensions is that of failing to conform with the LWH rule laid down in SM A3(a), requiring a dimension sequence of length, width, and height. Although this clause expressly refers only to the order of dimensions in the descriptions, convention extends this usage to the dimension column also and if it is not adhered to therein confusion will tend to arise—not least in the mind of the taker-off himself, when trying to identify his previous dimensions for conversion to subsequent measurements involving associated work.

It is also desirable that, where possible, the pattern of dimensions should represent the actual work configuration, since this is an aid to interpretation of the taking-off. Two triangular gable-ends, for example, each 6·00 metres at base by 3·00 metres high should not be denoted simply by taking 6·00 by 3·00 for the area of the two, although the result would be mathematically correct. Instead, the following should be written:

$$2/\tfrac{1}{2}/6{\cdot}00$$
$$\underline{3{\cdot}00}$$

since this represents two triangles, rather than one rectangle.

Again, a pair of double doors, each leaf size 0·75 metres by 1·90 metres should be set down as:

$$2/0{\cdot}75$$
$$\underline{1{\cdot}90}$$

and not as 1·50 by 1·90, although this would give the same area, and perhaps save the trouble of a preliminary waste calculation dividing the 1·50 figured dimension by 2.

A special word of warning is necessary with regard to timesing in general, and this is in connexion with 'dotting-on', i.e., the convention for adding instead of multiplying in the timesing column. It should be noted that any figure dotted-on to a timesing factor increases the timesing factor, and therefore multiplies instead of adding. Thus, for example, if it is required to add *one* rafter (extra for a hipped end) to twenty-four rafters which have already been twiced for both roof-slopes, care must be taken not to write:

$$1{\cdot}2/24/\underline{3{\cdot}75}$$

since the effect here is to give (1 + 2) × 24 = 72 rafters; while

$$2/1{\cdot}24/\underline{3{\cdot}75}$$

would give 2(1 + 24) = 50 rafters. In fact, however, *forty-nine* rafters only are required here, and once the original twenty-four have been timesed, single rafters can no longer be added by dotting-on; the only correct method being to add an individual one separately thus:

$$\begin{array}{r} 2/24/\underline{3{\cdot}75} \\ \underline{3{\cdot}75} \end{array}$$

Whenever the operation of timesing and dotting-on has been carried out, the taker-off should always check back the work in the timesing column to make sure that the correct number of units results, or serious errors may arise in the subsequent quantities.

Descriptions

The first line of each description should commence on a level with its first dimension. When traditional methods of working-up are used the taker-off writes customary abbreviations when framing his descriptions, and these are expanded by the draft biller. Modern systems of working-up may, however, require the taker-off to write his descriptions in full, or to use special abbreviations peculiar to the system. These changes have tended to render the old system of customary abbreviations somewhat obsolete. The use of such abbreviations does not save so much time as might be thought, since the time taken to take-off a set of dimensions is governed by thinking-speed rather than writing-speed. For these reasons the student of taking-off is recommended to write his descriptions in full, avoiding the use of any abbreviations which might be ambiguous.

The detailed framing of descriptions will be dealt with later, and it is chiefly their appearance on the page, rather than their content, which concerns us here. Points to be noted are as follows:

CHANGE OF COLUMN. Try to avoid breaking a description at the bottom of a page or column and continuing overleaf, or at the top of the next column. This tends to divorce the description from its associated dimensions. Anticipate the amount of space a description will take, and if doubtful, start the description on a fresh column.

Should the dimension-string out-run the column, do not simply write 'Ditto' at the head of the next column, since this will be rendered invalid if an additional item should be inserted at a later date. Instead, repeat the first phrase of the description, followed by '. . . all as before'.

ANDING-ON. Where two descriptions apply to one dimension or string of dimensions, they are separated by an ampersand (&) in the centre of the description column. Any number of descriptions may be 'anded-on' in this way, but not so as to extend over more than one column of

52 dimensions, or cohesion will be lost. When a new column is required, the dimensions should be written down afresh against the descriptions remaining to be anded-on. Examples of anding-on appear later.

REPEAT DESCRIPTIONS. When referring back, it must be made clear which description is intended. The description *100 × 25 mm skirting as before* will not be sufficient if two kinds of 100 × 25 mm skirting have previously been described. The phrase '. . . as last described' or '. . . as first described' should be used in this case, instead of '. . as before', which would be ambiguous. When in doubt, write the description in full, but when doing so make certain that the wording exactly matches the previous wording, or the items may not be correctly merged at working-up stage.

DEDUCTIONS. Each item to be deducted requires the description to commence with the word *Deduct* (underlined), often abbreviated to *Ddt* (still underlined). The word *Deduct* on its own is not sufficient to denote a deduction from a foregoing item; a description must always follow, even if it only consists of the word 'ditto' or clarity will be lost. It is customary to prefix the next positive description with the word *Add* (underlined) in order to emphasize the change from deductions to additions.

Timesing factors in the dimension column

Where two descriptions are anded together for reference to the same dimensions, it is possible to effect a change of measurement for one description only, by means of a timesing factor in the description column, written just below the relevant description, and heavily underlined, with a space for insertion of the subsequent squaring-result, thus:

50·00	100 × 25 *mm Softwood skirting plugged, including mitres and ends.*
	&
	Deduct Two coats emulsion paint on walls a.b.[1]
	Super × 0·10 =

If this method of timesing is used too frequently, a confused set of dimensions may tend to result, and its use should, therefore, be confined to cases such as the above, where common usage will have rendered it customary to the working-up department.

Unit statement

In certain cases the SMM requires the number of units to be stated, in

[1] *a.b.* is the customary abbreviation for 'as before'.

addition to the measured quantity. Thus certain types of casements must be given in square metres, stating the number (SM P27(a)).

The practice is to state the number of units in brackets just below the description, thus:

3/0·50	50 *mm Softwood rebated*
1·25	*and moulded casement.*
	(in No. 3)

If each dimension (untimesed) represents one unit (as above) the form (*in No. 1 each*) is sometimes used, instead of (*in No. 3*) or as the case may be. This is because if more casements of the same description were to follow later, for example, the total number of units would be more than three, and the (*in No. 3*) would then not be strictly correct.

This is rather an academic point, however, since the worker-up will read the (*in No. 3*) as applying to the one group of dimensions only, and will collect the total number of units together on the abstract. The form as illustrated above, therefore, may be considered suitable for all cases.

Bracketing

A bracket should be used wherever (a) more than one dimension applies to a description, or (b) more than one description applies to a dimension. The bracket comprises a straight vertical line overruling the printed line dividing the description column from the squaring column. The bracket is terminated by a horizontal seriph (i) at the top, just above the first line of the first description and (ii) at the level of the last dimension, or line of description, whichever is the lower.

Dimensions gain in neatness if the brackets are ruled, and if the paper is turned sideways at the completion of a page, the appropriate brackets can be quickly ruled-in and seriphed-off with little loss of time, and this method is recommended for examination purposes particularly, unless the student is possessed of a very steady hand when working under pressure.

It is the practice in some offices to bracket dimensions representing additions to those representing deductions, so that a net cast for the item can be written on the dimension-sheet at squaring stage. This saves carrying the deduction on to the abstract, and is a valid method provided the squaring staff are suitably briefed. It is a practice best avoided by the student, however, who should aim at clarity of presentation by making all his deductions separate items from the additions, particularly in view of the fact that this is in any case essential with some working-up systems.

54 **Nilling**

The standard conventions are as follows:

(a) INDIVIDUAL DIMENSIONS. In order to delete, the word *nil* is written in the squaring column opposite the lowest figure in the offending dimension, and is then underlined.

(b) STRINGS OF DIMENSIONS. The above convention is extended vertically by arrows terminating in horizontal lines thus:

0·75	
0·35	
1·05	↑
1·35	
2·65	*nil*
2·50	
1·85	
0·43	↓
3·15	
2·50	

The first and last dimensions in the above string still stand. The reason the squaring column is used for this purpose is to help prevent accidental squaring of the redundant dimensions.

(c) DESCRIPTIONS. These are nilled by simply nilling the associated dimensions. When a description is anded-on to others requiring the dimensions to be retained, it is nilled by crossing through with an oblique line, the word *nil* being written along the line, at the lower end. Alternatively it may be run through with a vertical wavy line, terminating in seriphs, with the word *nil* written against the lower end.

(d) WASTE CALCULATIONS. *Nil* is written beside figures forming the answer to any waste calculations requiring deletion.

(e) SIDE-NOTES. Incorrect words in side-notes or headings are run through with a horizontal stroke.

Correct nilling is very important, since any attempt to delete or alter, which does not follow the standard convention may lead to subsequent confusion, followed by an error occurring in the printed bill of quantities. This could have very serious financial consequences indeed, and must not be allowed to happen.

Even if, therefore, one figure only in the squaring column is redundant, it should not be crossed out or altered, since this could lead to ambiguity. The entire dimension should be nilled, and rewritten in the correct form.

Annotation

Since dimensions are a means of communication, it is important to communicate the meaning of every figure, for subsequent interpretation. This is achieved by annotation, which means the provision of side-notes to explain the derivation of the figures. These side-notes are of two kinds, (i) those annotating waste calculations and (ii) those annotating dimensions.

(i) WASTE ANNOTATION. Short or abbreviated words are all that is necessary, but the aim should be to annotate every figure, as far as is appropriate. The examples of waste calculations given in this book will serve as an indication of correct waste annotation procedure.

(ii) DIMENSION ANNOTATION. Every single dimension must be annotated for location, except in the following cases:

(a) Where its location is obvious (e.g., where the main area for surface strip is obviously over the whole extent of the building).
(b) Where it is obviously derived from previous (annotated) dimensions.
(c) Where it forms part of the measurement of a section of work covered by the last side-note to the dimensions in the same string.
(d) Where its exact location is implied in the description.

Dimension side-notes should also be used to explain any special method of measurement by which the dimensions have been arrived at, or any assumption made therein.

Side-notes to the dimensions require a single bracket on the *left-hand side only*, as indicated below. They should be located on the right-hand side of the description column, and not be allowed to stray over towards the squaring column, or they will tend to be read as descriptions rather than side-notes. If necessary the dimensions should be spaced out to accommodate their annotation, as below.

35·00 26·00	*Carefully strip existing turf average 75 mm deep, stack on site where directed and preserve for subsequent re-use.* *(Boiler house*
29·65 13·40	*(Annex*
20·33 16·65	*(Garage*

56 Abstracting instructions

It may occasionally be necessary for the taker-off to write a side-note which is an instruction rather than an explanation. For example, when measuring painting to be carried out by steelwork fabricators under SM Q23, it is customary for the quantities to be inserted at abstracting stage, after the total weight of steelwork has been calculated on the abstract. The item-description is written by the taker-off and the dimension column left blank; a side-note being necessary to instruct the worker-up to insert the correct quantity in due course.

Instructions of this kind are written as side-notes, preceded by the word ABSTRACTOR in block capitals, as follows:

[*ABSTRACTOR: please*
insert total weight
of steelwork.

Headings

Headings are a vitally important part of the taking-off, and the correct use of headings should be carefully studied by the student. They can be divided into two main kinds: (a) those necessary to enable the taker-off and subsequent users to find their way about the dimensions, and (b) those which require abstracting. The former might be called *fingerpost headings*, and the latter *transfer headings*, since they require transferring to the bill.

Fingerpost headings are of four main kinds, viz.:

(i) JOB HEADINGS. At the commencement of the taking-off, the job heading is written across the top of the first page as a title to the taking-off. When the dimensions are complete, this title is transferred to a title-page, tagged to the front of the complete set. The usual office practice is also to make up a rubber stamp containing the job title, and this is stamped on each sheet vertically in the binding-margin as the taking-off proceeds, to ensure recognition of any sheet which may become detached.

In examination procedure, the question number is written across the top of the first column, taking the place of the job heading.

(ii) SECTION HEADINGS. Each section of the taking-off is commenced with a heading across the top of the column, e.g., *Substructures*, *Internal Doors*, etc. Each section or group should start on a fresh sheet, and in some offices it is the practice to label each sheet 'Substructures 1', 'Floors 7', etc., in sequence, since this allows a numbering system to operate independently of sequential page numbering, and is desirable where more than one taker-off is employed on the same project.

(iii) SUB-SECTION HEADINGS. The commencement of each sub-section (e.g., *opening adjustments*, *roof coverings*) should be denoted by a suitable heading written across the column.

(iv) SUB-HEADINGS. These are written across the waste column only, and should be generously employed by the taker-off to explain which particular group of items is being dealt with at the time of writing. This will not only assist in the interpretation of the dimensions, but will also help the taker-off himself to arrange the work into a logical sequence of items, and help to ensure that nothing is overlooked. Unless the foundation work is very simple, for example, it will be helpful to insert the sub-headings *trenches*, *concrete*, *brickwork*, etc., in the dimensions, and this should always be done unless the relevant items are too few in number to justify sub-heading in this way.

Transfer headings are those with which the worker-up will have to deal, either by way of direct transfer to the bill as bill sub-headings, or as devices for the grouping of items together, prior to billing. The two main kinds are as follows:

(a) SPECIAL BILL HEADINGS. The SMM requires certain work to be grouped under separate headings in the bill instead of being distributed in accordance with the SMM sub-section headings. Joinery fittings, for example, under SM P40(c) are required to each be given under an appropriate heading, stating the overall size.

Where such special headings are required, they will need to be written at taking-off stage. Like the grouping headings which follow, they are distinguished from fingerpost headings by use of the phrase 'The following in . . .' at commencement of the wording; e.g.,

The following in sink unit size
1500 × 500 × 900 *mm high overall,*
all in softwood unless
otherwise described.

(b) GROUP HEADINGS. These take the same form in the dimensions as the special bill headings, but their purpose is to group together work which is required to be measured in detail prior to incorporation into a single bill description at working-up stage.

An example would be steel stancheons, required to be given in kilogrammes under SM Q6(a). Here it is necessary to measure the components of the stancheon in detail, each with a separate description to enable its weight to be calculated. One description only is required in the bill, however, describing the complete stancheon, against the total quantity in kilogrammes. The device employed by the taker-off in order to achieve this is a heading for the complete stancheon, followed by dimensions and descriptions for the components. The actual heading itself will finally become the bill description, the following descriptions being merely devices for achieving a total weight of steelwork.

Transfer headings should always be 'closed' by writing (*end of sink unit*) or as appropriate, in brackets and underlined, across the column at the end of the section of taking-off to which the heading refers.

Headings of any kind *must always be underlined* by the taker-off, or they may be confused with side-notes, special preambles, lump-sum items, and the like. Except for job headings and sub-headings, they are usually written across the dimension and description columns. As in the case of bracketing, students will be advised to rule the lines involved, and should note that the custom with headings is that *every line is underlined*, in order to make them stand out prominently from the remainder of the work.

Lump sum items

These are items not requiring quantities, and do not arise very frequently in intermediate work. They will be mainly P.C. items and 'spot' items, describing alteration work. Like special preambles, they are written commencing in the dimension column and extending across the description column.

Diagrams

Any diagrams or thumb-nail sketches drawn to assist the taking-off should be preserved as part of the waste calculations. The rule governing diagrams is: plenty in the waste column but *none in the bill*. A diagram which is necessary to explain a description to the estimator may be regarded as an admission of failure on the part of the quantity surveyor to measure and describe. Only refer a description to a diagram in extreme cases. These are not likely to occur in intermediate work.

General layout

A common error is to cramp the dimensions. This may cause difficulty at abstracting stage, and make direct-billing difficult due to inadequate space for any necessary collections. A good rule is to allow at least 20 mm between the end of one description and the beginning of the next (assuming no waste calculations). Where descriptions are anded-on, allow at least 20 mm between the end of one description and and and and and the beginning of the next. The fact that this sentence requires five identical consecutive words may help to jog the student's memory on this point.

Also remember to number the pages. Dimension folio numbers are necessary for reference purposes on the abstract. Some surveyors prefer to number the description columns instead, but there is no particular advantage in doing so, since the eye can easily scan a whole page when referring back.

Neatness is essential. A good-quality ball-point pen should be used; sketches only should be in pencil and all other work in ink, as a permanent record. If an error is made, avoid the reflex action of crossing-out. Instead, 'nil' in the correct manner any work which is faulty, except for individual words in a description or side-note, which may be neatly run through if redundant.

Indexing

The completed set of dimensions, after the sheets have been strung together with a treasury-tag through holes punched in the top left-hand corner, and provided with a title-page, is often supplied with an index. This consists of a list of all sections and sub-sections, with folio reference numbers, typed on foolscap or dimension paper and bound with the taking-off. The quickest way of tracing a bill item in the original dimensions, however, will usually be to refer back to the abstract. This will give the items in bill order, with dimension folio reference numbers displayed in each case.

Chapter Six

Measuring Procedure

The group system of taking-off is so called because the work is grouped into sections and sub-sections in a sequence which is designed to facilitate the speedy and accurate measurement of both small and large projects. A standardized order of taking-off has been established which allows the taker-off to concentrate his thought on one problem at a time, and then directs his mind to the next, ensuring that no single item is missed in the vast complexity of cost-components which go to make up a large-scale construction project.

The system depends on the taker-off measuring his items in the correct order, and although this order is to some extent flexible, the exact sequence depending on the job in hand, it will be necessary for the student to memorize what may be considered as the Standard Order of Taking-off so as to make use of the system.

Taking-off sequence

The main groups into which a simple building is divided for the purpose of taking-off are as follows:

Structure

1. Foundations
2. Brickwork, partitions and facework
3. Fires and vents
4. Floors
5. Roofs

Finishings

6. Internal finishings
7. Doors
8. Windows
9. Fittings and staircases

Services

10. Sanitary plumbing
11. Drainage
12. Engineering services
13. External works
14. Outbuildings.

It will be seen that these main groups fall under three headings. It will also be apparent that in certain cases, additional groups may be necessary. A steel-framed building would require the group of *structural steelwork*, for example, and framed buildings generally the group of *cladding*. As a basis for taking-off, however, the above list may be considered of general application for intermediate purposes, and is the general sequence to be followed when taking-off a complete building.

The detailed sequence of structure groups is dealt with in the following chapters, and in this connexion it should be noted that although the student may take a copy of the SMM into the examination-room, it will not help with taking-off sequence, since it is in bill sequence only. In order therefore, that the student knows (a) where to start, and (b) what to measure next, it would be wise for him to commit the above list to memory, as well as the detailed sequences which follow later.

The Standard Method of Measurement

In addition to following the standard sequence of taking-off, applied with such flexibility as the occasion demands, the taker-off must also follow the Standard Method of Measurement. No such flexibility is allowed in the use of this document, however, for although the introduction to the SMM provides for more detail to be given in certain circumstances, these are unlikely to arise in intermediate work, and for all practical purposes the intermediate student should regard the SMM as a rigid document, the exact wording of which he departs from at his peril.

It will be appreciated that the SMM is incorporated in the standard form of building contract, Clause 12 of the private edition with quantities stating that:

> *. . . the Contract Bills . . . shall be deemed to have been prepared in accordance with the principles of the Standard Method of Measurement of Building Works . . .*

Since this standard form of contract is in general use, it follows that under normal circumstances the SMM has the effect of a binding document as far as the quantity surveyor is concerned especially since, not only is he named personally in the contract articles of agreement, but his bills of quantities rank as one of the Contract Documents under the standard form.

It is essential for the student to acquire his own personal copy of the SMM before attempting to take-off quantities. He will require it at his elbow when studying this subject, and should read it in conjunction with this book. Needless to say, only the very latest editions of both should be used.

Drawing preparation

Before commencement of taking-off, the drawing should be carefully studied, together with any specification notes provided. Detailed advice will be given in the following chapters, where appropriate, on the subject of drawing preparation, but it should be noted here that, in general, it will usually pay to colour the drawing before commencing taking-off, as this will be a useful aid to readability when measuring. Brickwork should be lightly coloured-in with a red pencil, and concrete with light yellow, as this colour will enable any reinforcement to show through clearly.

62 The time taken to colour a drawing in this way is negligible, since this is done during the preliminary study of the drawing, and will be found to expedite rather than delay, its elucidation.

It might be thought unnecessary to point out that the scale shown on the drawing should be carefully noted, for use in case any measurements require to be scaled off the drawing. The possibility of accidental use of the wrong scale by the taker-off makes this warning necessary, however, and special care is needed to ensure that this does not happen, since the consequences of such an error could be grave.

Accuracy

In connexion with the use of a scale, it should be noted that, as a guiding principle, dimensions should be selected in the following order of preference:

(1) calculated in preference to figured,
(2) figured in preference to scaled,
(3) scaled only in the last resort.

Before selecting (1) however, make sure that your calculated dimension is correct and the figured dimension erroneous, and not vice versa! Also take immediate steps to see that the figured dimension on the drawing is amended.

Never scale a dimension which can be calculated from figured dimensions, unless the calculations involved would be obviously too laborious to justify the gain in accuracy, which is rarely the case.

Accuracy is a relative term. A sense of proportion must be maintained, or accuracy may degenerate into a fastidiousness bordering on procrastination.

Having said this, however, it is necessary to point out that 1 metre undermeasured in a bill of quantities for a small house could become 500 metres undermeasured if this bill is subsequently used as a 'type' bill for a housing scheme of 500 houses. Again, 0·1 m error in the width of a roof may result in a significant shortage of quantities of roof-coverings when multiplied by a length of 50 metres of factory roof.

Except in certain well-defined circumstances therefore, discussed in the following chapters, the only safe rule to adopt is that of making the measurements *exact* in every case, subject to the rule for rounding-off to the nearest centimetre in the dimension column. For it must be remembered that under the standard form of contract, the quantities form part of the contract, and any error subsequently discovered must be put right. If the error is an undermeasurement, this will add money to the Contract

Sum. This money must be paid by the building-owner. If he is by any chance unable to pay it, he may be made bankrupt. If the error is an overmeasurement, it will be deducted from the Contract Sum, and payment to the contractor will be reduced by the amount involved. Again, serious consequences could follow, and the surveyor will realize that the accuracy of his bills of quantities is no mere academic exercise, but a matter of great practical importance.

Precise accuracy coupled with strict adherence to the SMM are the keynotes of good taking-off.

Framing descriptions

In order to help decide whether an item-description is basically sound, the following seven questions should be asked.

1. Is the grammar faultless?
2. Does it 'scan' well?
3. Are the technical terms correct?
4. Is it in accordance with standard practice?
5. Does it comply with the SMM?
6. Is it as concise as possible?
7. Can the estimator price it accurately?

A description should be as euphonious as possible, contain no grammatical errors, and, unless very long indeed, should comprise one complete sentence. A complete sentence may, nevertheless, consist of one word only, viz., *Bends*, written 'short' after a description of drain pipe. Written-short items will be explained later.

TERMINOLOGY. The terms used must comply with current technical usage in the construction industry, and thus be readily and exactly understood by the estimator. Wherever possible use the wording which usage has rendered customary for bill descriptions. The majority of intermediate-stage descriptions are more or less standardized in form. Memorize the standard wording and always use it, amending as necessary to suit the circumstances. This will help to ease the task of the estimator.

BREVITY. Descriptions must be concise, or the estimator will bog down in a morass of wordage. Be brief, therefore. Every word must count, and each description must be examined for unnecessary words, which should be ruthlessly crossed through before passing as satisfactory.

PRICEABILITY. The true test of any bill description is: can the estimator price it quickly and accurately? Put another way; if your own money were at stake, would you be prepared to price your own description on its merits, and stand to lose if the price were too high, or the return inadequate?

Only by thus placing oneself in the position of the estimator is it possible to be certain that each taking-off description gives complete coverage of the work involved. Each description the taker-off writes is primarily addressed to the estimator, and only by mentally placing himself on the receiving-end in this way will the taker-off master the art of framing descriptions with confidence and precision.

ARRANGEMENT OF MATERIAL. Certain conventions exist regarding the arrangement of words within descriptions for common building components, and a study of these conventions will help to ensure that a description 'comes out right'.

For example; when describing linear items made of timber, the general arrangement of wording is (a) cross-sectional size, (b) material, (c) labours, (d) function, (e) method of fixing. Again, in the case of a glazing description the order would be: (a) thickness, (b) type of glass, (c) type of sash (wood or metal), (d) type of glazing-compound.

Attention will be drawn to such standard wording-arrangements where applicable in the chapters which follow.

When in doubt, a description should first be framed-up on scrap paper. It can then be scanned for errors, improved and polished up until perfect, and then transferred to the description column. Even highly experienced takers-off find it desirable to adopt this procedure on occasion, when dealing with a lengthy and unusual description. The student will find this method helpful whenever he is in any difficulty as to the outcome of any description he is trying to put together mentally, before committing it to the dimension paper.

Written-short items

It will be recalled from Chapter 3 that certain numbered items are billed immediately preceding their relevant parent linear items. The descriptions relating to written-short items are in truncated form in the bill (e.g., 'Mitres' after a parent item of skirting), and may be so written in the dimensions, provided they immediately precede the parent item in the taking-off. Later examples will clarify the instances where writing-short is customary.

Speed of working

This will come with practice and on no account should be forced by the student, or errors are likely to arise in the taking-off. Only by continual taking-off under office practice conditions can any real speed be attained, and the student should not worry if he finds himself to be a slow taker-off, for only painstaking care will ensure error-free dimensions, and painstaking care is time-consuming. Some people are by nature quicker at taking-off than others, but quick taking-off is not necessarily good taking-off, and, although speed is often necessary in office practice, especially when a 'rush' job comes in, it is achieved rather by sustained hard work than by undue haste. Let *more haste less speed* be the student's motto therefore, when taking-off quantities. Once the principles have been grasped and formed into reflex thinking-habits by constant practice, speed will look after itself.

The slower and more carefully the intermediate student takes-off his dimensions during study periods, the faster and more confident will be his performance in the examination-room.

Chapter Seven

Method of Study

The following methods of study are currently available to the student of quantity surveying:

(a) Full-time degree course,
(b) Full-time diploma course,
(c) Part-time day course,
(d) Evening course,
(e) Correspondence course.

In general, a full-time course will be found the most satisfactory. Students who study part-time, on the learn-while-you-earn basis, will be advised to take a part-time day, rather than an evening course. A correspondence course may prove a useful supplement to a part-time course, if taken concurrently, but a second-rate correspondence course is a false economy, and only a course of the highest reputation will prove to be value for money in the long run.

Practical work

Whichever method of study is adopted, practice in taking-off is the key to success in any examination in quantities. This is because a large proportion of the marks is normally awarded in respect of questions demanding the taking-off of quantities from set drawings.

In conjunction with a study of the theoretical principles of taking-off therefore, the student must ensure that adequate practice is gained, and this is only possible by means of actually taking-off quantities from drawings, and submitting the resultant dimensions to a tutor for marking. A study of worked solutions in textbooks, or of those taken down in the lecture-room is a necessary part of the student's training, but there is no substitute for the mental processes involved in preparing original dimensions from a drawing and specification notes, and the student should ensure that he obtains adequate taking-off practice in each constructional group, as well as mastering the theoretical principles discussed in this book.

Worked solutions

These will be found in textbooks, and others will be recorded by the student from black-board demonstrations in the lecture-room. When revising such notes, the student will find it advisable not merely to read them through, but when in doubt to put the worked solution aside and re-take-off from the drawing, comparing his own taking-off with the worked solution afterwards, and noting any discrepancies. In this way his mind will be activated to re-think the problem in depth, and useful revision will have been achieved.

Notes

In addition to worked solutions and the student's own marked exercises, a notebook should be kept containing a record of all notes which require memorizing. This notebook should contain a list of relevant formulae, and also the detailed taking-off sequence for each constructional group, as well as the general order of taking-off, previously mentioned. Under a heading for each group, 'general notes regarding taking-off procedure', 'description-framing', etc., should be recorded, together with any other points the student thinks may have slipped his memory by the time the examination becomes due.

Constructional elements

The main taking-off groups into which a simple building is divided have already been noted (page 60). Within each of these groups, however, a variety of constructional forms is possible. Thus *foundations* may be strip foundations, or of raft construction; and *roofs* may be of timber, or of concrete. Theoretically each main taking-off group may be thought of as encompassing an infinite variety of constructional forms, but it is possible to limit these to the main forms most likely to be encountered under normal circumstances. In this way, if the student works through the main forms most likely to arise in the examination, he should be able to extrapolate if any intermediate form occurs. These main forms are referred to as constructional elements.

Each of the following chapters of this book should be read concurrently with the appropriate constructional element being studied in the lecture-room, depending on the particular curriculum the student is following, chapters being skipped and referred back to later as appropriate. The general outline of the book will probably be found to follow most curricula, and the chapter on strip foundations should in any case be read first, since it lays down many of the fundamentals.

The candidate should check that all necessary constructional elements have been covered before sitting an examination. These are not at present displayed as such in current examination syllabuses, and a check-list of constructional elements is included in an appendix to this book to assist both students and lecturers in this respect.

This check-list does not claim to be exhaustive, and the constructional elements dealt with therein must be regarded as a guide only to the type of construction likely to be met with in examinations in this subject at the present time. For it will be realized that new constructional forms are continually evolving, and any attempt to codify them must be regarded as provisional, and should be periodically updated by the student, from the latest information provided by his construction technology lecturer.

part 3

Substructures

Chapter Eight

General Procedure for Strip Foundations

Definition

A simple structure, such as a dwelling-house, traditionally comprises load-bearing walls which support the upper floors and roof. Such walls require continuous support below ground-floor level, and this is achieved by a strip of concrete laid in a trench. Strip foundations of this traditional kind may also be used in connexion with framed structures, to support the ground-floor walls between column bases. In this case, however, they are best dealt with under the heading 'column base foundations', an advanced constructional group with which we are not at present concerned.

On a sloping site it may be necessary to introduce steps into strip foundations, so that the foundations may follow the slope of the ground. Such cases are more conveniently dealt with under the heading of stepped foundations (Chapter 13) so that we may define *strip foundations* for taking-off purposes as follows:

> *foundations on a flat site comprising load-bearing walls resting on a continuous strip of concrete and including any necessary piers or footings.*

Order of taking-off

Foundation work is taken-off in the general order in which the work is executed. It is merely necessary to visualize the work being carried out on the site and measure the items accordingly. This is not the case with most constructional groups, since the order of taking-off is designed to facilitate efficient measurement, which is not the same thing as efficient erection of the structure. In the case of foundation work the two coincide, however, and the difficulty lies not in remembering the order so much as in deciding where to stop. In other words, in determining where the foundations end and the superstructure begins.

Extent of the work

The normal practice is to include all work below general damp-proof course level under the heading of foundations (often called *substructures*) including the damp-proof course itself. There are some exceptions to this general rule, however, which may operate in certain circumstances, and before proceeding further it is necessary to examine these in detail.

(a) GROUND-FLOOR CONSTRUCTION. This will probably extend below damp-proof course level but is normally dealt with in the 'floors' section of the taking-off (Chapter 23) and not measured with the foundations. Any *excavation* necessary in connexion with ground-floor construction is normally taken with the foundations, however.

(b) EXTERNAL FACINGS. Although extending below the damp-proof course these are more conveniently taken with the superstructure facings, as they will be the same girth, and probably the same specification.

(c) EXTERNAL STEPS. These are more conveniently taken with the external doors, together with any necessary excavation.

(d) EXTERNAL PAVING. This is measured in the 'external works' section, not with foundations.

In considering the above, it must be borne in mind that there is no 'Foundation' or 'Substructure' section in the SMM, and because the presentation of the bill will normally follow the SMM work-sections, there will be no 'Foundation' section in the bill either, in the normal course of events. The items will be distributed into their various trades at abstracting stage.

In practice, however, it is sometimes necessary to separate the substructure from the superstructure in the bill of quantities, especially in the case of large projects. There are two main reasons for this, which may affect the taking-off.

(i) REMEASUREMENT OF SUBSTRUCTURES. Owing to uncertainties at design stage regarding the nature of the sub-strata it may be necessary to increase the foundation depth or otherwise alter the design after the trenches have been dug and the sub-strata examined. Careful site investigation at design stage may prevent such variations occurring, but if it is thought likely that, owing to the nature of the ground, substantial variations may arise, a separate substructure bill will facilitate the adjustment of such variations at post-contract stage.

Since floors, facings, etc., are unlikely to be affected by these variations, however, a separate substructure bill solely for the purpose of facilitating remeasurement may not affect the actual taking-off.

(ii) ALLOCATION OF FUNDS. It may be desirable to separate the cost of the substructure from that of the superstructure. In the case of a public authority, for example, it could happen that the funds allocated for substructures come under a different expenditure vote, or attract a higher subsidy grant, or must be separated for some similar reason connected with financial or cost considerations.

In this case it may be necessary to keep *all* structural work below DPC level separate in the bill, including floors and facings, and external steps as well. These items are then measured in the foundation section of the taking-off.

Examination procedure

The question now arises: what should the student do when asked to take-off 'the foundations' in an examination question? Should the ground floor, facings, etc., be included?

If, after a very careful study of the exact wording of the question (and of the drawing and specification notes) the candidate is still uncertain, he would probably be wise to adopt one of the following alternatives:

(a) IF TIME ALLOWS. 'Play safe' and include the facings, ground floor, and steps (if any) at the end of the taking-off.

(b) IF TIME IS SHORT. Include the facings and make a note in the dimensions that the ground floor is assumed measured in the floors section elsewhere, and the steps in the doors section.

If in doubt, it is always wise to include the facings, since if the common brickwork specification below DPC differs from that above, these facings will have to be kept separate in any case (if they are extra over commons), and it then becomes more logical to deal with them in the foundation section.

In examinations the rule is: study the question carefully and decide what the examiner expects you to do. When studying the principles of measurement, however, it will be more convenient to deal with floors, facings, and steps separately from foundations.

Chapter Nine

Site Preparation

SURFACE STRIP

Careful consideration must be given to preparing the site in readiness for erecting the building, and it is usually desirable for the quantity surveyor to visit the site in order to see for himself what requires to be measured under the category of site preparation work. This is dealt with under sub-section of the SMM; Clauses D2 to D5.

It will first be necessary to measure the removal of any trees, undergrowth or other obstructions in accordance with SM D5 and then to consider what preparation is necessary to the actual surface of the ground itself, over the area of the building, before construction can start. This is governed by two factors,

(i) turf or other vegetable matter (weeds, etc.) must first be removed,
(ii) ground-floor construction will normally extend below original ground level.

These factors must now be considered in turn.

Removal of vegetable matter

Virgin ground of almost any kind is soon covered with weeds, grass or similar vegetable matter which will have a deleterious effect if left growing underneath the building. This must, therefore, be removed over the whole area of the proposed structure, and the normal practice is for this to be done *before* trench excavation commences (a) to facilitate trench excavation and (b) to facilitate preservation of the vegetable matter in cases where this is in the form of re-usable turf.

It therefore follows that this *surface strip* as it is usually called, must be measured over the extreme plan area of the building *including projection of concrete foundations*, as shown on the cross-section illustration in Fig. 1.

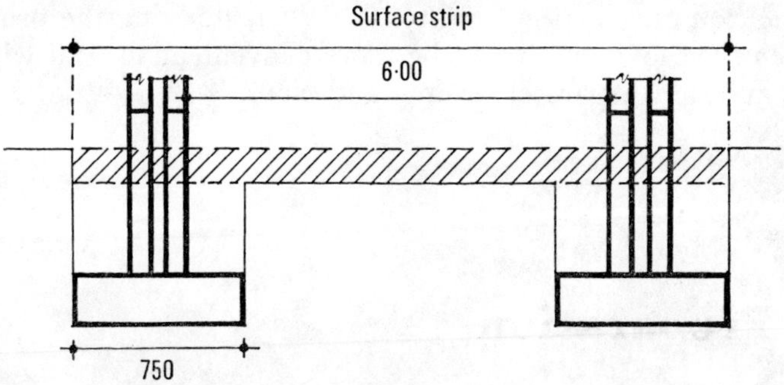

Fig. 1

This area must also include an addition for any piers or chimney-breasts projecting on the outside of the building, the measurement of which is discussed later.

There may, of course, be cases in practice where a site visit will confirm that the measurement of surface strip is not required since no vegetable matter exists. In the absence of evidence to the contrary, however, surface strip may be presumed necessary, and should be measured. Under examination conditions no question of a site visit will arise, and the student must carefully scrutinize the specification notes for details of the surface strip. If no surface strip is mentioned, then the above rule still applies, viz., *in the absence of evidence to the contrary, surface strip may be presumed necessary, and should be measured.*

Ground-floor construction

This will normally comprise a bed of concrete spread and levelled over the site in preparation for (a) floor finishings, or (b) timber floor construction. This ground-floor construction will probably include a layer of hardcore underneath the concrete bed. The level of the underside of this layer of hardcore (or of the concrete bed, if no hardcore) will almost certainly be below original ground level, even if only to allow for depth of surface strip, and when this is the case, this level is termed *formation level.* This is the level to which the ground will have to be reduced over the area of the building before floor construction can proceed, and the excavation necessary to achieve this is termed *reduced level excavation*, referred to in SM D7.

As previously stated, the actual ground-floor construction itself is normally measured with the 'floors' section of the taking-off, but a study of the following considerations will show why it is desirable to think about this reduced level excavation when taking-off the foundations.

Depth of surface strip

We are now in a position to examine the theoretical considerations governing the depth of surface strip, and it will be seen from Fig. 2 that

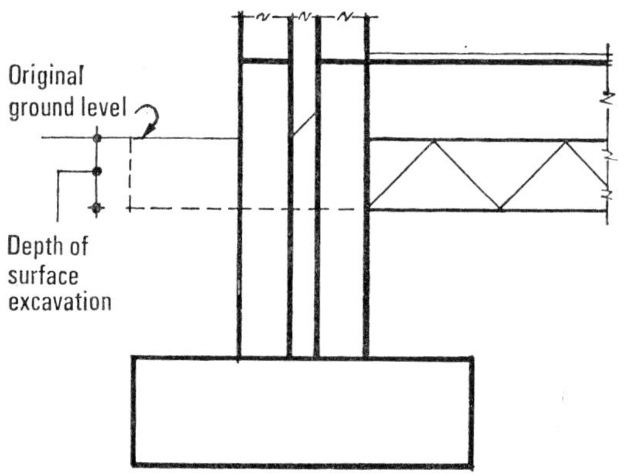

Fig. 2

 the total depth of any surface excavation is equal to the difference between original ground level and formation level.

Now it will be apparent that if the depth of excavation required to remove the vegetable matter is equal to the difference between original ground level and formation level, no reduced level excavation will be required, since formation level will have been achieved by the process of removing the vegetable matter. If it is possible to avoid reduced level excavation in this way, it is a reasonable assumption that the contractor will do so, and we must therefore measure accordingly. Before we can decide whether this will be the case, we must now ask the question, how thick will the surface strip be?

It is usual to assume that the removal of vegetable matter on an average building site will reduce the level of the ground by about 150 mm. It is possible, however, to strip turfs 50 mm thick or less; while, on the other hand, depths up to 300 mm may be achieved, if necessary, in what may be considered for this purpose as one operation. It is likely, therefore, that the contractor will attempt to achieve formation level with his surface strip, provided that this does not exceed 300 mm thick. The rules for determining depth of surface strip may therefore be summarized as follows:

(a) Determine the difference between original ground level and formation level. Call this difference *surface excavation.*
(b) If surface excavation does not exceed 300 mm thick, describe it as *surface strip*, giving the total depth as an average.
(c) If surface excavation exceeds 300 mm thick, take the first 150 mm as surface strip, and the remainder as reduced level excavation under SM D7.

Description of surface strip

Before framing the description it will be necessary to consider the method of disposal. Is the vegetable matter to be removed from the site, or preserved for re-use? The specification notes must be examined to decide this, but if no indication is given, the presumption is that the vegetable matter is not required for any special purpose, and it then becomes classifiable under the heading of *surplus spoil* and is dealt with under SM D16(f), the first sentence of which reads:

Surplus spoil shall be given in cubic metres.

Since surface strip is measured as a superficial item and surplus spoil as a cubic one, the two must be separated. Disposal must be anded-on as a separate item (cubed by the depth) and *not* included in the surface strip description. Typical descriptions are as follows:

super	*Strip surface soil average* 150 *mm deep and remove all vegetable matter.* & *Remove surplus excavated material from site.* *Cube* × ·15 =

Preserving turf

If the vegetable material consists of re-usable turf, the specification may call for its preservation on the site, with a view either to its future use by the building owner, or its re-laying by the contractor as part of the contract. This is covered by SM D3 which states:

> *Lifting turf which is to be preserved shall be given in square metres stating the method of preservation and disposal.*

If the specification notes expressly mention the removal and preservation of turf, then the surface strip item is best framed in the following (or similar) terms:

> *Carefully strip existing turf average* 75 *mm deep, stack on site where directed and preserve for subsequent re-use.*

For this operation a depth of 75 mm rather than 150 mm is more probable, to facilitate re-laying, and any further depth required should then be measured as reduced level excavation. It will also be noted that the actual re-laying of the turf is not included in this description. The reason for this is that *turfing* (i.e., the laying of turfs) is covered under SM D18 as a separate operation, and in any case is more conveniently measured in the 'external works' section of the taking-off, not with foundations. This will apply both to the laying of new turfs and the re-laying of existing turfs.

The student may notice that no mention of 'wheeling a distance not exceeding 50 metres' or such-like phrase appears in the above, as might perhaps be expected, and unless wheeling is mentioned in the specification notes (suggesting that the examiner expects to see it in the description) it is best omitted altogether. SM D3 does not demand a wheeling distance to be given, and if the analogy of SM D4 is invoked (preserving vegetable *soil*) the *average* distance must be stated, and this may not be easy to decide upon. It should be noted also that in the instance of lifting turf, wheeling might be deemed to be included under SM A3(b)(i), should any claim arise, so the simplest solution is also a safe one, viz.: wheeling need not normally be mentioned in this item.

It is interesting to note that while lifting turf which is to be preserved is fully covered under SM D3, stripping the site of vegetable matter not required to be preserved does not appear to be catered for in the Standard Method, unless we assume it to be included in 'bushes, scrub, undergrowth and the like' referred to in SM D5(c). If we assume this clause to apply, no average depth of excavation is required to be stated, and there is no mention of disposal. From this we might infer that SM D5(c) refers to preliminary site clearance rather than to removal of vegetable matter requiring actual excavation.

When referring to SM D3, it will be observed that no average depth is mentioned here either, and the student may wonder why this should be stated at all in connexion with surface strip or turfing, when the Standard Method does not seem to require it. The answer is that where this is measured over the area of the building, this depth directly affects the measurement of reduced level excavation, as we have seen. An exact depth must therefore be decided upon by the taker-off, and the assumption made known to the estimator, who can then price accordingly. This situation is envisaged by the Introduction to the Standard Method, which states: '. . . more detailed information than is demanded by this document should be given where necessary in order to define the precise nature and extent of the required work.'

Area of surface strip

As previously mentioned, surface strip is measured over the extreme area of the building, including concrete foundations, and unless a foundation plan is provided, the measurements will have to be taken from the $\frac{1}{100}$ ground-floor plan. This represents a plan as at 1 metre above floor level, and cannot therefore be expected to show the outline of the concrete foundations. A typical cross-section through foundations will have to be inspected, and the amount of concrete projection from the external wall face calculated therefrom, twice this amount being added to the overall brickwork dimension on plan to achieve a figure representing the extent of surface strip, extending over the concrete projection at each end.

The concrete projection should be carefully calculated *on waste* by subtracting the wall thickness from the width of concrete foundations, and dividing the result by two, for one projection. This assumes the wall to be centrally over the foundations, as is normal in simple types of buildings covered by the intermediate examination syllabus. Thus the waste calculation representing extent of surface strip for the building shown in Fig. 1 for example, would be as follows:

		length	6·00
	walls	2/·27	·54
strip width	·75		
less wall	·27		
	2)·48		
conc. proj.	·24	2/·24	·48
Extent of surface strip =			7·02

A similar calculation is then required for the width of the building, for timesing by this length in the dimension column.

Projecting piers, etc.

These must be bracketed-in after the main area, and each one carefully side-noted. A common error among beginners is to include piers (or chimney breasts) projecting *internally*, and care must be taken to measure only those projecting from the outside face of the external walls of the building.

It must also be remembered that any such projections shown on the ground-floor plan will represent brickwork at 1 metre above floor level, whereas the measurement required is that of the concrete foundations, not normally shown, even on section, in the case of projecting piers. A waste calculation will therefore be necessary in each case, to convert the brickwork dimensions to those of the concrete, and the following rules will be found of assistance in this connexion.

(a) The amount by which concrete foundations project from each face of an attached pier is equal to the amount by which they project from the main wall face as originally calculated.

(b) It follows that the extra *width* of projecting concrete is always equal to the extra width of projecting brickwork as shown on the plan.

(c) It also follows that the extra *length* of projecting concrete is always equal to the length of projecting brickwork *plus twice the original concrete projection.*

It is important that these simple rules (a more formal statement of which appears on page 95) be carefully studied by the student, as it will be found that a firm grasp of them will be of aid to speed and accuracy in measurement of projections, both for surface strip and (as will be seen later) trench excavation and concrete also.

80 **Corner piers**

External brickwork projections for piers, chimney-breasts, etc., usually occur along the length of the wall, but the above rules also apply in principle to corner piers of the type illustrated in Fig. 3, should these

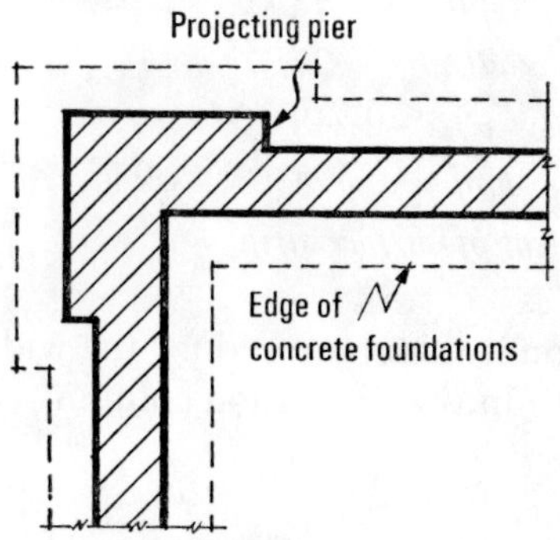

Fig. 3

occur. Since they may not project equally on both returns, however, difficulty may arise regarding the calculation of length, and it is therefore best to regard a corner pier as two separate piers, one on each wall face. One pier should be measured first, applying the rules as usual, but the second pier must have its length adjusted for overlap. This is illustrated

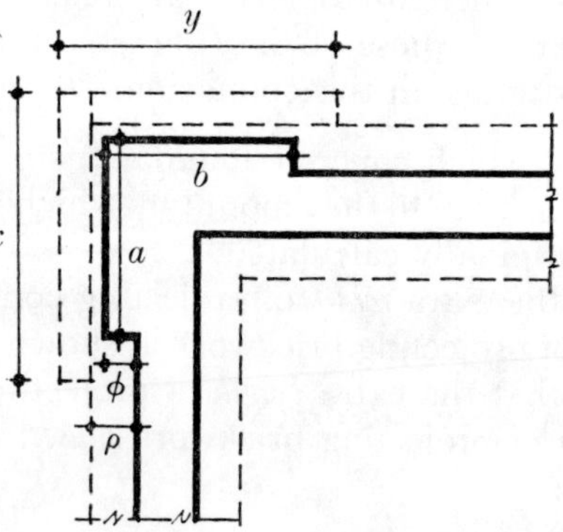

Fig. 4

in Fig. 4 in which *x* and *y* represent the lengths of concrete projection required to be measured in respect of a corner pier of total length

$a + b$. If we call the original concrete projection p, it will be apparent that

$$x = a + 2p$$

$$y = b + 2p - d$$

where d equals the distance which pier a projects from the wall face.

x is now supered by d in the dimension column, and y supered by the distance which pier b projects, and the correct additional area of surface strip for the corner pier has now been arrived at.

Deductions from total area

When the shape of the building is other than a simple rectangle, it will usually be found easier to measure overall and deduct, rather than to measure each section piecemeal. The resultant deductions will be of two general types, (i) corner deductions and (ii) side deductions, as indicated by C and S in Fig. 5. There may, of course, be combinations of both, but the following basic rules will still apply.

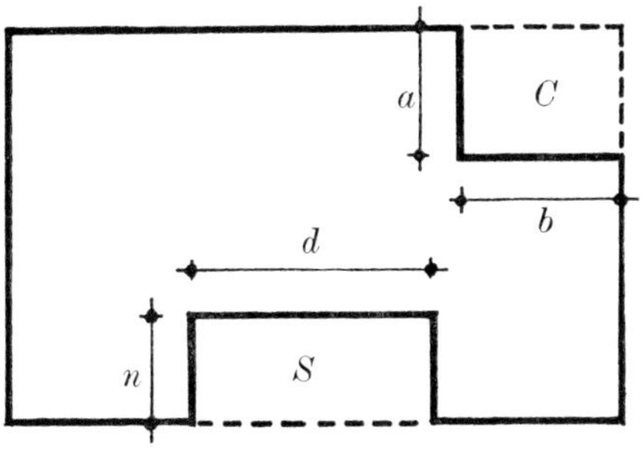

Fig. 5

It must be remembered that deductions are being made from a calculated plan area to outside of foundations, not the plan area shown by the ground-floor plan from which the taker-off will be working (unless a special foundation plan is supplied). It will therefore be found useful to remember the following:

(a) CORNER DEDUCTIONS. Since the concrete projection is constant throughout, it may be ignored. The brickwork dimensions may therefore be taken for surface strip deduction.

(b) SIDE DEDUCTIONS. The above is true, but for only one dimension (n, Fig. 5). The other dimension (d) is always equal to the external brickwork dimension less *twice the concrete projection.*

 If, for example, we regard Fig. 5 as being the outline of external face of brickwork, as shown on the ground-floor plan, then the areas for surface strip deduction would be

$$Area\ C = a \times b$$

$$Area\ S = n(d - 2p)$$

where p is the original concrete projection.

We have assumed here that the concrete projection is constant, and this will normally be the case provided the external walls are of a constant thickness, which again is normal practice in traditional house construction. Should the external walls vary in thickness, however, the cross-sections must be examined to see whether the concrete projection also varies, in which case the dimensions must be adjusted accordingly.

Irregular areas

Shapes other than rectangular must be treated on their merits, and are sometimes best measured net, rather than overall, with a deduction. An example is shown in Fig. 6, which represents the ground-floor plan of a

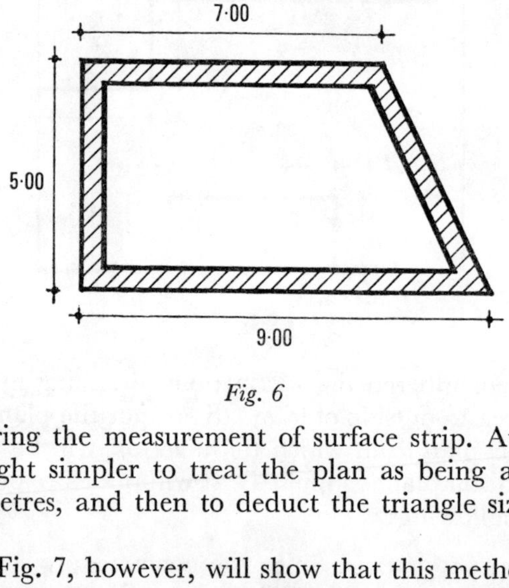

Fig. 6

building requiring the measurement of surface strip. At first glance it might be thought simpler to treat the plan as being a rectangle size 9·00 × 5·00 metres, and then to deduct the triangle size 2·00 × 5·00 metres.

A glance at Fig. 7, however, will show that this method of approach presents difficulties, for when the dotted lines showing edge of concrete foundations are plotted on the ground-floor plan, it will be seen that the required deduction is represented by the triangle *ABC*, the area of which must be calculated having regard to the projection p. If a plan is available showing the edge of concrete foundations, this triangle may be

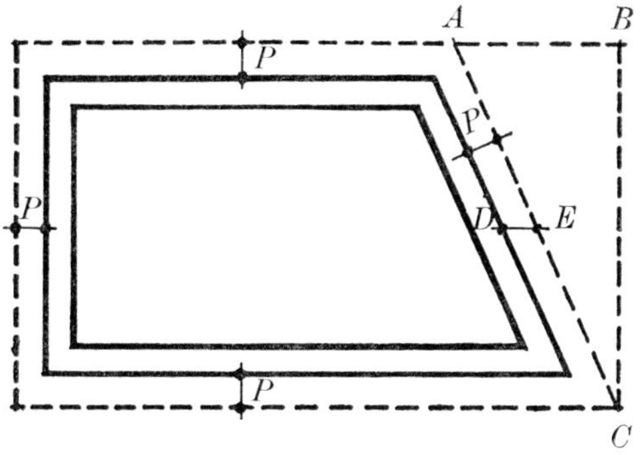

Fig. 7

scaled off the drawing, but when working from the ground-floor plan, the following method will be found useful when dealing with shapes such as the one shown:

Area = total mean length × total width.
Find mean length of brickwork and add concrete projection thus:

	7·00
	9·00
	2)16·00
Mean length bwk.	8·00
Add conc. proj. p (Fig. 7)	·30
	8·30
Add conc. proj. DE (scaled)	·35
Total mean length	8·65

Then calculate the total width as usual, thus:

	5·00
Add two conc. projs. 2/·30	·60
Total width	5·60

The area then = 8·65 × 5·60 *for transfer to dimension column.*

It is, of course, possible to calculate the projection *DE*, but this is hardly worthwhile in view of the fact that a short pencil line plotted on the plan at distance *p* from, and parallel to, outer face of brickwork along the line *AC* is the work of a moment for the taker-off, and scaling the horizontal projection from this simple plot will give results of sufficient accuracy for all normal purposes.[1]

[1] For planking and strutting to surface strip see page 103.

84 **Excavating vegetable soil**

This may be defined as the layer of soil immediately below the vegetable matter. We have seen that any further excavation necessary after removal of vegetable matter is measurable as reduced level excavation, so vegetable *soil* (as opposed to vegetable *matter*) may be regarded for our purposes as part of the reduced level excavation, unless the contractor is expected to sort out the vegetable soil from the top layer of surface strip, which is unlikely.

Now SM D4 deals with the excavation of vegetable soil, but only, it should be noted, with the excavation of vegetable soil *which is to be preserved.* The inference is that unless the vegetable soil is to be preserved for future use, there is no purpose in keeping it separate from ordinary reduced level excavation. Vegetable soil may be valuable for horticultural purposes, and where *soiling* is specified to garden beds, etc., (measurable in external works under Clause D18) the preservation of any vegetable soil met with in surface excavation (or trench excavation, for that matter) will save the cost of buying vegetable soil and importing it to the site for soiling purposes. In such a case the following item might be measured with the foundations, after the surface strip (to which it might be anded-on) and before taking any reduced level excavation:

Excavate average 150 *mm deep to remove*
vegetable soil, wheel a distance
average 50 *metres and spread and level*
on site where directed.

Note here that the *average* (not maximum) wheeling distance must be given (SM D4) and the specification notes must be examined to determine (a) the average depth of vegetable soil and (b) whether it is to be spread and levelled as described above, or merely deposited on site in permanent spoil heaps, in which case the description would be:

Excavate average 150 *mm deep to remove*
vegetable soil, wheel a distance
average 50 *metres and deposit.*

This item would be measured in cases where the vegetable soil was required to be left on site for future use by the building owner. When it is to be spread and levelled by the contractor as part of the contract, *temporary* spoil heaps may, of course, be required, until such time as the contractor is ready to spread and level the soil in the requisite areas. The formation of temporary spoil heaps to suit the contractor's convenience, however, constitutes *multiple-handling*, and is deemed to be included under

SM D16(a), and should therefore not be mentioned in the first description above, unless particularly specified.

It is true that the actual process of soiling (i.e., spreading and levelling the soil) is included, like turfing, as a separate operation under SM D18, and it might therefore be thought necessary to keep this operation entirely separate from the excavation, describing this as being deposited in temporary spoil heaps. However, SM D4 states that soil spread on site shall be so described, presumably to give the estimator an opportunity of pricing on the basis of wheeling direct to the spreading area, whereas turfs will normally require stacking, prior to careful re-laying after preparation of the ground. Spreading and levelling of existing vegetable soil is therefore included in the excavation item, whereas re-laying existing turf is more appropriately measured separately under SM D18.

Use of vegetable soil by contractor

Where there is no mention in the specification of the preservation of vegetable soil, its existence, as such, will be ignored by the taker-off. Any vegetable soil then met with in the excavation will be carted away by the contractor with the remainder of the surplus spoil, measured under SM D16(f). Supposing this vegetable soil is valuable to the contractor, and he sells it at a profit. Can the building-owner recover the money?

If the building-owner was entitled to the saving involved, the procedure would be to adjust the bill of quantities by means of a variation order at post-contract stage, for it would be difficult for the taker-off to measure in advance the quantity of vegetable soil met with in the excavation which may be of economic value to the contractor at a later stage. The Standard Method does not provide for the measurement of vegetable soil not specified as being 'preserved', however, and the contractor may therefore be presumed to have made allowance for the credit value of vegetable soil in his rates for removing surplus spoil, after inspecting the trial holes mentioned in SM B2(b), so that in the absence of any special restrictions imposed upon removal of vegetable soil in the preliminaries, the building-owner would not be entitled to a saving, and no adjustment would be required to the bill of quantities. To repeat, therefore: where there is no mention in the specification of the preservation of vegetable matter, its existence, as such, may be ignored by the taker-off.

Depth of vegetable soil

Where vegetable soil exists, its thickness varies depending on the type of sub-strata, and the specification must be examined to find the thickness of the layer of vegetable soil required to be preserved. In the absence of instructions, 150 mm might perhaps be taken as a reasonable average

86 thickness to which vegetable soil may be expected to extend below the initial 150-mm surface strip, but the average wheeling distance, (which must always be stated in the description of this item, when it is measured at all) can only be estimated from an inspection of the plan, failing any distance being mentioned in the specification.

REDUCED LEVEL EXCAVATION

Any further excavation necessary to achieve formation level, after removal of vegetable matter (and excavating vegetable soil to be preserved, if required) is measured as reduced level excavation under SM D7. Like surface strip and removal of vegetable soil, reduced level excavation will probably be carried out *before* commencement of the trench excavation, and for this reason is again measured over the complete plan area, *including projecting concrete foundations*. The same area, in fact, as the surface strip (assuming formation level to be constant over the area of the building). If not exceeding 300 mm deep, reduced level excavation is measured super, and might therefore be anded-on to surface strip, or if over 300 mm deep, anded-on and cubed by the depth. A typical superficial item would read as follows:

super	*Excavate average* 200 *mm deep to reduce levels*
	&
	Remove surplus, a.b.
	Cube × ·20 =

It is, of course, conceivable that the trenches may be excavated first, leaving the 'dumpling' in the middle (as the remaining reduced level excavation between trenches is called) to be dug out afterwards. Since trench excavation is normally more expensive than reduced level excavation, however, it is reasonable to assume that the contractor will do the reduced level excavation first, thus reducing the volume of trench excavation, with its planking and strutting both sides, to a minimum. It is therefore the usual practice to measure it in this way, and if it subsequently transpires that the contractor excavates the trenches first, the onus will be upon him to show why an apparently uneconomical method of construction was adopted, before a variation order can be issued at post-contract stage.

This argument illustrates a particular case of the following general principle, which may be held to apply in all normal circumstances:

If two or more methods of construction are equally practicable, measure the most economical.

Note that this principle refers to methods of construction, not to details of specification. It applies only where the method of construction is not specified, but is left to the discretion of the contractor, and must not be confused with the following principle which applies when the taker-off is left to decide not a method of construction, but a *detail of specification* to be confirmed by the architect:

Where no specification exists, assume a reasonably good quality construction. If two or more specifications are equally reasonable, measure the dearest.

This last rule only applies when the necessary specification details are not available at time of taking-off, and efforts to obtain the architect's decision on the point in question have for some reason been unsuccessful. In such circumstances, the rule is a kind of fail-safe mechanism. In the event of confirmation from the architect not being available before the bill goes out to tender, it will ensure financial coverage. And in the event of a candidate having to assume his own specification, it will help to ensure that the result meets with the examiner's approval.

It must be stressed, however, that when the above rule is used in practice, (a) confirmation from the architect must be obtained as soon as possible, and (b) steps must be taken to ensure that the specification and contract drawings comply with the decision taken. Otherwise an error in quantities might subsequently be held to have arisen.

Depth of reduced level excavation

SM D7 states: 'Excavating surfaces over 300 millimetres deep to reduce levels shall be given in cubic metres.' Could this mean over 300 mm deep *from original ground level*? If so, only the first 150 mm would be measured super, assuming a 150-mm surface strip to begin with.

The depth of foundation trenches and pits is related (in SM D10 and 11) to 'starting level', and it would not seem unreasonable, therefore, to assume that the word 'deep' in SM D7 means the depth from starting level, which is not necessarily original ground level. We may thus presume that SM D7 intends us to measure the first 300 mm of reduced level excavation super, and the remainder (if any) cube.[1]

[1] For planking and strutting to reduced level excavation, see page 101.

Foundation Trenches

TRENCH EXCAVATION

Strip foundations may be expected to occur under all external walls of the building, but internal load-bearing partition walls may also extend down to foundation level, and the cross-sections must be carefully examined to determine if and where such internal foundations are required. Light non-load-bearing partitions are normally put up off the floors, and can therefore be ignored when measuring foundations, but all brick internal walls must be carefully considered, to decide whether foundations require measuring to these.

External walls

Since foundation trenches are measured cube (SM D10) it will first be necessary to find the total length, which can then be transferred to the dimension column for timesing by the width and depth. Assuming a constant wall thickness, and therefore trench width (and depth), the length along the centre-line of the walls will be the total length required. This centre-line length around the perimeter of the building, is referred to as the *mean girth*, and this mean girth of the external walls can be used later, in calculating the quantities for planking and strutting, concrete, damp-proof course, etc., and the walls themselves, as well as the foundation trenches.

It is therefore necessary to take special care in ensuring that the mean girth (M) of the external walls is calculated correctly in the first place. The following formula, which is correct for a building of any given closed rectangular shape having external walls of constant thickness, should be used for this purpose:

$$M = 2(L + W) \pm 4T + 2n_1 + 2n_2 + \ldots$$

where L = overall length of the building,
W = overall width of the building,
T = thickness of wall,
and n = net projection of each re-entrant.

Note 1. $4T$ is plus or minus according to whether the overall length and width of the building is measured between

(a) internal face of walls (+) or
(b) external face of walls (−),

depending on whether internal or external dimensions are figured on the plan.

Note 2. A *re-entrant* is a section of the perimeter which enters the interior of the plan shape and returns outwards again, thus forming a 'side deduction' as *s* in Fig. 5.

The *net projection n* of any re-entrant is the length of the face of one return wall; or where the return walls are unequal, the length of the face of the *shorter* return wall, as shown in Fig. 8.

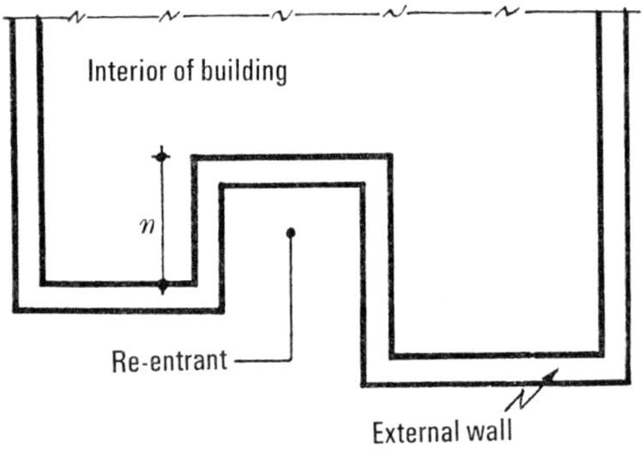

Fig. 8

Example

PROBLEM

Find the mean girth of external walls for the building shown in Fig. 9.

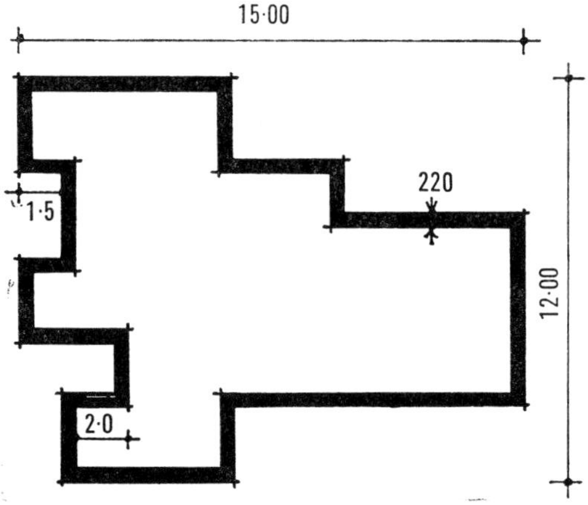

Fig. 9

SOLUTION

In the formula $M = (L + W) \pm 4T + 2n_1 + 2n_2 + \ldots$

$$L = 15{\cdot}00$$
$$W = 12{\cdot}00$$
$$T = 0{\cdot}22$$
$$n_1 = 1{\cdot}50$$
$$n_2 = 2{\cdot}00$$

Substituting we obtain:

$$M = 2(15{\cdot}00 + 12{\cdot}00) - 4(0{\cdot}22) + 2(1{\cdot}50) + 2(2{\cdot}00)$$
$$= 54{\cdot}00 - 0{\cdot}88 + 3{\cdot}00 + 4{\cdot}00 = \underline{60{\cdot}12 \quad \text{ans.}}$$

The actual waste calculation would be set out on dimension paper as follows:

	15·00
	12·00
	27·00
2/ =	54·00
−4/·22	·88
	53·12
Re-entrant 1. 2/1·50	3·00
” 2. 2/2·00	4·00
Mean girth of walls =	60·12

The advantages of using the above method will be very apparent with buildings of any complexity of shape, for it will be seen that in the above solution all breaks in the perimeter have been completely ignored, with the exception of the two re-entrants. The reason for this is that in any rectangular closed figure, the sum of the external angles is always greater than the sum of the internal angles by four. This is why we add (or subtract) four times the wall thickness in the formula, the extra girth represented by one angle being the equivalent of the wall thickness.

Internal walls

The length of foundation trenches to any internal walls extending below floor level must now be calculated, for timesing by their width and depth in the dimension column. If these are of the same width and depth as the external trenches, it would be valid to add their lengths to the mean girth of external walls on waste, and thus have only one dimension for squaring in the dimension column. Since the girth of external walls *only* may be required later, however, (when measuring facings, for example) reference back will be facilitated by keeping this entirely separate, as will the annotation of the dimensions. It is considered preferable therefore to bracket-in the internal trenches as separate dimensions from the external ones, rather than adding on waste to the mean external girth.

It is important to note, however, that internal wall foundations should not be measured separately from external ones, in a separate section of the taking-off, as this will make for needless repetition of descriptions and consequent loss of time, particularly valuable in the examination-room. The taker-off must remember therefore to bracket-in all internal wall foundations with each description as he goes along, annotating his dimensions with suitable side-notes.

No comprehensive formula exists for calculating the total length of internal trenches, and these must be measured piecemeal, each one being dealt with on its merits. It must be particularly noted, however, that the length of a cross-wall *is greater than the length of its foundation trench.* This may be expressed by the following formula:

$$l = c - 2p$$

where l = length of foundation trench,
c = cross-wall length,
and p = concrete projection on external walls.

The above becomes obvious when the concrete foundations are superimposed on the ground-floor plan as in Fig. 10, or when a foundation

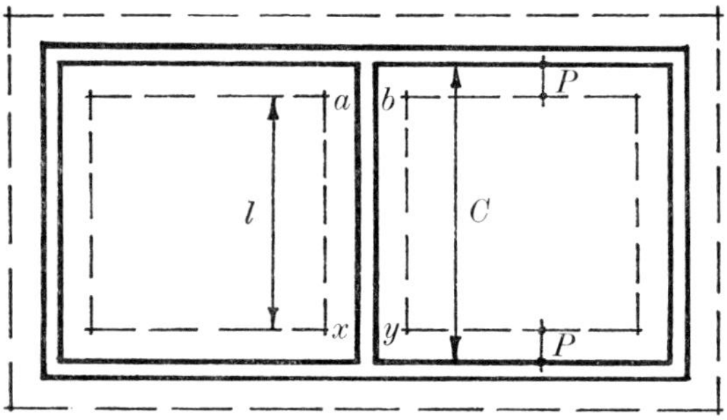

Fig. 10

plan is provided, but is easily overlooked by the student when measuring from floor plans showing brickwork only.

Walls of differing thickness

The thickness of internal walls may well differ from that of the external walls, and since the wall thickness tends to govern the width of concrete foundations, the trench width will then differ also, and possibly the depth as well. Again, internal walls may carry less load, and therefore require lighter foundations than the external walls. It is important therefore to keep an eye on the cross-sections when measuring trenches for these walls, and adjust the widths and depths accordingly.

In the case of external walls, any variation in thickness will again denote a probable variation in trench width, and will in any event upset the parameter T in the formula $2(L + W) \pm 4T$, and thus render the formula invalid. Where the variation is confined to one straight section of wall it is usually best ignored in the first instance, and the girthing formula applied as usual, with T representing the thickness of the remaining walls. The 'rogue' section should then be deducted from the girth *on waste*, and a final net girth arrived at, for transference to the dimension column, to be followed by dimensions of the rogue section, suitably annotated.

Making the adjustment on waste in this way will (a) avoid a separate deduction item in the dimensions and (b) leave the correct net girth of the bulk of the external walls available for future reference back when measuring the concrete, brickwork, etc. A word of warning is necessary here, however, for the net girth for brickwork will not necessarily be the same as that for the trenches, as the plan in Fig. 11 shows, and a suitable adjustment will usually have to be made.

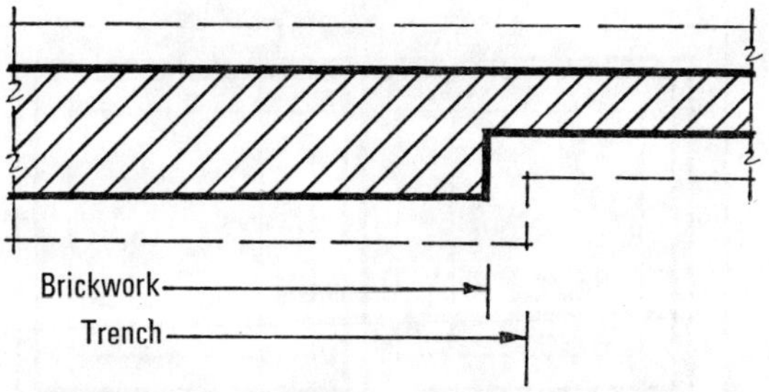

Fig. 11

Where the variation in external wall thickness is at all extensive, it will usually be convenient to discard the girthing formula and *measure each section of wall separately*, collecting the lengths of any similar sections on waste, and transferring the totals to the dimension column, for timesing by their relevant widths and depths. Special care must be taken to establish the points where the trench widths (as opposed to the wall thicknesses) change, and when in any doubt, the line of trench should be plotted on the plan in pencil by the taker-off, to assist in clarity of thought, and constitute a graphical record of what has been measured. Each section of trench should be ticked off on the drawing as it is measured, to avoid duplication, and the dimensions side-noted wherever possible.

Plan forms of irregular shape

If the building contains squint quoins (i.e., angles other than right angles) it will often be found most convenient to apply the girthing formula in the first instance, and then to adjust the girth *on waste* to allow for any sections of wall which are not perpendicular on plan. A simple example is shown in Fig. 12. In this instance no re-entrants exist, and the girthing formula can therefore be reduced to the basic form,

$$M = 2(L + W) \pm 4T.$$

The waste calculations would be as follows:

		10·50
		9·00
		19·50
	2/ =	39·00
	+ 4/·44	1·76
		40·76
less AB (scaled)	2·30	
,, *BC (scaled)*	2·30	4·60
		36·16
Add AC (scaled)		2·75
Mean girth		38·91

It is necessary to plot the centre-line *XACY* of the wall in pencil on the drawing, and also the notional centre-line *ABC*, to be deducted. *AB*, *BC*, and *AC* will then have to be scaled off the drawing, since it is unlikely that figured dimensions relating to the *centre-line* of the wall will be available.

A section of wall *curved on plan* would be dealt with in the same way, the notional centre-line being deducted on waste. The addition, however, would need to be added as a separate item in the dimension column, instead of on waste, since curved trenches require to be separately described under SM D10(b). The description

Ditto but curved on plan

would be sufficient here.

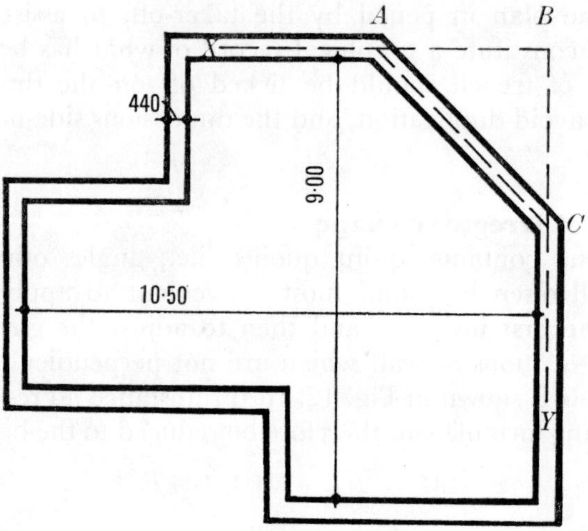

Fig. 12

If squint quoins or curved walls predominate, it will probably be found convenient not to use the girthing formula, but to calculate the length of each section of wall separately from the start. In general, however, the use of the girthing formula where possible will be found to simplify measurement, save time, and help to ensure accuracy.

Trench excavation for projecting piers

Projections from the walls forming attached piers or chimney-breasts will normally extend down to the same foundation level as the wall from which they project, and will thus require what is, in effect, a mere widening of the foundation trench in order to accommodate the extra width of foundations required. This is why the only mention of 'piers' in the excavation section of the SMM is in respect of *isolated piers* in SM D11(a), for piers and other projections attached to walls are simply added in, for excavation, with the trenches: it is this type of pier which is most likely to occur in simple unframed buildings.

For our present purpose, then, we may define the word 'pier' to mean a projection from the wall face (on plan) to form a pilaster, chimney-breast, etc., requiring foundations. These are best dealt with as follows:

(i) First measure the external and internal trenches, ignoring all piers in the first instance, measuring the walls right through.
(ii) Bracket-in the dimensions representing the extra width of trench required for each pier, on to the end of the trench dimensions, annotating each dimension.

The first two dimensions for each pier will, of course, be the length and width; the third one (depth) being in all probability the same as the depth of the main trench. The problem of arriving at the length and width will be a repetition of that met with in measuring surface strip, when dealing with externally projecting piers.

Once again it is the extra width of *concrete foundations* which is required, not that of the brickwork shown on the drawing, and it might be as well, therefore, to repeat the rules previously cited in this connexion, bearing in mind that they now apply to both external and internal projections. For trench excavation it will be convenient to restate these rules in the following form:

$$TW = BW$$
$$TL = BL + 2p,$$

where TW = extra trench width,
BW = extra brickwork width (pier projection),
TL = extra trench length,
BL = extra brickwork length (pier length),
and p = original concrete projection.

The above equations will be readily understood if reference is made to Fig. 13, and may be worth bearing in mind until the student becomes accustomed to measuring invisible foundations from a ground-floor plan.

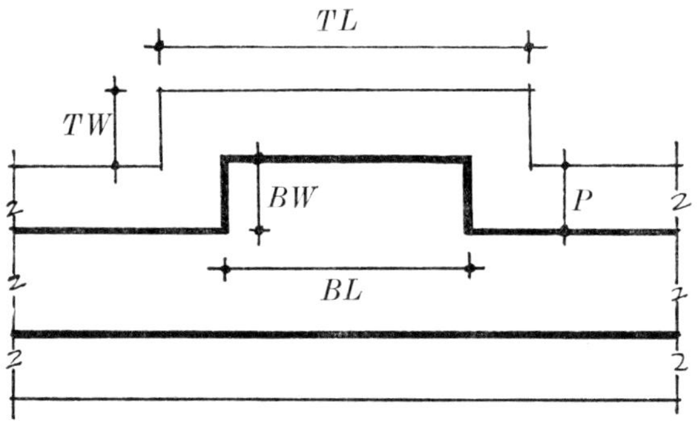

Fig. 13

The additional area required for corner piers, for timesing by the depth, again follows the same pattern as for surface strip (Fig. 4 refers). Those on the outside of the building may be dealt with by simply copying the relevant surface strip dimensions straight from the dimension column, and this applies to non-corner external piers as well. Piers on the inside of the building will need to be worked afresh on waste, and those on external *angles* are best dealt with by the previously mentioned equations:

$$x = a + 2p$$
$$y = b + 2p - d,$$

treating the two returns as separate piers.

Most internal corner piers, however (and some external ones too), will be on internal *angles*, following the configuration shown in Fig. 14. It is

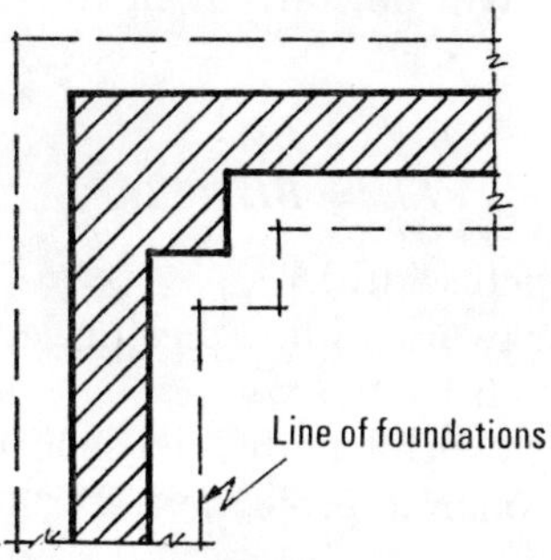

Fig. 14

important to note that, since the concrete projection from each pier face is equal to the concrete projection from the main wall face, the additional area required for trench excavation is the same as the area of the projecting brick pier. The figured (or scaled) dimensions of the brickwork can therefore be set down direct in the dimension column as additional trench excavation in respect of all corner piers occurring in internal angles of walls.

Depth of trenches

This must be calculated on waste, with reference to the cross-sections showing foundation depths. The required depth will be the difference between original ground level and bottom of concrete foundations *less* the depth of surface excavation already measured. Should this be greater than 1·5 metres, the trench excavation must be split into separate items, one for each successive 1·5 metres in depth, in accordance with SM D6(f) as follows:

Excavate foundation trench starting
at formation level and not exceeding
1·50 *metres deep.*

Ditto over 1·50 *but not exceeding* 3 *metres deep.*

Ditto over 3 *but not exceeding* 4·50 *metres deep.*

although trench depths of over 1·5 metres are the exception rather than the rule in the case of small dwelling houses on flat sites.

Description of trenches

It will be seen from SM D10(a) that the 'starting level' must be given in the description. Unless no surface excavation has been measured, this starting level will not be ground level, but formation level, as previously explained. Where reduced level excavation has been measured, however, it is the usual practice to describe the trenches as 'starting at underside of reduced level excavation',[1] especially when the reduced level excavation is more than 300 mm or so deep. This indicates to the estimator that the work has been measured on the assumption that trench excavation will not commence until reduced level excavation has been completed, so that he can price accordingly. A good rule of thumb is to use the phrase 'starting at formation level' unless *cube* reduced level excavation has been measured, in which case say: 'starting at underside of reduced level excavation'.

Many surveyors still use the word 'commencing' instead of 'starting', but when in doubt the student will probably be wisest to adopt the terminology of the latest revised edition of the SMM in this respect.

It should be noted that trenches are measured the width of the foundations (SM D10(a)), and that no allowance is made for any extra space required to accommodate planking and strutting (SM D6(b)). Also, getting out the excavated material is deemed to be included (SM D6(c)). These two factors will be automatically allowed for by the estimator in his price, and need not be mentioned in the description.

RETURN FILL AND RAM

Although the 'getting out' (of the trench) of excavated material is deemed to be included, its subsequent disposal must be measured. Some of this material will require to be tipped back into the trench again, to fill the space between the brickwork and the trench side above the level of the concrete strip foundation. It is common practice to measure the *whole* of the trench disposal as 'return fill and ram' in the first instance.

[1] Some offices make a practice of stating the actual depth below ground level.

The volume of earth *not* required to be returned is measured later, with the concrete and brickwork which will displace it. This will be found a very convenient method (except in those cases where no return fill and ram is required at all), and allows the return fill and ram to be anded-on to the trench excavation, since both are measured cube (see SM D16(c)), as follows:

cube	*Excavate foundation trench starting at formation level and not exceeding* 1·50 *metres deep*
	&
	Return fill and ram surplus excavated material around foundations.

LEVEL AND RAM

Treating the bottom of excavation is a measurable item under SM D17(a), and it may be assumed that the treatment required at bottom of trench excavation is that of levelling and ramming to receive the foundations. This must therefore be measured next, as a superficial item, and the area required will be that represented by the first two figures of each dimension for trench excavation (the third being the depth in each case).

The student will now appreciate the convenience of the length-width-depth rule (SM A3(a)), for providing this has been followed it will not be necessary to refer to the drawing, but simply to copy down the first two components of each trench dimension direct into the dimension column in the form of supers, and the result will be the complete string of dimensions ready for bracketing against a description worded 'level and ram bottom of excavation'.

Description of level and ram

Treating the bottom of excavation will be required not only to foundation trenches, but to other excavation as well, notably to surface excavation down to underside of the ground floor of the building. Treatment to this last will probably also consist of levelling and ramming, as will treatment required to any other the bottom of any other excavation calling for measurement in connexion with the project.

Now if this treatment is the same in each case, a description which is common to any location will tend to minimize the number of bill items, and should therefore be adopted. Use the form 'level and ram bottom of excavation' therefore, rather than 'level and ram bottoms of *trenches*' or 'level and ram bottom of excavation *to receive concrete*'. Unless, of course, you are certain that the additional wording will not result in unnecessary bill items caused by taking-off descriptions of identical items which cannot be merged at working-up stage because the wording differs.

Extent of level and ram

Since level and ram is required to be measured to bottom of surface excavation, it might be thought that this could be anded-on to the surface strip item. Measurement in this way will not always give a strictly accurate result, however, because the area over the trenches not requiring treatment will not necessarily compensate for levelling and ramming bottoms of trenches. A typical example is shown in the cross-section in Fig. 15, where the trench-bottom *AB* can be seen to overlap the level and

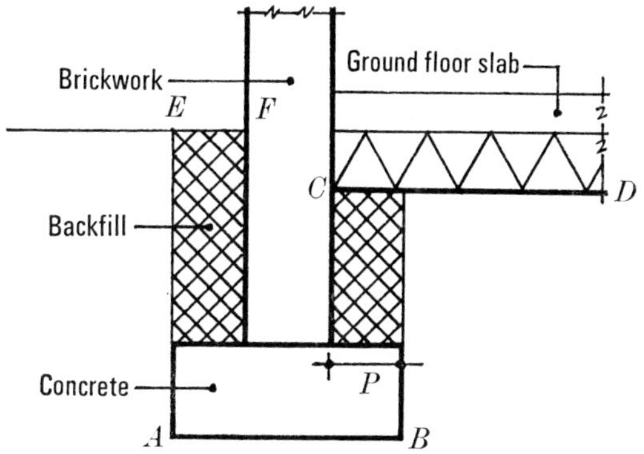

Fig. 15

ram at formation level *CD* by the amount *p*. It may be argued that since the shaded portion is return fill and ram, no ramming need be measured to its surface, since this is included. The question of levelling still remains, however, and SM D17(a) specifically mentions this; 'Treating . . . the surface of filling . . . (e.g., levelling . . .)'. The question arises: should the surface of backfilling under the floor be measured for levelling only, and if so, does this apply to the external backfilling at surface *EF* also?

The top surface of external backfilling must be left reasonably even, but the operation of careful levelling sufficient to justify a measured item is not generally considered necessary. The top surface of any backfilling under floors, on the other hand, must be prepared to receive the hardcore bed, or concrete, as the case may be, and this may necessitate additional ramming as well as levelling. It is therefore desirable to measure level and ram to the overlap *p* in Fig. 15, but this is not usually measured to the surface *EF*.

PLANKING AND STRUTTING TO TRENCHES

Planking and strutting to sides of trenches is measured superficially under SM D21(a)(iv), and this is achieved by referring back once again to the trench excavation dimensions, and converting these to supers representing the area of the planking and strutting. In this case the first and last component of each dimension (length and depth respectively) is written direct into the dimension column, *twiced for both sides of the trench*, and the result is a string of dimensions which can now be bracketed against a description for planking and strutting.[1]

super — *Planking and strutting to sides of foundation trenches starting at formation level and not exceeding* 1·50 *metres deep.*

Any curved trenches are dealt with in the same way and separately described as 'ditto but curved on plan' in accordance with SM D20(d). Certain adjustments may be necessary, however, and the following points must be borne in mind:

Pier adjustments for trench P & S

The importance of having annoted the piers at excavation stage will now be apparent, for the dimensions of these are best ignored in the above process, and dealt with separately, to avoid confusion. The additional planking and strutting for piers should, however, be bracketed-in to the same item as the trench planking and strutting, since piers are regarded as a widening of the trench. The additional length of planking and strutting required for each attached pier will be found to be equal to the following:

(i) SIDE PIERS. Twice the amount of brickwork projection.

(ii) EXTERNAL ANGLE CORNER PIERS. Twice the *sum* of both brickwork projections.

(iii) INTERNAL ANGLE CORNER PIERS. Nil.

The student should satisfy himself as to the validity of the above three statements by sketching foundation plans for the three types of piers concerned, and marking on his sketches the amount of planking and strutting

[1] If P & S to surface strip is to be included, note that each side of trench will then have a different starting level and girth—see pages 103–6.

accounted for in trench measurement. The remaining lengths can then be checked against the above.

Passings

Where internal trenches exist, a deduction will have to be made where an external trench passes the end of a cross-wall. Thus in Fig. 10 the dotted line representing the planking and strutting on the inside of the external trench is broken at *ab* and *xy*, whilst the external planking and strutting is continuous. A separate 'deduct' item will be necessary here, to account for each end of any internal walls, the dimensions required being the internal trench width by its depth, twiced for both ends as necessary. The dimensions should be annotated 'passings', and care should be taken to ensure the correct depth for this adjustment which is not necessarily the full depth of the *external* trenches.

Depth of planking and strutting

It will be recalled that trench *excavation* is measured in *successive* stages of 1·5 metres in depth (SM D6(f)), each stage requiring a separate item. Planking and strutting to trenches, however, merely requires the depth to be stated in multiples of 1·5 metres, as SM D21(a), and does not require measuring in separate successive stages.

The reason for this difference in treatment is that the cost of hand excavation is directly proportional to the height to which the earth has to be lifted, 1·5 metres being the maximum vertical distance a labourer is expected to throw with a shovel on to the temporary staging above. Each successive stage downwards therefore is more expensive by reason of additional handling, and must be kept separate. The cost of planking and strutting, on the other hand, may depend on the total depth at any one point, since deeper timbering may affect the timbering nearer the surface. Thus the total depth must be stated and the planking and strutting measured the full depth in one item.

PLANKING AND STRUTTING TO REDUCED LEVEL EXCAVATION

Where shallow reduced level excavation has been measured to flat sites of the type under consideration, the area required for planking and strutting, measurable under SM D21(a)(i), will be the perimeter of the building by the depth of the excavation to reduce levels. The required perimeter is the girth of the external strip foundations measured at extreme edge of concrete projection, as shown by line *A* in Fig. 16.

Now this required girth bears a relation to the original mean girth of

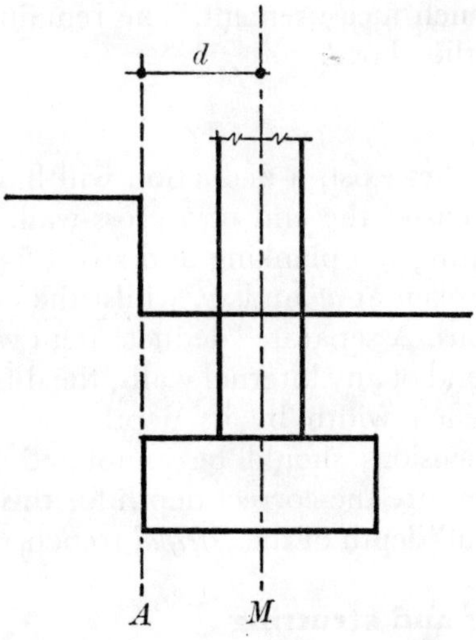

Fig. 16

external trenches on the centre-line *M*, and it is for this reason that the item is best measured at this stage, rather than with reduced level excavation, before the original mean girth has been calculated. By dealing with it following trench planking and strutting in this way, the mean girth for external trenches may be used as a basis for measurement.

In the case of a building of very simple plan-shape, it might be thought convenient to take these measurements from the reduced level excavation dimensions, but for complex plan-shapes the following method will be found invaluable, and its simplicity is such that the student would do well to make a habit of applying it to both simple and complex shapes, thus achieving uniformity of procedure, by means of the following formula:

$$A = M \pm 8d$$

where A = adjusted girth,
M = original mean girth,
and d = cross-sectional distance from M to A (Fig. 16 refers).

Example

PROBLEM

The mean girth of external walls for the building shown in Fig. 9 has been calculated as 60·12. Assuming a trench width of 1·00, find the length (girth) required for planking and strutting to side of reduced level excavation.

SOLUTION

In the formula $A = M \pm 8d$

$$
\begin{aligned}
M &= 60{\cdot}12 \\
d &= \text{distance from } M \text{ to } A \\
&= \text{distance from centre-line to edge of concrete} \\
&= \tfrac{1}{2} \times 1{\cdot}00 \\
&= 0{\cdot}50.
\end{aligned}
$$

Substituting we obtain:

$$
\begin{aligned}
A &= 60{\cdot}12 + 8 \times 0{\cdot}50 \\
&= 60{\cdot}12 + 4{\cdot}00 \\
&= \underline{64{\cdot}12} \quad \textit{ans.}
\end{aligned}
$$

The actual waste calculation would be set out on dimension paper as follows:

Mean girth	60·12
+ 8/½/1·00	4·00
	64·12

From this it can be seen that only a very simple waste calculation is required in order to shift the line of the original mean girth in this way, for any re-entrants included in the mean girth are unaffected by the shift of centre-line, and the girth adjusting formula will remain true for a building of any given rectangular closed shape.

It should be noted that $8d$ is plus or minus according to whether the girth is to be shifted outwards (+) or inwards (−) towards the interior of the building. In the case of planking and strutting to reduced level excavation the shift will normally be outwards, and, therefore +.

Pier adjustments for reduced level P & S

As in the case of trench planking and strutting, the reduced level planking and strutting will require adjustment for piers, where these exist, but only for those piers projecting on the outside face of the walls, as for surface strip. The adjustment will take the form of bracketing-in the additional lengths of planking and strutting arrived at by means of the rules laid down under pier adjustments for trench planking and strutting on page 100. Although it may be possible to pick these out from previous dimensions, a fresh look at the drawing is probably desirable at this stage, to ensure that only external piers are included in this item.

Planking and strutting to surface strip

If this is measurable, it is dealt with at this stage, and the girth arrived at as described above for reduced level planking and strutting. First, however, the taker-off must address himself to the question, is this item measurable and if so, under what circumstances?

The difficulty arises from SM D20(c), which states:

> *Planking and strutting shall not be measured to any excavation which does not exceed* 300 *millimetres in depth* . . .

Now we have already seen that surface strip (as opposed to reduced level excavation) will not be expected to exceed 300 mm in depth, and as such, therefore, does not qualify for any planking and strutting measurement under the above clause. It might be held, however, that the *total* depth of excavation where the planking and strutting occurs includes the trench depth as well, and is therefore exceeding 300 mm in depth. Indeed, SM D21(a) states that the depth of any trench immediately below the side of a basement shall be included in the total depth of that side. Might not this principle be held to apply to any trench immediately below the side of surface strip excavation?

Again, while the above extract from the SMM might be held to presume that planking and strutting is not normally required to such shallow excavation, is it equally reasonable to suppose that a contractor would stop back his timbering 300 mm below ground level when excavating, say, a deep basement in unfirm ground?

On the other hand, no surface strip classification is mentioned in SM D21(a), and this tends towards the view that none is intended to be measured. However, the classification of 'excavation to reduce levels over site' in SM D21(a)(i) presumably refers to excavation under Clause D7, which is headed 'surface excavation', which might be held for this purpose to include surface strip, in the absence of any other SMM reference to this item. On this reasoning, planking and strutting to edge of surface strip should be included in with that to sides of reduced level excavation, where such exists. Alternatively, with that to sides of trenches (classification D21(a)(iv)) on the 'side of basement' analogy, above.

In practice, when timbering a trench, one side of which is higher than the other by virtue of surface strip excavation (or for any other reason), would the operative see any reason for treating the two sides differently? After all, SM D20(b)(ii) states that planking and strutting shall be measured to both sides of trenches which are over 300 mm deep. To the full depth, presumably. And supposing the contractor elects to slope back the trench side, in lieu of planking and strutting, which is measurable whether or not any is in fact required (SM D20(b)). The cost of this would certainly relate to the *total* depth, and would need to be equated with a total depth of notional planking and strutting.

If it is held that no planking and strutting is measurable to surface strip by reason of SM D20(c), the same argument would hold good in respect of any reduced level excavation not exceeding 300 mm deep which occurred *below* the surface strip. If a 300-mm deep trench occurred below *this* level, followed by another 300 mm of (say) stancheon-base excavation, a total depth of 1200 mm would be reached, with no planking and strutting measured!

An interesting point arises in connexion with the starting level, which is required to be stated in all cases under SM D21(a). The starting level of any planking and strutting to edge of surface strip will be ground level; that to reduced level excavation strip level; and that to trenches formation level. If surface strip planking and strutting is included as (say) trench planking and strutting therefore, the outside face of external trenches would need to be kept separate from the inside face, since the starting levels would differ. In such a case, if the planking and strutting to the outside of the trench was described as '. . . to sides of foundation trenches starting at ground level', the estimator may wonder why no trenches starting at ground level have been measured to go with this item.

The above arguments for and against this relatively unimportant item have been set down in some detail to illustrate to the student the variety of factors which may be involved in even the most elementary quantity surveying problem. Problems of this kind can only be solved by thought processes based on a logical analysis of the SMM, coupled with accurate visualization of the site operations involved. If study of the relevant work-section clauses of the SMM fails to provide a solution, Section A (General Rules) may help to solve the problem. In the case under discussion, for example, SM A1 might perhaps be invoked. The first sentence of this clause states:

> *Bills of quantities shall fully describe and accurately represent the works to be executed.*

In the final analysis, therefore, and failing any other express SMM guidance, if the contractor executes the planking and strutting in question, he has at least a *prima facie* case for having it measured.

Meanwhile, at taking-off stage, unless the quantity surveyor suspects the ground to be exceptionally unfirm, he would be making a reasonable assumption if he concluded that this planking and strutting to edge of shallow surface strip above trench level would not in fact be executed, and he would therefore be justified under the SMM in not measuring it.

Under examination conditions the student, after carefully reading the specification notes and drawing, should decide whether any evidence exists to suggest that this item is measurable. If not, a side-note should be made in the dimensions as follows:

P & S to surface strip assumed not measurable by reason of SM D20(c)

and one can reasonably assume that the examiner will be satisfied.

However, many offices normally measure this item, including it with the trench P & S, and giving each side a separate starting level (and girth).

Should there be no other excavation immediately below the side of that which is not exceeding 300 mm deep, SM D20(c) will always apply of course, and planking and strutting must not then be measured in any case.

Chapter Eleven

Foundation Concrete and Footings

CONCRETE FOUNDATIONS

These are measured with reference to the trench excavation dimensions, which will be identical except for the depth. The cross-sections must be examined to check on the thickness of concrete, care being taken to check for the possibility of thinner sections in the case of internal walls, which may be carrying a lighter load. It should be noted that SM F3(a) requires the thickness to be stated in three stages (0–150, 150–300, and over 300 mm thick), and where the thicknesses vary, the sections must be kept separate accordingly. With this proviso, the trench dimensions may be converted to concrete dimensions, and entered direct into the dimension column, including piers and other projections.

The volume of concrete will displace an equal volume of earth not required to be backfilled into the trench, and since the whole of trench excavation disposal has been measured initially as returned filled and rammed (see page 98), an adjustment will now be necessary. This takes the form of anding-on a description of '*deduct* return fill and ram', and then anding-on another description of '*add* remove surplus' to the concrete description, thus:

cube	*Concrete* (1:2:4) *over* 150 *and not exceeding* 300 *mm thick in foundations in trenches.*
	&
	Deduct R.F.R. *a.b.*
	&
	Add *Remove surplus a.b.*

Description of foundation concrete

The SMM divides in-situ concrete into two main kinds, viz.: (i) plain, and (ii) reinforced. These are required by Clause F2 to be given under separate headings in the bill, and this is achieved by the taker-off using a different description for each of the two. The usual practice is to stipulate if the concrete is reinforced, all concrete not so described being assumed plain concrete at working-up stage.

Although any type of strip foundations may be reinforced with rods or fabric, plain concrete will be assumed in this section, and reinforcement dealt with later (see Chapter 15).

SM F1(a) stipulates four groups of particulars which must be given in respect of concrete, but the first three refer to details of materials and tests which may be assumed covered by preambles, and only the fourth requires incorporating in the item description. This refers to 'composition and mix'.

Composition of concrete

Concrete is composed of cement and aggregate, mixed together with water. The quality of these ingredients will be covered in the preambles, leaving only the *type* of cement and the *size* of aggregate to be stated in the measured item description, as follows.

TYPE OF CEMENT. The two main types in use are (i) Portland cement and (ii) high alumina cement. Portland cement is the most usual, and the term *concrete*, unless qualified, will be taken as meaning Portland cement concrete, but where any doubt is thought to exist, the letters P.C.C. are often used by the taker-off, in lieu of the word 'concrete'.

Concrete made with high alumina cement should be described as 'high alumina cement concrete'. Concrete containing other types of cement should be similarly described, e.g., 'rapid hardening cement concrete'.

SIZE OF AGGREGATE. Aggregate used in plain foundation concrete may be specified as being of a size to pass through a 38 mm mesh sieve (or through a 38 mm ring). That for reinforced concrete through a 19 mm ring. These sizes are fairly standardized, and if not more than one size applies to any given item description, the sizes may be considered sufficiently dealt with in preambles, or covered by standard bill headings, inserted at working-up stage. When in doubt, however, the taker-off should insert the aggregate size after the mix, thus: 'Concrete (1:2:4) (19 mm ring) in foundations a.b.'

Concrete mix

Concrete may be regarded as a mixture of cement and aggregate, but the aggregate may take two forms (i) a mixture of sand and coarse aggregate, or, alternatively (ii) 'all-in' aggregate, which contains its own sand.

(i) CEMENT, SAND AND COARSE AGGREGATE. The mix is denoted by the proportions in brackets after the word 'concrete', thus: 'Concrete (1:2:4) in foundations a.b.'
(ii) CEMENT AND ALL-IN AGGREGATE. Two numbers in the brackets, instead of three, will be deemed to imply an all-in aggregate, as 'Concrete (1:12) 150 mm thick, spread and levelled over site.'

The taker-off should make a habit of always following the word 'concrete' by brackets containing the mix in this way.

Surface finish to concrete

Treating the surface of unset concrete is dealt with under SM F14(a), and if the question of 'grading' is ignored (since foundation concrete requires a level surface) the possible types of treatment normally met with are:

(i) SPADE FACE. The surface is levelled with the spade, possibly with the help of wood rules between pegs.

(ii) FLOATED FINISH. A wood float is used to achieve an accurate level surface, the finished texture of which is rough.

(iii) TROWELLED FINISH. A steel trowel is used to work the surface smooth. Trowel marks are not easy to conceal, and surface cracks may show, but a trowelled finish is necessary if a really smooth surface is required.

(iv) TAMPED FINISH. A tamping-board is used, which leaves a rough patterned surface. Often used for roads and pavings, or where a rough surface to form a key for floor coverings is desired.

Unless otherwise specified, the taker-off may assume that a spade face is all that is necessary as a finish to concrete foundations. Spade face is not mentioned in SM F14(a) however. Neither is floated finish. Are we to assume that these two treatments are not measurable?

'Treating the surface' of unset concrete may reasonably be held to refer to treatments which require a separate operation, and floating, trowelling and tamping cannot normally be executed until the surface is fairly free of water, after the concrete has begun to dry out. Spade face, however, is executed during deposition of the concrete and cannot really be regarded as a separate operation. Floating, although not specifically mentioned in SM F14(a), may thus be held to come under 'e.g.' in the first sentence of that clause. Spade face, however, may be considered included in the deposition of the concrete, and therefore not measurable.

Since spade face only will be required to concrete strip foundations, no surface treatment of concrete is measurable, and this item may therefore be ignored.

Curved foundations

Excavating curved trenches, as we have seen, is measured separately from straight trench excavation, and this applies to the planking and strutting as well. It does *not* apply to the concrete, however, for it may be assumed that no difference exists between the cost of tipping concrete into a straight trench, and that of tipping it into a curved one. Although SM F1(e) provides for the separate measurement of curved *labours* on concrete, therefore (unlikely to be required in foundations), no provision is made for curved in-situ concrete, which is bracketed-in to the same description as the straight concrete.

Formwork

This may be defined as any temporary work necessary to support the surface of concrete during setting, and is measurable under SM F20, etc. The requirement is that the actual area of *concrete surface needing support* shall be measured, and the student should visualize the concrete as being in liquid form, and measured as formwork any surfaces necessary to contain the liquid concrete.

In the case of foundations the concrete will, of course, be contained by the trenches, and no formwork will therefore be necessary, except in certain circumstances such as the following.

110 Consider, for example, Fig. 17 showing the junction between an external wall and a cross-wall. Should the foundation depths differ, then formwork may be necessary, as in the present case, where the main wall is deeper than the cross-wall. The bottom surface (soffit) of cross-wall

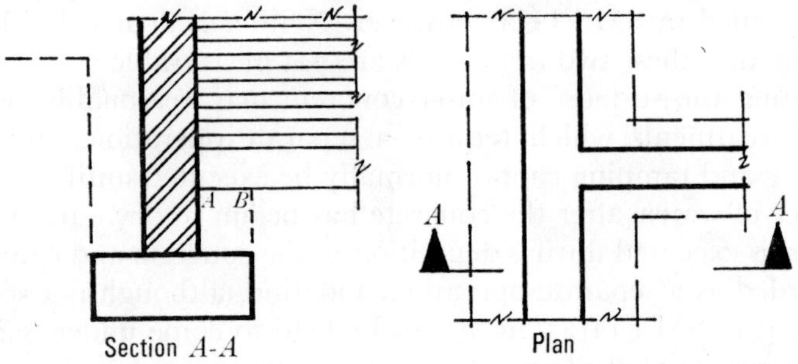

Fig. 17

concrete is unsupported by its trench over the distance *A–B*, and the sides are similarly unsupported at this point. Formwork must therefore be measured to the sides and soffit respectively, and the areas concerned entered against the following descriptions.

(i) Sawn formwork to vertical sides of foundations.
(ii) Ditto to horizontal soffit of do.

When measuring foundation concrete an eye must be kept open for any such unsupported surfaces, which are sometimes not easy to visualize, and formwork must then be measured following on the concrete dimensions.

BRICK FOOTINGS

The term 'footings' appears in SM G4(i) and refers to a widening of the base of a wall by means of brick steppings as shown in section in Fig. 18(a). This method of spreading the load is not often used nowadays, but may still be met with occasionally, and is dealt with before commencing the actual wall measurement.

Footings are measured *beyond the wall face* as shown in Fig. 18(b), and described as 'brickwork in projections' in accordance with the SMM, which requires them to be measured super and reduced to one-brick in thickness (Clause G4). The reduction to one-brick thickness is carried out at working-up stage, the *actual* thickness being given by the taker-off. Since this varies with each course, the *average* thickness must be

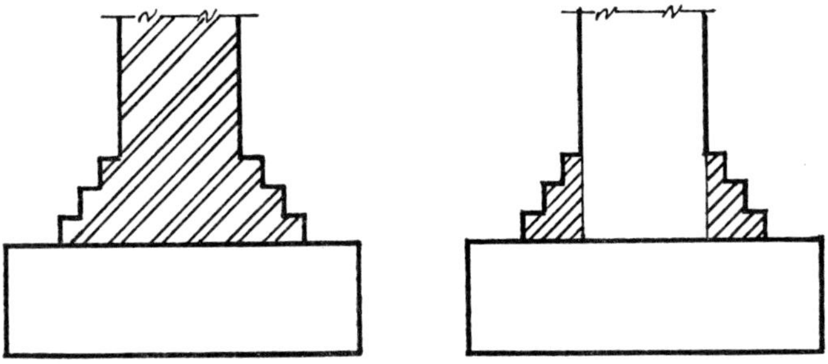

(a) (b)

Fig. 18

calculated for this purpose, bearing in mind that the standard amount of projection is normally quarter-brick in thickness for every course in height.

Projection of footings

The average width of footings is calculated on waste, and the best procedure is to start at the top course and write down the thickness of each successive course in terms of the one-brick thickness to which final reduction is required, before dividing by the number of courses to arrive at an average. If *B* represents one-brick thickness therefore, the waste calculation for Fig. 18 would be as follows:

$$\begin{array}{r} \tfrac{1}{2}B \\ 1\ B \\ 1\tfrac{1}{2}B \\ \hline 3)3 \quad (1B \textit{ average.} \end{array}$$

whilst that for five courses of footings would be:

$$\begin{array}{r} \tfrac{1}{2}B \\ 1\ B \\ 1\tfrac{1}{2}B \\ 2\ B \\ 2\tfrac{1}{2}B \\ \hline 5)7\tfrac{1}{2} \quad (1\tfrac{1}{2}B \textit{ average.} \end{array}$$

The result is always an average between the first and last courses, and in the case of an even number of courses, always equivalent to the width of the middle course; but since the above process must be gone through in order to establish the width of the middle and lower courses in the first place, little is gained by these short cuts. It is best to simply write down

112 the first course as $\frac{1}{2}B$ and increase by $\frac{1}{2}B$ for each successive course, taking the average as above noting especially that the result is the total average thickness for *both sides of the wall.*

Length of footings

Having established the average width of the footings, for insertion after the description, the height and length will be required for the superficial dimensions, representing the area on elevation. The height will be 75 mm by the number of courses, but the length may not be so obvious.

EXTERNAL WALLS. The mean girth, used for concrete measurement, will be correct for footings also, since these are centrally placed over the concrete.

INTERNAL WALLS. The figured dimensions of brickwork will require adjustment to allow for projecting footings already measured to external walls. Thus the correct length for footings to the cross-wall shown in Fig. 19 will be an average between the top and bottom courses shown in section *A–A*.

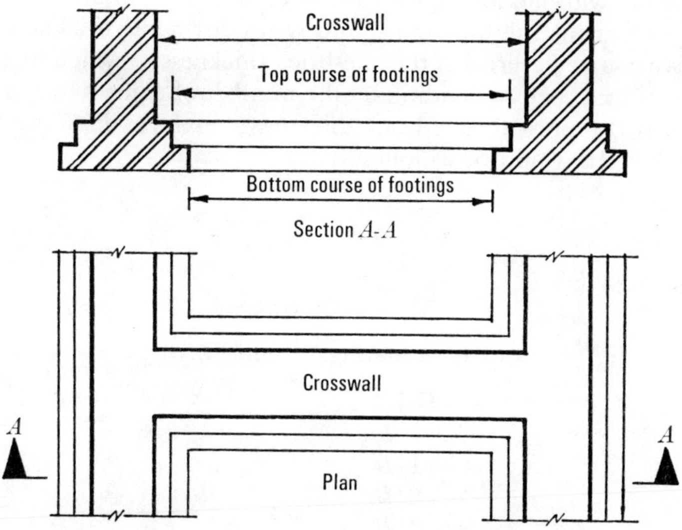

Fig. 19

Strictly speaking, an average length is not mathematically correct, since the width has also been averaged, and each course should be measured separately if precise results are required. The error is so small, however, that it may be considered negligible for this purpose, and the average length regarded as satisfactory.

SIDE PIERS. Footings may be assumed to return round the sides and front of all projections, and these must be added in. It will be seen from the cross-section in Fig. 20 that the footings to the *front* of the projection

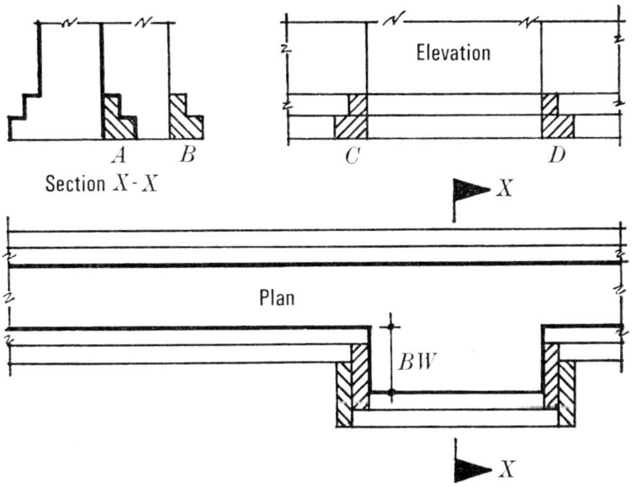

Fig. 20

at *B* are equal to the footings already measured through on the main wall at *A*, and which do not actually exist where the pier occurs. It therefore follows that *all footings to the front of piers must be ignored.*

Footings to the side of these piers are extra, however, as shown at *C* and *D* on the elevation in Fig. 20, and a glance at the plan will show that the length of each course (hatchured) is the same as the length *BW*, which is the figured dimension for brickwork width. It therefore follows that the extra length of footings required in respect of side piers is always *equal to the brickwork projection of the pier.*

Note especially that since each side of the pier is equivalent in footings to one side of the wall, the footings on both sides of the pier are equal to those on both sides of the wall. Because, therefore, the average width previously calculated was for *both* sides of the wall in the first place, the extra length of footings required is equal to *once* the pier projection, and must not be twiced.

CORNER PIERS. As in the case of planking and strutting, the extra footings required for *internal* angle piers is: nil. Piers to *external* angles are also best treated on the planking and strutting analogy, since the extra length on face has already been measured through on the main wall, leaving only the returns as a measurable extra. Twice the sum of both brickwork

114 projections, as given for planking and strutting, however, would give each projection separately, whereas the projection on *both* sides of the pier will accord with the average width previously calculated. The extra length of footings required in respect of external corner piers, therefore, is *equal to the sum of both brickwork projections of the pier.*

Example

PROBLEM

Take off the quantities for the footings to the building shown in Fig. 21.

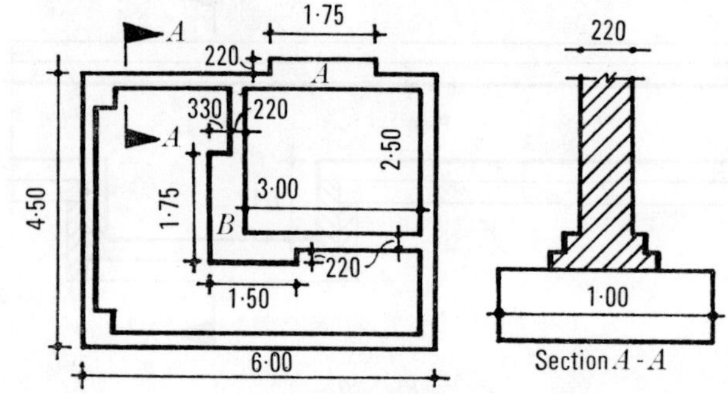

Fig. 21

SOLUTION

	4·50	½B
	6·00	1B
	10·50	2)1½B(¾B
2/ =	21·00	
− 4/·22	·88	
Mean girth =	20·12	
	3·00	
	2·50	
	5·50	
corner	·22	
int. wall	5·72	
2/·055	·11	
top cos.	5·83	5·83
	·11	
bot cos.	5·72	5·72
		2)11·55
average		5·775

20·12	*Reduced brickwork in commons in cement mortar* (1:3) *in projections.*	
0·15		
		(*footings*
	× $\frac{3}{4}$*B* =	
5·78		[*internal walls*
0·15		
0·22		
0·15		(*pier A*
	·22	
	·33	
	·55	
0·55		
0·15		(*pier B*

Curved footings

Any footings to curved walls must be kept separate under SM G1(d), and the mean radius stated. This SMM clause also calls for rough cutting to be included in the case of curved brickwork, so that a description similar to the following will be required for any curved footings:

> *Ditto but curved on plan to a mean radius of* 3 *metres, including rough cutting.*

Description of footings

The framing of brickwork descriptions is dealt with on page 117, but in the case of footings it should be noted that the word 'footings' does not actually appear in the description for this item. This is because SM G4(i) requires footings to be grouped together with other types of projection, and all projection items must therefore be merged at working-up stage into one item, viz., 'brickwork in projections'. The use of the word 'footings' in the description would cause this item to be separated out from other items dealing with projections, and the word 'footings' must therefore be relegated to a side-note, not forming part of the description itself.

The term 'reduced brickwork' will be taken to mean brickwork reduced to one-brick thickness, as required by SM G4, and the average thickness of the footings must be given *after* the description, as shown, to allow this reduction to be achieved at working-up stage.

116 Rough cutting to footings

Clause G10(a) stipulates that 'rough cutting' shall be measured to 'projections which are not a multiple of half-a-brick in thickness . . .'. If the usual quarter-brick projection for each course is assumed, then *every alternate course of footings projects a distance from the wall-face which is not a multiple of half-a-brick.*

At first glance it would certainly seem that every alternate course of footings requires measurement of rough cutting in accordance with SM G10(a). If the actual site operation is visualized, however, it will be apparent that no rough cutting exists, for each course of footings goes right through the wall; the whole purpose of a quarter-brick projection on each side being to avoid the necessity for rough cutting (assuming the wall itself to be a multiple of half-a-brick). SM G10(a) does not specifically mention footings, and it is reasonable to assume that the SMM does not intend non-existent rough cutting to be measured. SM A1 may therefore be invoked here, and the quantities made to represent the work executed. No cutting therefore is normally measurable to footings, except in the following instance.

Squint angle cutting to footings

Where squint quoins occur, rough cutting is measurable under SM G10(a) . . . 'rough cutting at squint or birdsmouth angles'. The convention is to take the cross-sectional area of footings for this purpose, once for every squint quoin. Thus, the item required for one squint quoin, where footings average $\frac{3}{4}B$ in thickness would be:

		$\frac{3}{4}B$ = 165 *mm*
		height = *(say)* 150 *mm*
0·17	*Rough cutting.*	
0·15		
		(squint.

Note that the word 'squint' does not form part of the description itself, since this cutting must be grouped in the bill with other types, under the omnibus description of 'rough cutting' in accordance with SM G10(a).

Foundation Walls and Damp-proof Courses

BRICK WALLS

These must now be measured, from top of concrete up to damp-proof course level, together with any piers or other projections, excluding projections for footings (if any) which have already been measured. Brickwork above DPC level must not be included here, but with the superstructure, measured later, as previously noted.

External walls are measured first, using the original mean girth as calculated for the trenches. Any internal walls going down to foundation level are then bracketed-in to the description, or described separately if of different thickness from the external walls, care being taken to adjust their depths, if these vary.

Like trenches, all walls are measured through, ignoring any projecting piers, as these are dealt with under a separate description as projections, measured from the wall-face.

Brickwork descriptions

Unless over two-brick thick (unusual in present-day construction) the thickness of brick walls must be stated in the description, in accordance with SM G3(a). The thickness in millimetres may be stated, if preferred, but the more usual practice is to state the thickness in terms of the length of a brick, thus:

Half-brick wall . . .
One-brick wall . . .
One-and-a-half brick wall . . .

Particulars to be given are stipulated in SM G2(b), and relate to (i) kind, type and size of bricks, (ii) type of bond and (iii) composition and mix of mortar.

(i) KIND OF BRICKS. These are of two main kinds, (a) common bricks and (b) facing bricks. Since facing bricks are only used above ground level, these are more conveniently dealt with in the superstructure section, rather than with foundations, as previously discussed on page 72.

118 Common bricks only, therefore, will be dealt with at this stage, and since the specific type of common brick (e.g., flettons, wirecuts, etc.,) will normally be covered by a preamble, the term 'commons' may be sufficient to describe the kind of brick in the item-description, unless more than one type of common brick occurs in the project.

This last may well be the case, since the common bricks used below DPC often differ from those used above, when the term 'wirecuts' (e.g.) must be used instead of 'commons', to distinguish from (say) flettons. Under examination conditions it is always wise to describe the exact brick-type specified, if this is indicated, in order to be on the safe side, rather than to use the general term 'commons'.

Unless more than one size of brick is being used, or the size is other than normal, no mention of size need be made, as this will be covered by preambles.

(ii) TYPE OF BOND. With facings, this must be included in the description because it may have a marked effect on the price. In the case of commons, however, here under discussion, preamble coverage is all that is necessary, unless the bond varies. No mention of bond need normally be made in the taking-off description.

(iii) MORTAR. This will probably be either cement mortar (cement and sand) or gauged mortar (cement and sand gauged with lime). As with concrete, the mix is always stated in brackets following the word 'mortar'.

A 1:3 mix of cement and sand is denoted by: *cement mortar* (1:3). A 1:2:9 mix of cement, lime and sand, respectively, is denoted by: *gauged mortar* (1:2:9).

The description for a brick wall under two-brick thick must therefore contain the following, in the order stated:

(i) Thickness (G3(a)).
(ii) Class (i.e., 'wall') (G3(a)(i)).
(iii) Kind of brick (G2(b)(i)).
(iv) Kind of mortar and mix (G2(b)(iii)).

This will produce a description in the following form:

One-brick wall in commons
in cement mortar (1:3).

or

Half-brick wall in flettons
in gauged mortar (1:2:9).

Should the wall be two-brick thickness or over, it must be described as reduced brickwork in accordance with SM G3(a), with the thickness stated *after* the description, as in the case of footings, thus:

Reduced brickwork in walls in
commons in cement mortar (1:3).
$\times$ 2*B* =

Hollow walls

Although solid brickwork has been assumed in the foregoing, the external walls of dwelling-houses are usually of hollow construction, comprising two half-brick skins with a cavity between, tied together with wall ties. The inner skin may be of clinker blocks above DPC level, or may be thicker than half-brick, but the standard 270 mm hollow wall below ground level will be considered here, to which the same principles apply. Hollow walls do not have brick footings as a rule, but are built directly off the concrete, the cavity being filled in solid with concrete up to ground level.

The skins of hollow walls are measured under SM G3(a)(iii) and cavities under G8(a), and since both are measured super, it might be thought convenient to and-on the skins to the cavity. The following factors must first be taken into consideration, however:

(i) WHERE THE SKINS ARE IDENTICAL. This will normally be the case below ground level, and the skins must not then be separately described, but merged together as 'half-brick skins of hollow walls' under SM G3(a)(iii). Since the total area of the skins will be twice that of the cavity, anding-on cavity to skins will involve either writing the skin descriptions twice (once for each skin), or writing the dimensions twice, once for the cavity, and again (twiced) for the skins.

(ii) WHERE THE SKINS DIFFER. A difference of thickness or brick-type between the two skins will involve not only separate descriptions, but separate dimensions for each skin, owing to the difference between the areas of inner and outer skins, both of which differ in area from the cavity. Three separate items with three sets of dimensions will therefore be required, one for each skin, and one for the cavity.

In general, case (i) will apply below ground level and case (ii) above ground level, for here the outer skin will be of facing bricks, and therefore the skins will differ. The question of facings will be dealt with later, however, on the assumption that any facings between ground level and DPC will be adjusted with the superstructure, and if we confine ourselves to the standard 270 mm hollow wall below ground level (case (i),

120 above), it will be found that the following set of descriptions will be suitable for setting down against the mean girth, timesed by the height from top of concrete to damp-proof course:

<u>*super*</u>	*Half-brick skins of hollow walls.* *&* *Ditto as last item* *&* *Form cavity* 50 *mm wide between skins of hollow walls, including galvanized twisted wall ties, four per square metre.*

Projections on Hollow Walls

Projections are measured 'beyond the face of the wall' (SM G4(i)) and this will apply to hollow walls, as for solid walls. Projections for attached piers, etc., on hollow walls will constitute a thickening of one skin (usually the inner skin) since the cavity, to be effective, will normally run right through. Thus a 270-mm cavity wall with a 110-mm projection on one side could be regarded as a 380-mm cavity wall, comprising 110-mm and 220-mm skins, with a 50-mm cavity between. The question arises, therefore: what length (on plan) may such a projection be before it is reasonably regarded as constituting a wall?

Under SM G3 (a) (v), brick walls having a length on plan not exceeding four times their thickness are classed as isolated piers. Again, under SM F8(a), attached *concrete* piers having a length over four times the thickness are classed as walls.

There is no such restriction on the length of brick projections for 'attached piers, chimney-breasts . . . and the like' measurable under SM G4(i), however, and all such cases must therefore be measured as projections from the wall-face, and not as walls of greater thickness than the main wall.[1]

Under this system of measurement, with the main wall measured right through and the projection measured from the wall-face as a separate item, there will occur in the bill a section of (say) 'half-brick skins of hol-

[1] An exception would be made in the case of a projection of such length on plan, and in such a position, that its function could not be said to constitute a projection as defined in SM G4(i). In such a case it may be regarded as a wall of greater thickness than the main wall, and measured as such.

low walls' which does not in fact exist. Where the inner skin *A* passes through the pier at *B* in Fig. 22 it is no longer a half-brick skin, built in stretcher bond, but a section of (say) chimney-breast, abutting the cavity. Students are sometimes concerned (very properly) that any cost difference between the two should be reflected in the measurements, and wonder how the difficulty can be overcome.

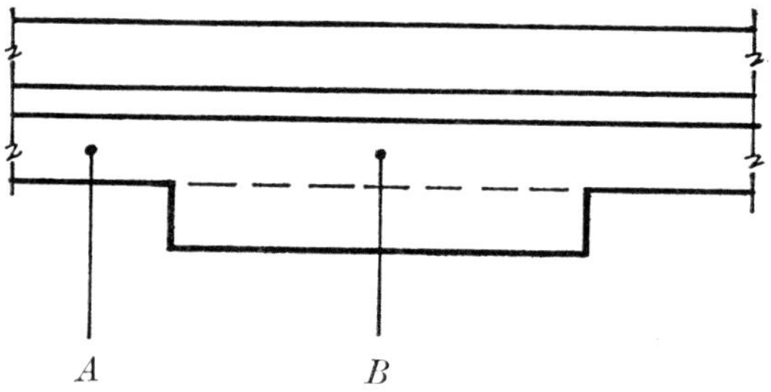

Fig. 22

This situation is, however, inherent in any method requiring projections to be measured from the wall-face, and applies equally to projections on solid walls. Thus a pier projecting from a one-brick wall measured in accordance with SM G4(i) as previously discussed, leaves the section of brickwork measured through the pier included as 'one-brick wall', when in fact the brickwork is thicker at that point.

The estimator knows (from the SMM) that the brickwork has been measured in this way, however, and is theoretically able to make some kind of adjustment in his price for brickwork on this account if he considers it worthwhile. Thus the specific directions of the SMM may override SM A1, and the bill of quantities made to accurately represent the intentions of the SMM rather than the works to be executed. This is only valid in cases such as the present one, where the intention of the SMM is clear. In instances where the clause wording is loose, or appears illogical (as in the previous case of the non-existent rough cutting) then the general principle invoked by SM A1 must prevail.

For further discussion of projections on walls, see superstructure brickwork, Chapter 16.

Curved walls

These must be kept separate under SM G1(d), and the girth can usually be taken from previous measurements of the curved footings, or curved trenches. As with curved footings, the mean radius must be stated, and rough cutting included in the description.

122 In the case of curved cavity walls, the two skins and the cavity must be described as curved, but where the two skins are identical (as is probable below ground level) the term 'mean radius' in SM G1(d) may be construed as referring to the mean radius of the complete cavity wall, and it is not considered necessary to calculate three separate radii. The measurements for 270-mm cavity walls 1 metre high, beneath a semi-circular bay window of 3 metres mean radius, for example, would be as follows:

Radius = 3·00
Dia = 2/3·00 = 6·00

$\frac{1}{2}/3\frac{1}{7}/6{\cdot}00$ 1·00	*Half-brick skins of hollow walls, curved on plan to 3 metres mean radius including rough cutting.* & *Ditto as last item.* & *Form cavity* 50 *mm wide all as last described, but curved on plan to* 3 *metres radius.*

Squint angle cutting to walls

As in the case of footings, previously described, rough cutting is measurable to walls at squint angles, should these occur, under SM G10(a). In the case of walls, the convention is to take the cross-sectional area of the wall (once for each squint quoin) on a cross-section *through the angle* as shown at *A–A* in Fig. 23. The distance *B–B* is scaled from the drawing

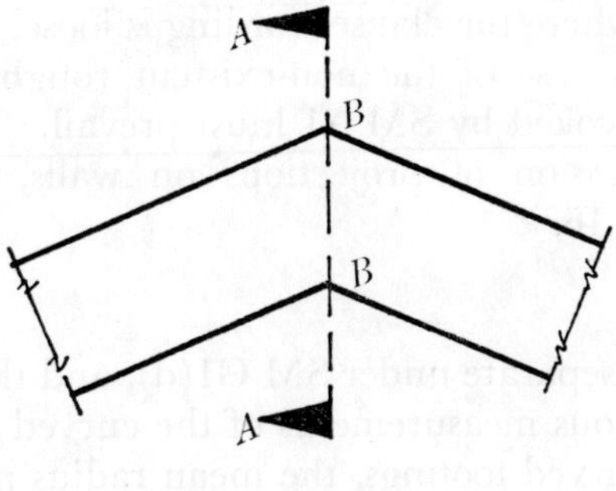

Fig. 23

and supered by the height as 'rough cutting'. This represents a notional amount of rough cutting sufficient to cover the extra cutting involved in bonding a squint quoin as opposed to a square quoin, the cutting for which is deemed to be included in the price of normal brickwork.

Cavity filling

This refers not to the quantity surveyor's dental appointment, but to concrete filling to hollow walls measurable under SM F8(d). The cavity is always filled solid up to ground level (*not* DPC level) with a splayed top edge to direct condensation from the cavity through open vertical brick joints (not measurable) in the outer skin at ground level. This is measured super, taking the outer girth of the cavity previously calculated by the height of the filling; a suitable description being:

Concrete (1:10) *in filling*
to 50 *mm cavity of hollow wall.*

The splaying to top edge is deemed to be included under SM F8(d), and should not be mentioned in the description.

On the analogy of concrete foundations, curved cavity filling is not kept separate from straight, but forming the cavity itself may involve extra labour if curved, and should therefore be kept separate as illustrated above.

DISPOSAL ADJUSTMENT FOR BRICKWORK

As in the case of the foundation concrete previously discussed, the volume of brickwork will displace a volume of earth not required to be backfilled into the trench, and now that the brickwork has been measured, a suitable adjustment can be made.

Since disposal is measured cube, it could simply be anded-on in the case of the concrete, which was also measured cube. It is a common occurrence for students to mistakenly and-on the disposal to the brickwork descriptions, but unless the device of (e.g.) *cube by* 0·22 *equals* is employed, this is obviously invalid.

Even this device is suspect, however, for the total volume of brickwork will not in fact be equal to the total volume of disposal, owing to the fact that return fill and ram has been measured to *top of trench excavation only*. It is best therefore to leave disposal adjustment until after the brickwork has been measured, and not to attempt any short cuts which might lead to inaccuracies.

The procedure is to refer back to the brickwork dimensions and convert these to cubes. The first component of each dimension is transferred direct from the dimension column, but the second component (width) must be taken from the width (e.g., $\frac{3}{4}B$ = 0·17 (rounded)) in the description column. The third component (depth) may be taken direct, but for *footings only*. All other projections, and all walls, must first be adjusted for height to allow for the difference between wall height (to DPC) and trench depth.

Since this will be a constant, however (a flat site being assumed) only one very simple waste calculation is required in order to reduce the constant height of all walls and piers to the correct dimensions for trench depth (less concrete), as follows:

Example

PROBLEM

Calculate on waste the correct depth of disposal adjustment for brickwork to the wall shown in Fig. 24.

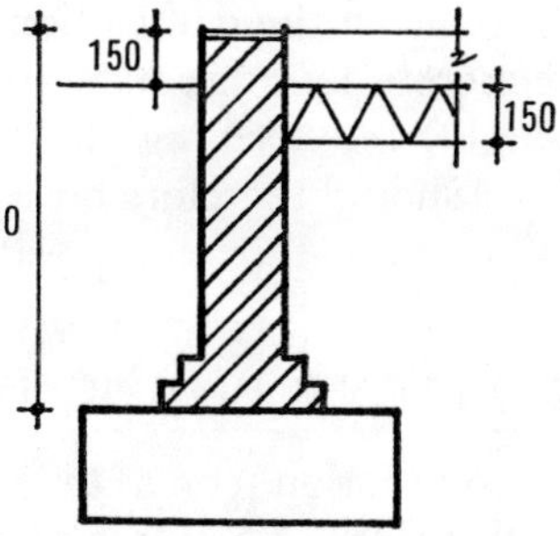

Fig. 24

SOLUTION

Height of bwk.		1·10
to g. l.	·15	
strip	·15	·30
Disposal depth =		0·80

DAMP-PROOF COURSES

The horizontal damp-proof course to walls at ground-floor level will generally be one of the two main types mentioned in SM G44(a), viz.: either (i) slate or (ii) sheet. In either case the method of measurement is the same, i.e., not exceeding one-brick wide linear; exceeding one-brick wide super.

The girths for DPC measurement can be picked out from the brickwork dimensions and transferred direct to the dimension column, since DPC's will be required to all walls previously measured. An eye must be kept on the brickwork thickness in the description column, any change in this requiring a fresh DPC description, and any walls thicker than one-brick requiring a superficial measurement. All very straightforward, but there is one special snag to watch out for in the process of copying from previous brickwork dimensions, and this arises in connexion with projections.

DPC to projections

Brickwork projections, as we have seen, are measured from the wall-face. This does not apply to DPC, however, and care must be taken therefore not to include any brickwork *projection* dimensions in the above process of transference. The DPC may be assumed to be required to all projections, however, and when all the DPC to walls has been measured, an adjustment for the projections (if any) must then be made.

In most cases, this adjustment will not take the form of simply adding-in an extra, because the additional width required will change the DPC description, and may perhaps involve a change from linear to super, as in the case of pier *A* in Fig. 25.

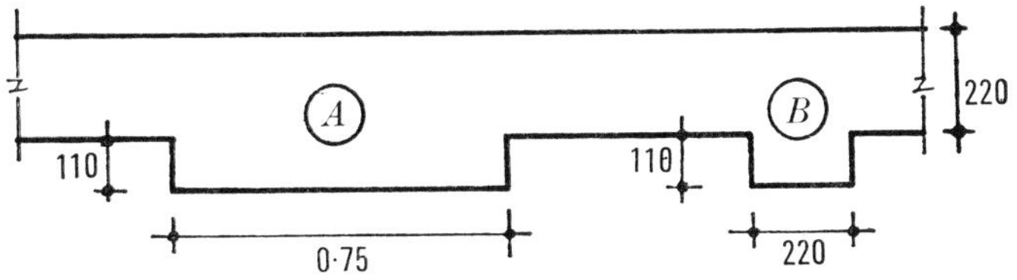

Fig. 25

Here, the DPC to the wall will have been measured through the pier (as taken from brickwork dimensions) and described as 220 mm wide, but where the pier occurs it will in fact be 220 + 110 = 330 mm wide. The following adjustment will therefore have to be made for this pier:

	·22 ·11 ·33	
0·75 0·33	*Horizontal DPC as described*	
		(pier A
0·75	*Deduct* Ditto but 220 mm wide a.b.	
		(wall

126 In cases such as pier *B* (Fig. 25) where the pier *length* is the same as the wall *width*, the pier DPC may be regarded as simply an extention of the wall DPC, to which it should be bracketed-in, or measured as:

0·11 *Horizontal DPC* 220 *mm wide a.b.*

(pier B

there being no differentiation in the SMM between walls and projections in the case of damp-proof courses.

In order to avoid confusion it is better to measure the DPC's to projections from the drawing, considering each one on its merits, rather than taking these from previous brickwork dimensions. The important thing is to decide in each case what the actual physical width of the DPC is at that particular position, and describe accordingly.

DPC descriptions

For a slate DPC, the following traditional standard description may be considered suitable for all normal circumstances (adjusted if necessary to suit any special requirements) and should be memorized by the student:

Horizontal DPC of two courses of stout whole Welsh slates laid breaking-joint and bedded in cement mortar (1:3).[1]

Note that pointing exposed edges is deemed to be included under G44(c), and there should not therefore be mentioned in the description.

Although sheet lead and copper are mentioned in SM G44(a), the usual alternative to slates in domestic work is that of a layer of bituminous felt (also referred to in this clause), and this may take two forms:

(i) BITUMINOUS FELT DPC. BS 743:1951 lists three varieties:

(a) hessian base,
(b) fibre base,
(c) asbestos base.

Since the cost may differ, the description should state which of these three is required.

(ii) LEAD-CORED BITUMINOUS FELT DPC. This material is as described for (i) above, but incorporates a layer or core of sheet lead within the thickness of the bituminous felt, which again may be either hessian, fibre or asbestos.

[1] The width being stated (as with felt DPC's) if measured linear.

Under SM G44(a) no allowance in measurement shall be made for laps, and '*this shall be stated in the description*'. The number of *layers* must be given in the description under SM G43(iii), but the substance, or weight (SM G43(ii)) is laid down in the British Standard for each material, and, like the extent of lap, may be considered as covered by preambles. Thus the following (or similar) taking-off description will be found to evolve:

Horizontal layer of hessian
base bituminous felt DPC,
bedded in cement mortar (1:3)
(no allowance for laps).

Note that the composition and mix of bedding material is required under SM G43(iv), but no requirement is called for regarding *location* of the DPC.

This last point is worth noting, since the intention of the SMM is (presumably) to group all horizontal DPC's together in the bill. Any mention of 'bedded on brick *walls* . . .' in the description would have the effect of separating out DPC's to chimney-stacks for example, measured later with superstructure. The 'no deductions under 0·5 square metre' (e.g., for flues) in SM G44(a) is generally considered to be framed for the purpose of avoiding such separations, and to prevent needless repetition of items in the bill, therefore, a description (applicable to stacks, parapet DPC's, etc.,) in the above form should be used for all horizontal damp-proof courses.

DPC to curved walls

Curved work must be kept separate under SM G44(a), and can usually be taken from the previous dimensions for the curved walls. Cutting must be given in the description, and the requirement to state the mean radius of curved work under SM G1(d) is usually held to apply to damp-proof courses as well as brickwork. The measurement of DPC to the semi-circular bay window brickwork measured on page 122, would be as follows:

$2/\frac{1}{2}/3\frac{1}{7}/\underline{6{\cdot}00}$	*Horizontal DPC* 110 *mm wide* *all a.b. but curved on* *plan to* 3 *metres mean radius,* *including cutting.*

Note that the term 'rough' cutting is a bricklayers' term referring to the cutting of bricks, not DPC.

128 **Raking cutting to DPC**

This is not mentioned in the SMM, but the usual practice is to take a linear measurement where required, under SM A1 and the 'more detailed information' rule of the introduction to the SMM.

Squint quoins are a case where this item is considered measurable, and the convention is to take a linear measurement across the angle (*B–B* in Fig. 23) and describe as:

Raking cutting to horizontal
slate DPC.

or as the case may be.

DISPOSAL ADJUSTMENT FOR SURFACE STRIP EXCAVATION

When the foundation work so far measured has in fact been executed, there will remain a small trench extending round the perimeter of the building adjacent to external face of brickwork caused by the difference in level between formation level and original ground level. This small trench is shown at *T* in Fig. 26(a) and will require backfilling, since the backfilling hitherto measured extends only to top of foundation trench level, at *E*.

The earth necessary to complete the remaining backfill may be regarded as being the surplus spoil from surface excavation, previously measured as removed from site. It is therefore necessary to calculate the volume of the trench *T*, and describe as:

Deduct *Remove surplus a.b.*

&

Add *RFR a.b.*

This adjustment can be made following on the disposal adjustment for brickwork if desired, but the convenience of measuring the DPC as soon as possible after the brickwork to which it is related favours taking surface excavation adjustment at this later stage.

Calculation of surface excavation adjustment

The width required for this item will be the original concrete projection *p* (Fig. 26(a)), the depth being the total depth of surface excavation originally measured (surface strip plus any reduced level excavation). The required length will be the mean girth of the trench *T*, and care is needed to calculate this accurately, an incorrect girth for this item being a common source of error among beginners.

The best procedure is to use the girth adjustment formula in order to shift the original mean girth to a position on the centre-line of the trench *T*. Then add twice the brickwork projection in respect of any attached piers or chimney-breasts occurring on the outside face of the walls.

It will be recalled that the girth adjustment formula was used (page 102) to establish the length of planking and strutting to reduced level excavation, and when used for surface excavation adjustment purposes, special care is needed to ensure that the correct value is given to d in the formula

$$A = M \pm 8d.$$

It will be apparent from Fig. 26(b) that the value of d for this purpose

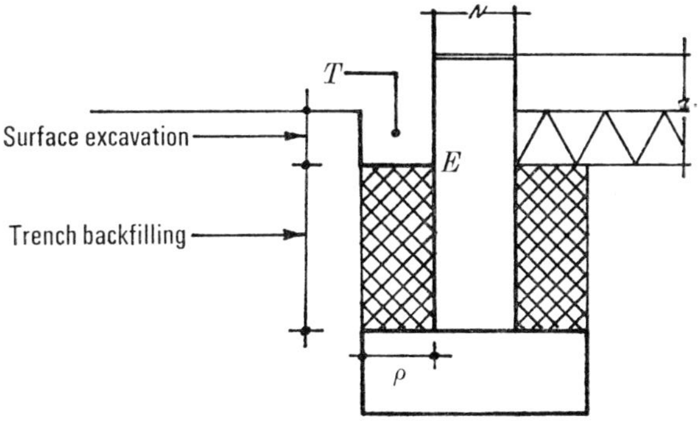

*Fig. 26*a

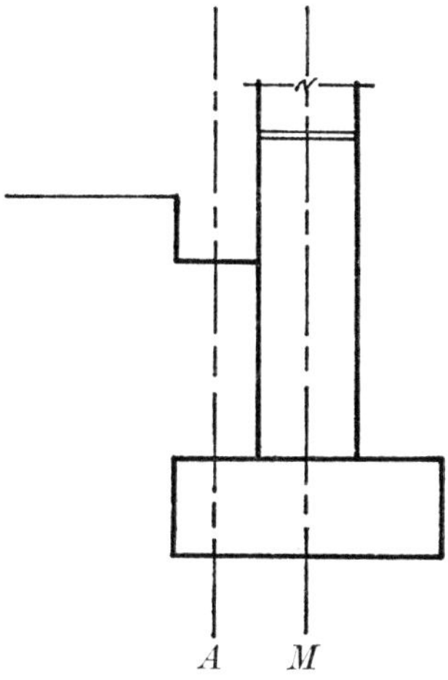

*Fig. 26*b

 is: *half* the wall thickness plus *half* the original concrete projection. Put another way:

$$d = \tfrac{1}{2}(T + p)$$

and the formula needed to express the final length *L*, required for surface excavation adjustment would be:

$$L = M + 8\left(\frac{T + p}{2}\right) + 2BW_1 + 2BW_2 + \ldots$$

where M = original mean girth,
T = thickness of wall,
p = concrete projection,
and BW = brickwork width of any externally projecting pier.

As in the case of pier adjustments for reduced level planking and strutting, however, the student may well find it desirable to take a fresh look at the drawing before including any pier adjustments for the item in question, as this will remind him to treat any external angle corner piers as two separate piers each, and to ignore any internal angle corner piers altogether. It is perhaps better therefore to calculate the mean girth *G* of trench *T* from the formula:

$$G = M + 8\left(\frac{T + p}{2}\right)$$

and then to bracket-in the additional volume required for any piers afterwards. This additional volume will be a length of trench arrived at by means of the rules laid down under pier adjustments for trench planking and strutting, which are repeated here for convenience:

(i) SIDE PIERS. Twice the amount of brickwork projection.

(ii) EXTERNAL ANGLE CORNER PIERS. Twice the *sum* of both brickwork projections.

(iii) INTERNAL ANGLE CORNER PIERS. Nil.

Example

PROBLEM

Take off the quantities for surface strip adjustment to the building shown in Fig. 21. Assume surface excavation 150 mm deep has been measured.

SOLUTION

$$\text{in } G = M + 8\left(\frac{T + p}{2}\right)$$

M (from page 114) = 20·12
T = ·22
concrete 1·00
less wall ·22
p = 2)0·78(·39
$\frac{1}{2}(T + p)$ = 2)·61
8/ ·305 2·44
G = 22·56

22·56	*Deduct* *Remove surplus a.b.*
0·39	[*to replace*
0·15	*surface soil*
	&
	Add *RFR a.b.*
2/0·22	(*pier A*
0·39	
0·15	

It will be remembered that any re-entrants included in the mean girth are unaffected by a centre-line shift. These may therefore be ignored in the above process, since the formula $A = M \pm 8d$, from which $G = M + 8[(T + p)/2]$ is derived, is valid for a building of any given closed rectangular shape. Any extra required will normally be in respect of external projections only therefore, as illustrated above.

Alternative formula for surface excavation adjustment

In the above example it was necessary to calculate the value of p on waste, by taking the wall thickness from the concrete spread and dividing by two. When it is convenient to search back through the dimensions and pick up the value of p from previous waste calculations, this will not be necessary. In cases where it is thought desirable to recalculate p at this stage, however, the waste calculation will be simplified by incorporating the concrete spread S in the formula, and substituting $\frac{1}{2}(S - T)$ in place of p, thus:

$$G = M + 8\left[\frac{T + \frac{1}{2}(S - T)}{2}\right]$$

from whence, by cancellation we obtain:

$$G = M + 2(T + S).$$

Using this formula in lieu of $G = M + 8[(T + p)/2]$ the waste calculations in respect of the above example are simplified to the following:

	$M =$	20·12
$T =$	·22	
$S =$	1·00	
	2/1·22	2·44
	$G =$	22·56

Whichever method is adopted it is important that the student should study the formulae concerned and ensure that the principles involved are fully understood. In this way errors arising from faulty adjustment of surface strip disposal can be avoided.

GIRTH ADJUSTMENT TO IRREGULAR PLAN-SHAPES

In the foregoing discussions on girth adjustment, both for planking and strutting and for surface strip disposal adjustment, it was assumed that the structure under consideration formed a closed rectangular plan-shape. Reference was made in the brickwork section, however, to the possibility of plan-shapes containing curved work, or squint quoins. Where this is the case, the following adjustments will be necessary in the case of any girth-shifting operations.

Curved work.

Should a section of external wall be curved on plan, it will have been deducted-out on waste and measured separately, leaving a net mean girth. If such a curved section were to occur at a *corner* of the building, it would replace an angle. The plan-shape must now be regarded as *unclosed* as far as girth shifting is concerned, since a portion is missing when the girth-adjusting formula is applied.

In the girth-adjusting formula $A = M \pm 8d$, the figure 8 represents $2d$ for each of the four external corners which are left over after internal angles have been deducted from external angles (the sum of external angles always being greater than the sum of internal angles by four). If one of the external angles has been eliminated from the mean girth M in the process of establishing a net mean girth, the figure 8, representing four corners timesed by two, will no longer hold good. It will then become 6 in fact, representing three corners timesed by two.

In an unclosed plan-shape therefore, the figure 8 must be regarded as a variable parameter, depending on the number of corners missing. It will still represent twice the number of external angles less the number of internal angles, however, and the girth adjustment formula will still apply, provided it is in the form:

$$A = M \pm 2(E - I)d$$

where E = sum of external angles contained in M,
I = sum of internal angles contained in M.

When using the above formula to adjust a net mean girth representing an unclosed plan-shape, the student must, of course, remember to add back the curved section as well, after calculating its adjusted girth by inspection of the drawing.

Squint quoins

As previously noted, straight sections of external wall which are not perpendicular on plan are best adjusted for on waste and included in the mean girth M (provided they are of the same thickness and depth as the other walls). Where this has been done, although M will represent a *closed* plan-shape, it will not represent a *rectangular* plan-shape, and once again, therefore, the girth adjustment formula will be invalid in its original form $A = M \pm 8d$. The reason for this is that squint angles *greater* than a right-angle (internally) have an adjustment-factor *less* than $2d$, while those smaller than a right-angle have a factor which is greater than $2d$.

Where the building is of a shape which has permitted a mean girth M to be arrived at, however, the best plan when adjusting this girth is to apply the formula $A = M \pm 8d$ in the first instance, and then check for any squint angles, adjusting the result accordingly. It will be found a relatively simple matter in the case of a squint angle to deduct $2d$ from the adjusted girth and scale-off the correct additional girth for the angle in question by plotting the line of girth at that point on the plan in pencil.

In general it will be found that where two squint angles form a splayed corner to a building, as in Fig. 12, the alteration required to the adjusted girth by reason of these two squint quoins replacing a square quoin is fairly marginal, consisting of a slight reduction. It is therefore usually ignored, as being negligible.

Missing walls and rogue sections

When a section of external wall is missing, to form an open side to the building, as in a car-port, for instance, or where the section is of different thickness or depth from the remainder, this section will have been deducted-out on waste, leaving a net mean girth. For purposes of girth

134 adjustment, the problem then falls into the same category as for curved sections of wall, described above, and is dealt with accordingly.

FORMULA SUMMARY FOR STRIP FOUNDATIONS

The following formulae in connexion with strip foundations, which the student may wish to memorize for examination purposes, are collected together for easy reference.

NOTES: p = concrete projection,
T = wall thickness,
M = mean girth of external walls.

(1) SIDE DEDUCTION FORMULA (page 82).

$$S = n(d - 2p)$$

where S = area of side deduction,
n = net projection of wall re-entrant,
d = dimension between re-entrant walls.

(2) CROSS-WALL FORMULA (page 91).

$$l = c - 2p$$

where l = length of foundation trench,
c = cross-wall length.

(3) SIDE PIER FORMULA (page 95).

$$TW = BW$$
$$TL = BL + 2p$$

where TW = extra trench width,
BW = extra brickwork width (pier projection),
TL = extra trench length,
BL = extra brickwork length (pier length).

(4) CORNER PIER FORMULA (page 81).

$$x = a + 2p$$
$$y = b + 2p - d$$

where x and y = extra lengths of strip/trenches/concrete required,
a and b = brickwork dimensions of each face,
d = distance face a projects from main wall.

(5) GIRTHING FORMULA (page 88).

$$M = 2(L + W) \pm 4T + 2n_1 + 2n_2 + \ldots$$

where L = overall length of building,
W = overall width of building,
n = net projection of each re-entrant.

(6) GIRTH ADJUSTMENT FORMULA (page 102).

$$A = M \pm 8d$$

where A = adjusted girth,
d = cross-sectional distance $M - A$.

(7) STRIP ADJUSTMENT FORMULA (page 130).

$$G = M + 8\left(\frac{T + p}{2}\right)$$

where G = girth of strip trench (less pier additions).

(8) ALTERNATIVE STRIP ADJUSTMENT FORMULA (page 131).

$$G = M + 2(T + S)$$

where S = spread of concrete.

(9) UNCLOSED GIRTH ADJUSTMENT FORMULA (page 133).

$$A = M \pm 2(E - I)d$$

where E = sum of external angles contained in M,
I = sum of internal angles contained in M.

Chapter Thirteen

Stepped Foundations

Definition

When a building is sited on sloping ground, both the floors and the foundations must maintain the horizontal, in order to ensure stability of the structure. In such circumstances, however, strip foundations, if maintained at one level throughout would tend to be either needlessly deep at one end, or dangerously shallow at the other. Steps must be inserted to avoid this, but any in the finished floor level are obviously undesirable, and the usual solution, where the slope of the ground is fairly gentle, is to keep the ground floor at a constant level and introduce steps at intervals in

136 the concrete foundations. In this way the foundations, although horizontal at any one point, are made to follow the contour of the ground, and the trenches can then be maintained at an economic depth throughout, as shown diagrammatically in Fig. 27(a).

When the slope of the ground is steep, however, such a solution would be uneconomical, owing to the difference between ground level and floor level at the lower end of the site; when, in order to reduce the amount of foundation brickwork, it becomes necessary to lower the floor level such that it comes *below* ground level at the upper end of the site, as in Fig. 27(b).

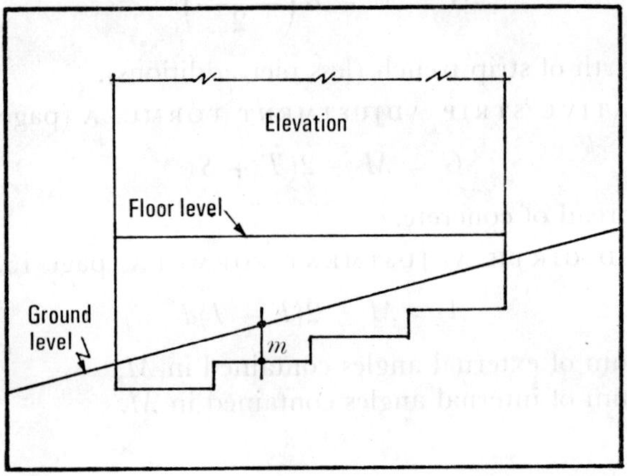

(a)

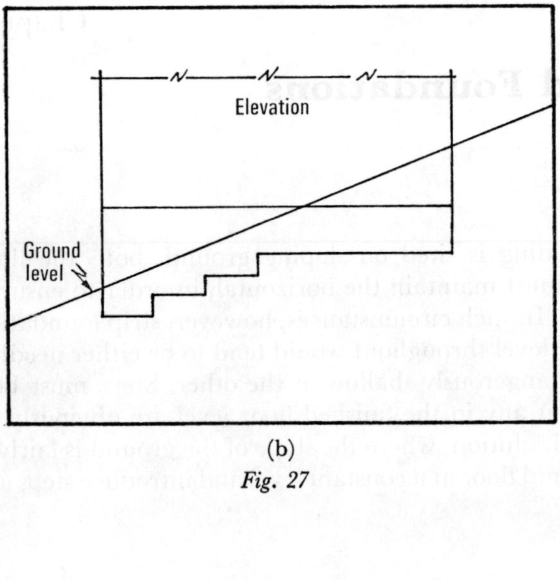

(b)

Fig. 27

In this last instance, although steps may still be used in the foundations, as shown, the taking-off is complicated by the fact that (a) these will not enable the trench depth to follow the contour of the ground at the upper end of the site, and (b) preliminary site excavation will be necessary to achieve formation level over certain parts of the building, possibly involving the formation of embankments at the top end of the site. The problem will be further complicated if the site is sloping in more than one direction.

For these reasons, foundations on sites on which the slope of the ground is steep enough to justify lowering the floor level below original ground level, are best regarded as 'sloping site foundations', an advanced constructional group not dealt with at this stage. We may therefore define 'stepped foundations' for taking-off purposes as:

> *Strip foundations on gently sloping sites, where floor level is above ground level and the foundations are stepped to follow the contour of the ground.*

Order of taking-off

The general order of taking-off is as for strip foundations, described in Chapter 8, and the typical item sequence will remain the same, subject to the following special factors to be considered in respect of stepped foundations.

SURFACE STRIP

In the case of stepped foundations, it will be apparent from Fig. 27(a) that although the finished floor is horizontal, the level at underside of surface excavation will have to follow the slope of the ground, or be stepped to conform with the foundations. Since the ground-floor construction does not carry a structural load, there is no objection to supporting it on a sloping surface, and the practice is therefore to fill in the space between ground-floor construction and formation level with a hardcore bed of varying thickness, as shown in Fig. 28, where the slope of the ground has been somewhat exaggerated in the interests of clarity.

Depth of surface excavation

It is not constructionally desirable that the thickness of the hardcore bed should die away to nothing at the shallow end, as the uneven bearing may cause settlement. It is therefore the practice to specify a *minimum*

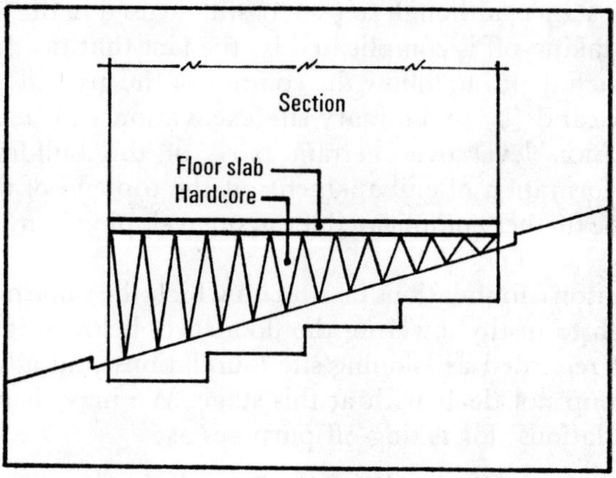

Fig. 28

thickness of hardcore in cases of this kind, and the depth of any surface excavation on sloping sites will be governed by this minimum value. The drawing must be carefully examined at the point where ground level is highest to determine the relation between ground level and the level of the underside of minimum-thickness hardcore, i.e., formation level at this point.

If the difference between the two levels is greater than the depth of surface strip required over the site, reduced level excavation may have to be measured to the line of intersection between strip level and formation level.

Figure 29 should make this clear. In this instance the difference $G - B$ between ground level and formation level is greater than the depth of surface strip, $G - A$. Hence reduced level excavation is required to the extent of the triangle *ABC*.

Unless the ground level is higher than floor level, however (which would disqualify the foundations from consideration under the present definition), it will be apparent from Fig. 29 that the amount of reduced level excavation required is unlikely to be extensive, and if the portion *XABY* is regarded as trench excavation, it is reduced in this instance to the negligible proportions of triangle *CXY*.

It will usually be found therefore that a suitable adjustment of the surface strip depth will eliminate measurement of reduced level excavation in the case of stepped foundations, and the student should examine the cross-sections to see if this solution can be achieved, bearing in mind that, as we have seen in Chapter 9, 300 mm is the maximum depth

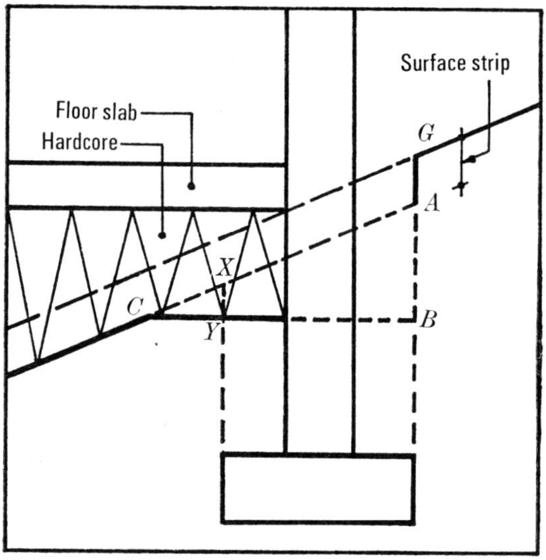

Fig. 29

which can reasonably be allowed for this item. Otherwise 150 mm plus a suitable volume of reduced level excavation will have to be measured.

Area of surface strip

The method of measurement is identical to that for ordinary strip foundations; surface strip must be measured over the extreme plan area of the building, including concrete projections, with additions for any side or corner piers, bracketed-in as described on page 79. Any corner deductions must be made, and side deductions measured in accordance with the formula $S = n(d - 2p)$, as hitherto explained.

The question sometimes raised by students in connexion with surface strip to sloping sites is whether an allowance should be made for the fact that the *actual* area differs from the *plan* area owing to the slope of the ground. This might be significant in the case of 'sloping site foundations', but in the case of ordinary stepped foundations, where by definition, the slope of the ground is gentle, this difference will probably be negligible.

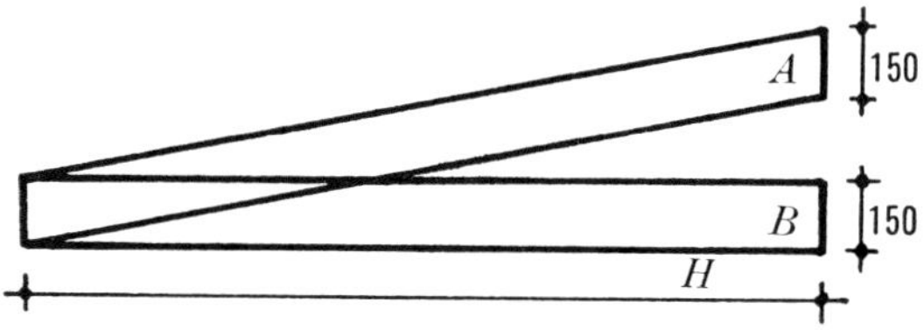

Fig. 30

In any case, however, the actual volume of earth removed will be unaffected by this particular factor, as the diagram in Fig. 30 shows. The sloping section of surface strip *A* will be seen to have the same cross-sectional area as the horizontal section *B*, the area of either one being $H \times 150$ mm. Provided the depth is measured vertically downwards therefore, the increase in volume for this item occasioned by the method of measurement will compensate for the shortage of area in the quantities, and no allowance need be made for this when taking-off.

FOUNDATION TRENCHES

Since, by definition, the foundations are stepped to follow the contour of the ground, the *average* distance between ground level and bottom of concrete foundations will be a constant. This constant will be the depth midway between any two steps, as at *m* in Fig. 27(a).

This average depth should be established by the taker-off from a careful inspection of the drawings. It will probably be necessary to plot a pencil-line on the elevation or section to represent original ground level, and from this line to scale off the average depth *m* at a point midway between two steps. From this depth, the thickness of surface strip already measured can then be deducted on waste, and the average depth for trench excavation arrived at.

With stepped foundations of this kind, the taker-off should aim at establishing a single average depth in this way for the whole of the foundation trenches, which can then be measured exactly as for ordinary strip foundations, ignoring all steps. In doing so, however, the following points must be borne in mind.

Variation in foundation depth

Certain sections of trench, particularly those to internal walls, may upset the average *m* by a depth-variation due to a difference in load-bearing requirement, or for other reasons. The depth of all foundations must be carefully examined in order to identify any such rogue sections, and these must be measured separately, with a separate average depth calculated in each case. Where such sections occur in external walls, it will usually be convenient to calculate the mean girth as usual, by using

$$M = 2(L + W) \pm 4T + 2n_1 + 2n_2 + \ldots$$

and timesing by the average depth *m*. An adjustment can then be made for any rogue sections by deducting these and adding back at a revised average depth.

On the rare occasions when depth-variation is such that a reasonable average *m* is impossible to arrive at over any major portion of the building, there remains nothing for it but to abandon the girthing formula and measure each section of trench individually, establishing the correct average depth for each section, and ticking-off each length on the drawing as it is measured. It will nearly always be found possible, however, to avoid this by establishing a reasonable average for the external walls, or at least by splitting the building into (say) two portions, with an appropriate average for each, and taking internal walls separately if necessary.

Changes of depth-range

It will be recalled that trench excavation is measured in successive stages of 1·5 metres in depth under SM D6(f), and on sloping sites, where the depth is constantly varying, a weather-eye must be kept open for any possibility of a portion of trench exceeding 1·5 metres deep, for although the *average* for any section may not exceed this depth, the *actual* depth may exceed 1·5 metres at the deeper end of the section, and this can easily pass unnoticed by the taker-off.

It would not be valid, for example, to take 1·45 as the average depth for a section of trench varying from 1·15 at the shallow end to 1·75 at the deep end, and describe this as 'excavate foundation trench . . . not exceeding 1·5 metres deep', for part of it would in fact be exceeding 1·5 metres deep, and must be so described.

The correct procedure in such a case, illustrated in Fig. 31, is to lay a

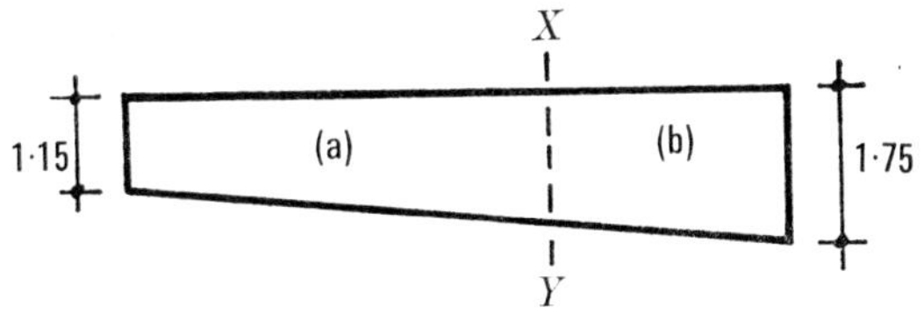

Fig. 31

scale down vertically on the drawing, and slide it along the section until 1·50 is read off, when a pencil-line *X–Y* can be drawn, dividing the section into part (a) not exceeding 1·5 metres deep, and part (b) over 1·5 metres deep. The horizontal length of each part can now be scaled off, and its average depth calculated, bearing in mind that part (b) will require splitting into two separate descriptions.

Projecting piers

Strictly speaking, each projection should be bracketed-in and timesed by its individual depth, but where it has been found possible to obtain an average depth *m* for the whole of the trenches, it is common practice to use this depth for projections also, provided these are few in number and more or less evenly distributed, especially if the slope is a very gentle one. Discretion must be used here, bearing in mind that excavation and disposal are reduced to the nearest cubic metre at abstracting stage, and that the volume of excavation and disposal to projecting piers is often a very small fraction of the whole.

PLANKING AND STRUTTING

Like level and ram, planking and strutting will follow the trench dimensions as for ordinary strip foundations, subject to the proviso regarding depth-range, as above discussed for trenches. Even this will tend to resolve itself automatically, since the taker-off will notice any changes in description when working through the trench dimensions, and this will remind him to keep separate any planking and strutting to sections of trench over 1·5 metres deep under SM D21(a).

It might be desirable to once again remind the student that, unlike trench excavation, planking and strutting is not measured in *successive* stages, but in one stage only, with the total depth stated in multiples of 1·5 metres. Any section of trench over 1·5 metres deep therefore must be grouped together with its corresponding section not exceeding 1·5 metres deep when converting the trench dimensions into planking and strutting dimensions.

Planking and strutting to steps

At the position where a step occurs in the foundations, the trench-bottom itself will, of course, require to be stepped. This step in the trench-

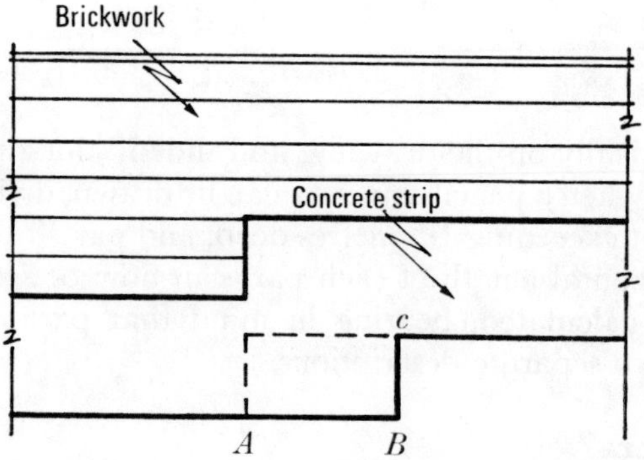

Fig. 32

bottom will not affect the volume of trench excavation, since this is taken care of in the average depth. It will affect the quantity of planking and strutting, however, for this will be required transversally across the trench in order to support the 'riser' to the step in the trench-bottom. This is shown at *B–C* in Fig. 32, which represents the longitudinal section through a typical foundation step.

Under SM D20(c), nothing will be measurable here unless the amount of step exceeds 300 mm in height, but where this amount is exceeded, the area of the riser to every such step should be bracketed-in to the trench planking and strutting, and annotated 'steps' by means of a side-note.

CONCRETE FOUNDATIONS

The concrete will follow exactly as for ordinary strip foundations, subject to an addition for extra concrete where each step occurs, and formwork to support the concrete riser.

Concrete to steps

The additional concrete is necessitated by reason of an overlap in the concrete strip, as shown by the area bounded by the dotted line in Fig. 32. This overlap must always be provided at each step to maintain adequate strength, and unless otherwise shown, may be taken as equal in length to the thickness of the concrete or 300 mm, whichever is the greater.

When measuring the concrete strip, the foundations are measured right through in the first instance, ignoring all steps, exactly as for ordinary strip foundations (care being taken to remember to and-on the disposal items). From Fig. 32 it can be seen that the extra volume of concrete required for each step is therefore the overlap *A–B* by the width of trench, timesed by the step height *B–C*. This additional volume for each step is bracketed-in with the rest of the foundation concrete, and side-noted 'steps', as in the case of the planking and strutting. There is one slight snag which may arise here, however, in connexion with the thickness of concrete.

It will be remembered that SM F3(a) requires concrete foundations in trenches to be given stating the thickness in stages: 0–150, 150–300, and over 300 mm thick. Now over the length *A–B* in the figure, the foundation concrete has undergone a change in total thickness, and this may well involve a change in thickness-range under the above clause. In this case it will be necessary to add an area of concrete timesed by its *total* depth at the step, under a fresh description, after deducting the length *A–B* from the original measurements.

144 **Formwork to steps**

Formwork will always be required to support the concrete riser, and the area involved will be the same as that for the transverse planking and strutting, previously discussed.

Formwork to 'vertical . . . sides of foundations' is a superficial measurement under SM F21(a)(iv). On the other hand, formwork to 'risers of steps' is linear under SM F22(a)(iv). Since 'risers of steps' is grouped with 'staircases' under the latter clause, it would seem to be more appropriate to measure the formwork to foundation steps under the former one, since this is less likely to mislead the estimator. This formwork should therefore be measured super, and may be bracketed-in with any other formwork required to vertical sides of foundations, as discussed in the previous chapter.

BRICKWORK BELOW DPC

We have seen how, in the case of stepped foundations, trench excavation is averaged for depth. Such is not the case with brickwork, however, for although the bottom course of brickwork may follow the contour of the ground, more or less, the top course remains constant at DPC level. Again, brickwork is an expensive item, in terms of cost, and must therefore be measured with extreme accuracy, averaging only being permissible where the degree of mathematical error involved is completely negligible.

Figure 33 represents in elevation the foundation brickwork to a building with stepped foundations, and it will be seen that all brickwork above

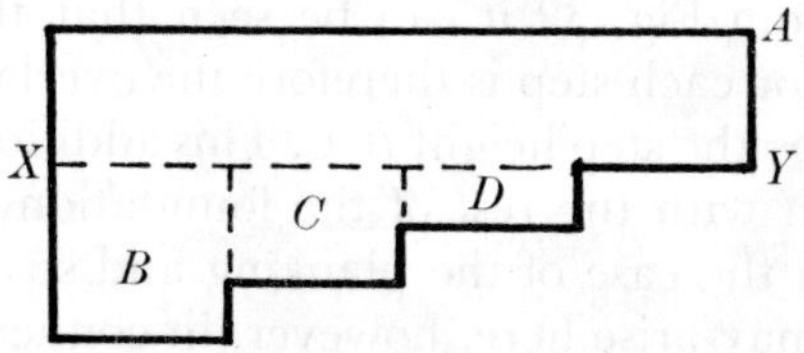

Fig. 33

the level *X–Y* may be measured without regard to the position of any of the steps. The remaining brickwork can then be dealt with on its merits, the procedure for brickwork measurement being as follows.

Brickwork above steps

First ascertain the minimum depth of brickwork, i.e., the difference in level between DPC and top of concrete at the *highest step* (*A–Y*, Fig. 33).

Now measure the whole of the brickwork to this depth all as for ordinary strip foundations, employing the trench girths as usual, and adding-in for any projecting piers.

Brickwork below steps

Each section of brickwork below the level *X–Y* must now be dealt with, and it is usually convenient to start with the deepest section *B*, as shown, and follow with *C*, *D*, etc., until all the brickwork below *X–Y* is accounted for.

For this operation it will usually be desirable to mark the position of the steps in pencil on the plan, if these are not already shown, and the calculated depth of brickwork for each section can then be marked on the drawing, prior to commencement. It will prove helpful to quickly colour in each section of brickwork on the plan with a coloured pencil, using a different colour for each depth. Careful preparation of the drawing in this way is well worth the time spent, for measurement can then proceed with speed and accuracy, each section being ticked-off on the plan as the taking-off proceeds, to ensure that none is missed, or measured twice. When measuring brickwork in this way, two further points might perhaps be worthy of mention:

(i) MEASUREMENT AT CORNERS. To ensure that these are not over-measured, it is a good plan to make a habit of measuring vertical sections net, and horizontal sections gross. Thus, when measuring a length of brickwork between two steps as shown in Fig. 34, for example, the vertical

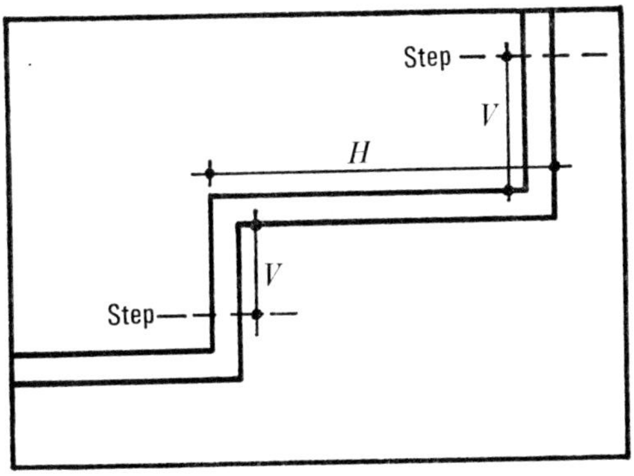

Fig. 34

sections *V* would be measured to the wall-face only of the horizontal section *H*, which would be measured the extreme length as shown.

(ii) ADDITIONS FOR PIERS. The additional depths of projections for attached piers or chimney-breasts are best dealt with separately at the end, since these will require a fresh description in any case. If the relevant depths for each section have been marked on the plan, as recommended above, it will be a simple matter to find the required depth for each pier by reference to the drawing.

Footings

Unlike the concrete strip, footings do not require any overlap where the steps occur. It follows that footings, should these occur, are completely unaffected by steps in the foundations, and are measured all as for ordinary strip foundations, previously described.

Chapter Fourteen

Deep Strip Foundations

When the sub-soil comprises firm shrinkable clay, a type of foundation is sometimes used which consists of a narrow strip of concrete underneath external walls, extending from ground level down to a metre or so in depth, as shown in Fig. 35. This type of construction, in which the

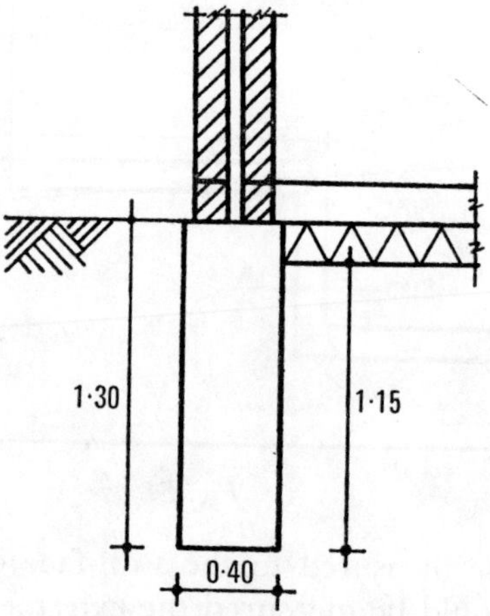

Fig. 35

thickness of the concrete strip is greater than its width, is known as *deep strip foundations*, and although the general method of measurement will obviously be the same as that for ordinary strip foundations, one or two points call for special attention, as being more likely to arise in connexion with deep strip foundations than with the type of foundation hitherto dealt with.

Adjustment of disposal

It will be apparent that if the foundation trench is completely filled with concrete up to original ground level, there will be no backfilling required. When measuring the trenches therefore, *Remove surplus excavated material from site* should be anded-on to the trench description, instead of the usual *Return fill and ram surplus excavated material around foundations*. There will then be no disposal adjustment descriptions to and-on to the concrete when this is measured.

Concrete abutting hardcore

Where hardcore abuts concrete, as shown in the cross-section, the surveyor must decide which is most likely to be executed first, the concrete or the hardcore, since this will determine whether or not any formwork is required.

In cases where the abutment is a *vertical* one, as illustrated, one might reasonably assume that the concrete would be poured first, and formwork would therefore be necessary to support the vertical face at the point above the inner trench-side, where the surface excavation will subsequently be backfilled with hardcore under the floor. SM F21(a)(iv) applies here, and a superficial measurement of *sawn formwork to vertical sides of foundations* meets the case.

Where it is suspected that the hardcore bed might be laid before the concrete is poured, an item of *hand-packing hardcore* should be measured under SM D22(f), in lieu of formwork. As a general rule, however, the concrete may be assumed poured first, except in cases there the type of abutment is such that hand-packing of hardcore would obviously be more convenient than the construction of formwork. This often applies in the case of raft construction, when the edge-beams have splayed internal angles, and is further discussed in the chapter on raft foundations.

Narrow trench excavation

In the foregoing paragraphs it has been assumed that the work is constructed as shown in Fig. 35, and that the foundation trench is dug to the exact width of the concrete, as in the case of strip foundations hitherto discussed. It will be observed, however, that the particular trench in question is 400 mm wide and 1·15 metres deep, and in order to appreciate the special significance of these dimensions we must now turn to SM D10(a) which states *inter alia*:

> *In the case of trenches over* 1 *metre deep the minimum width measured shall be* 0·75 *metres for the full depth and this minimum shall apply also to the concrete foundations therein.*

This particular rule applies to any type of foundation trench, not only to *deep strip* foundations, and a watchful eye must always be kept open when measuring ordinary strip and stepped foundations in case the trenches should fall into the above category. Since it is far more likely to arise with deep strip foundations, however, it is more conveniently dealt with at this stage. The following points should be noted in cases where this 'narrow trench rule' applies.

SURFACE STRIP. The importance of a careful routine preliminary study of the drawing is emphasized here, for without this the trench depths may not be noticed until after the surface excavation has been measured, in which case the surveyor will have to tear up his dimensions and start again!

Why is the surface strip affected? Under the above clause, could not the minimum width be achieved by widening the trench *inwards* and thus economizing on hardcore?

This solution would satisfy the requirements of the SMM, but if the structure was built in this way, the walls would be off-centre of the concrete foundations and the resultant overturning moments might be unacceptable from a structural viewpoint. Such a solution should not therefore be adopted without the consent of the designer, and unless this is forthcoming, the additional concrete spread must be taken into account when calculating the surface strip area, *the waste calculations being carefully annotated accordingly.*

If only a centimetre or two of additional width is involved, it would probably be acceptable to widen inwards in fact, but since the time saved in adjusting surface strip would be offset by the necessity of a girth adjustment as between foundations and external walls, the best rule is to widen the trench on its centre-line in all normal cases.

PROJECTIONS. The additional areas of trench (and surface strip) required for the projection of piers, chimney-breasts, etc., normally given by the equations:

$$TW = BW$$
$$TL = BL + 2p$$

will be affected, where these occur in sections of trench the width of which have been adjusted under the narrow trench rule.

The additional trench length *TL* will remain unaffected, but the additional width *TW* will no longer be equal to the brickwork width *BW*, and the equation $TW = BW$ will therefore no longer hold good.

A cross-section through a typical projection is shown in Fig. 36, *BW*

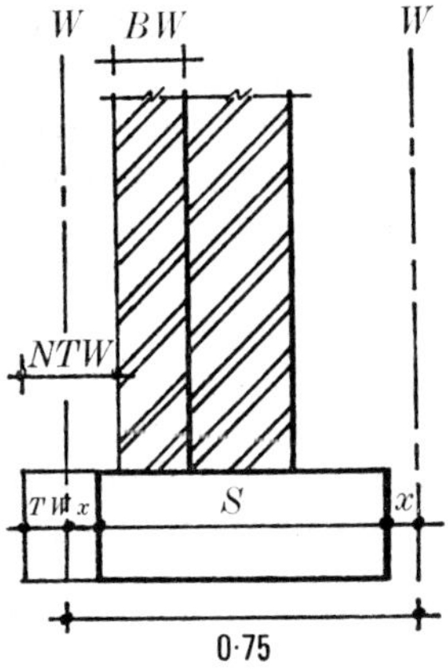

Fig. 36

being the projection of brickwork and *NTW* being the normal additional amount of trench width. It is assumed that the narrow trench rule applies, however, and the dotted lines *W* indicate the minimum width required to be measured, the concrete spread *S* being less than 0·75 metres.

Now since the trench has been widened by the amount *x*, it will be seen that the additional amount of trench width required is $NTW - x$. In fact:

$$TW = BW - x$$

The value of *x* will always be half the difference between the concrete spread shown on the drawing and the minimum width required to be measured, viz., 0·75 metres. In short:

$$TW = BW - \left(\frac{0{\cdot}75 - S}{2}\right),$$

which may be called the Narrow Trench Projection Formula.

It will be apparent that if the result of the above formula is zero or negative, no additional excavation (or concrete) is required for the projection, unless specially ordered for structural reasons. In this case therefore, such projections are normally ignored, until measurement of brickwork commences.

The case dealt with above represents a side pier, but any corner piers will, of course, require similar adjustment for width, while those occurring on the outside of the building will affect the surface strip as well as the actual trench excavation, etc.

A point of procedure arises in connexion with foundations which are *actually executed* in accordance with Fig. 35. Does SM D10(a) cover cases where the design of deep strip foundations is specifically intended to cater for narrow trenches mechanically excavated to the required width? This would seem an unreasonable interpretation, for it cannot be supposed that the SMM requires the contractor to be paid for twice as much concrete as he actually uses, which would be the case in this particular instance.

If it is subsequently found therefore that the work is executed in accordance with the drawing, instead of in accordance with SM D10(a), Clause A1 might be invoked as overriding D10(a), and the quantities adjusted at post-contract stage so as to accurately represent the work executed.

In the event of the taker-off suspecting that this might be the case, and the designer confirming that the drawing must be adhered to, a special preamble to this effect should be inserted in the *BQ*, and the work measured in accordance with Clause A1, which, after all, states that the bills shall accurately represent the works *to be* executed, and would therefore appear to cover such a case.

It should be stressed, however, that unless the taker-off has special reasons for supposing that the designer intends the work to be carried out exactly as shown on the drawing, the narrow width rule in SM D10(a) must always be applied in cases where a foundation trench over 1 metre deep is less than 0·75 metres wide.

Internal foundations

Load-bearing internal walls may be built up off the floor slab in certain cases, when the foundations to such walls may take the form shown in Fig. 37. Here the strip foundations may comprise what is, in effect, a thickening of the floor slab, and where this strip is very shallow, it is sometimes measured linear and described as extra over the floor slab.

The SSM does not expressly sanction this method of measurement, however, and unless the strip is only a few centimetres in depth (or is reinforced), it is best treated as ordinary foundation concrete up to the level of the underside of floor slab, and measured cube.

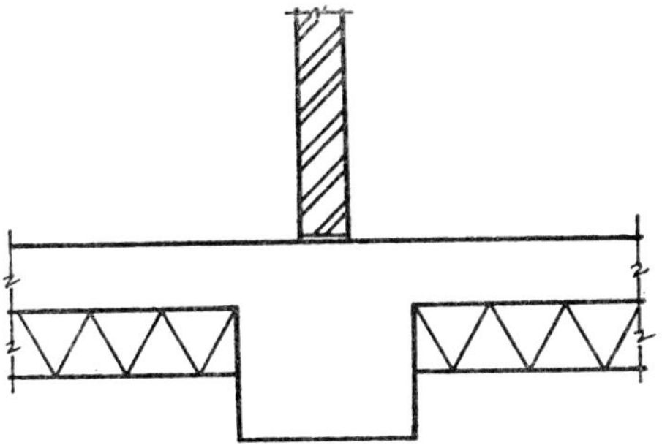

Fig. 37

Note especially that concrete projections of this kind, below floor slabs, must only be measured up to *underside of floor slab*, since this will be the actual method of construction; the floor slab being laid right across afterwards. This ensures that the joint between the floor slab and the foundation strip will be horizontal and not vertical, which might lead to settlement.

At first sight it may seem incorrect to describe the concrete immediately below the internal wall as (say) 150 mm thick, when the total thickness of concrete at that point is in fact 150 mm plus the thickness of the projecting foundations. From the constructional point of view, however (and therefore from the estimator's viewpoint), this method of measurement is correct, since the 150-mm slab will have to be laid as a more or less separate operation, as described above.

Planking and strutting

A point to bear in mind when measuring foundations of the type shown in Fig. 37 is in connexion with the measurement of planking and strutting. The foundation trench itself will be measurable from formation level downwards as usual, but these trenches are sometimes quite shallow, and the depth must be checked before measuring planking and strutting, which is not measurable to any excavation which does not exceed 300 mm in depth (SM D20(c)).

DISPOSAL OF WATER

This chapter on deep strip foundations provides a convenient point at which to introduce the student to the problem of water disposal. Like the application of the Narrow Trench Rule, the question of water disposal may apply to any type of foundation, but is more likely to be relevant in the case of deep rather than shallow foundations.

 Nevertheless, the term 'deep strip' foundations does not necessarily mean that these foundations will be any deeper than ordinary strip foundations; it merely refers to the type of construction, illustrated in Fig. 35.

General water

To enable work to proceed, it is necessary to keep all excavation reasonably dry and free from water. Various methods are available, but the employment of mechanical pumps is the one most usual in the case of small building sites.

Water caused by rain, etc., is called *general water*, which is defined by SM D19(b) as *all water except spring or running water and water below the water-level in the ground.*

The SMM requires that keeping the excavations free from general water '. . . shall be given as an item' (SM D19(b)), and since this will require to be done in the case of all projects involving excavation in the open, it is normally given as a standard preamble item in the bill, and is not therefore written at taking-off stage.

Spring or running water

A provisional sum must be given for keeping the excavations free from spring or running water under SM D19(d), which allows the alternative of measuring a provisional number of actual pumping hours instead. A provisional sum is the simplest solution, and suitable for small projects where it is suspected that running water might be met with in the excavations. For large projects, when extensive pumping is thought to be required, a provisional number of pumping hours in the bill enables the estimator to tender competitively for this item, and may therefore be preferable in this case.

Water below the water-level

Permanent underground water exists in the form of underground streams, etc., and the level of such water, which varies with the place and time, and is often known as the 'water table level' is referred to in the SMM as the *water-level.*

It frequently happens, in the case of small building projects with shallow foundations, that this water-level is known to be well below foundation depth, and in this case the taker-off will not be concerned with this class of water.

Where foundations are deep, however, or the water-level is known to be near the surface, provision must be made for dealing with this water, and the SMM requires that the following steps must be taken.

(a) ASCERTAINMENT OF WATER-LEVEL. When it is suspected that the water-level may be above the lowest point of the foundations, the water-level should be ascertained and stated in accordance with SM D1(a). Only where this information is not available may it be ascertained at post-contract stage.

(b) PROVISION OF BILL ITEM. Keeping the excavations free from water below the ascertained water-level must be given as an item under SM D19(c). Where lack of information has prevented the ascertainment of the water-level at taking-off stage, a provisional sum may be given (or provisional pumping hours measured).

(c) MEASUREMENT OF EXCAVATION. In addition to the above, the quantities of excavation below the water-level must be measured and described as extra over ordinary excavation, under SM D6(d).

It will be apparent from the foregoing that the taker-off will be faced with one of three situations regarding disposal of water. They are as follows.

(1) Water-level below foundations

When taking-off shallow foundations (not exceeding about 1·5 metres deep) the student will be justified in assuming that the water-level is below foundation level *provided that nothing to the contrary is indicated in the specification notes or on the drawing.*

Subject to the last proviso, the student will also be justified in assuming that (a) general water is deemed taken care of in preambles, and (b) running water is deemed unlikely and will, if necessary, be paid for out of the general provisional sum for contingencies allowed for in preliminaries (see SM B25).

(2) Water-level above foundations

Where the water-level is indicated on the drawing or in the specification notes as being above the lowest level of foundations, a 'special preamble' must be written by the taker-off on the following lines:

> *The water-level in the ground was measured on* 12*th August* 1965 *and found to be* 1·25 *metres below original ground level, which shall be deemed to be the normal water-level. The contractor is to allow for keeping the excavations free from all water below this level.*

154 Note that the date the water-level was ascertained must be given, in accordance with D1(a), and the practice is to include this (together with the actual level) in the preamble item requiring the contractor to keep the excavations free from water below this level. Thus the above clause is made to satisfy the provisions of both SM D1(a) and SM D19(c).

Unless the specification notes suggest otherwise, *general water* may normally be deemed included in preambles, but the question of *running water* still arises. If the drawing or specification notes suggest that this may be encountered, the following (or similar) provisional sum should be written by the taker-off, in order to comply with SM D19(d).

> *Allow the provisional sum of*
> £20.00 *for keeping the excavations*
> *free from spring or running water*
> *above the normal water-level,* 1·25 *metres*
> *below ground level.*

The general preamble covering *general* water would be as follows:

> *The contractor is to allow for*
> *keeping the excavations free of*
> *all general water (other than*
> *spring or running water, for which*
> *a provisional sum has been included)*
> *above the normal water-level,* 1·25 *metres*
> *below ground level.*

As previously stated, this preamble covering general water may normally be assumed included at working-up stage. Where the taker-off is given the normal water-level as being above foundation level, however, the general water clause is often drafted by the taker-off, in order to ensure that the wording is appropriate, and this practice is recommended to the student.

There remains the measurement of the extra over item for all trenches below normal water-level. This is probably best carried out after measurement of the trenches, which should then be cubed below normal water-level and described under SM D6(d) as:

> *Extra over ordinary excavation*
> *for excavating below the*
> *normal water-level.*

(3) Water-level suspected above foundations

In the case of deep foundations (3 metres deep or so) it would be reasonable to suspect that the normal water-level might be reached. Failing any specific information as to water-level, the student should proceed under the second sentence of SM D19(c), which states:

> *Where the water-level remains to be ascertained in accordance with the provisions of Clause D1(a) hereof, the work shall be given as a provisional . . . sum . . .*

It is usual to merge this provisional sum with the one for running water, thus:

> *Allow the provisional sum of* £20.00 *for keeping the excavations free from all water other than general water, i.e., from spring or running water or water below the normal water-level.*

This is accompanied in the bill by the normal preamble covering general water.

If the water-level subsequently proves to be above foundation level, the extra item required by SM D6(d) will, of course, need to be measured also, and it is desirable that provisional quantities should be taken-off to cover this in the first instance. Alternatively the provisional sum can be expanded to include any funds necessary to pay for this particular extra, in cases where the amount involved is likely to be a comparatively small one. The actual amount itself may be left blank by the taker-off, for later decision, unless included in the specification notes as a definite sum of money.

Running silt or sand, etc.

It will be observed that SM D6(d) requires not only excavation below normal water-level to be measured as extra over ordinary excavation, but also requires excavation in *running silt or running sand (grouped together)* to be similarly treated.

The taker-off need not normally mention the nature of the ground in his excavation descriptions, since this will be covered in preambles in accordance with the first two sentences of SM D1(a). Running silt or sand is an exception, however, as is rock under SM D6(e). If the specification indicates any excavation in running silt or running sand, this must be cubed and described as:

> *Extra over ordinary excavation for excavating in running silt or running sand.*
> *(Provisional)*

156 the normal practice being to mark the item 'provisional' as shown, in case adjustment to the quantity is necessary when operations actually commence.

The measurement of excavation in *rock* under SM D6(e) is also measured extra over excavation, but unlike water and sand, rock is measured extra over *each of the various descriptions* of excavation. Thus, instead of 'extra over ordinary excavation . . .' the description for rock would be:

Extra over foundation trench excavation not exceeding 1·50 *metres deep for excavating in rock.*

Ditto exceeding 1·50 *metres and not exceeding* 3 *metres deep for excavating in ditto.*

and so on, each type of excavation being kept separate. It is usual to mark these quantities as provisional also, since the extent of the rock is usually difficult to predict with certainty.

In the case of rock, the alternative method of including it in the actual excavation description is allowed under SM D6(e), but it will usually be found more convenient to measure it extra over, as described above, unless all or most of the excavation is in rock.

Running sand or rock below water-level

It will be noted that excavating in running silt or running sand *below the normal water-level*, should this occur, must be described, under SM D6(d). Where the general excavation has been measured for 'Extra over ordinary excavation for excavating below the normal water-level', it will be necessary to deduct this item when running sand occurs below water-level, since running sand is not 'ordinary' excavation, and the extra cost of excavating below water-level must in any case be accounted for once only. The following items will therefore be necessary:

Extra over ordinary excavation for excavating in running silt or running sand below the normal water-level.

&

Deduct *Extra over ordinary excavation for excavating below the normal water-level.*

The same procedure will be necessary in the case of rock. Rock below water-level is not specifically referred to in the SMM, but where rock is given as extra over each of the various *descriptions of excavation* under SM D6(e), it is reasonable to assume that a *description of excavation* is intended to include that for excavating below water-level. In order to avoid an extra over an extra over, therefore (which would occur if one wrote 'Extra over extra over ordinary excavation for excavating below the normal water-level for excavating in rock') the following descriptions would be appropriate, where rock occurs below water-level:

Extra over foundation trench excavation exceeding 1·50 *and not exceeding* 3 *metres deep for excavating in rock below the normal water-level.*

&

Deduct *Extra over ordinary excavation for excavating below the normal water-level.*

Planking and Strutting Below Water-level

Where work below normal water-level occurs, not only must an additional excavation item be measured (as well as a special preamble), but the planking and strutting description must also be adjusted, to conform with SM D21(d). This applies also to work in running silt and running sand *but not to rock (see SM D21(e)).*

Since planking and strutting is *not* measured in *successive* stages, as is excavation, any planking and strutting which extends wholly or partly below the normal water-level must be so described and measured from the starting level of the excavation to the full depth. The following is a typical description:

Planking and strutting to sides of foundation trenches starting at formation level and not exceeding 3 *metres deep and extending partly below the normal water-level.*

Raft Foundations

Where the bearing capacity of the ground is poor, or where the structure is a very light one, foundations may take the form of a solid slab of reinforced concrete laid over the area of the building at ground level, or below. This is known as a solid slab raft, and the edges of the raft are often turned down to form an edge beam, the chief purpose of which is to contain any hardcore underneath the raft, and to prevent erosion of the soil from around the edges. In certain cases cross beams may also be provided, to help stiffen the raft, or to help strengthen it under load-bearing internal walls.

Raft slabs are normally reinforced with steel in order that any stresses which may build up are satisfactorily distributed, and in some cases the edge beams may also be reinforced, as may the cross beams, if any. The introduction of steel reinforcement into concrete foundations is not necessarily confined to raft foundations; strip foundations may also be reinforced when the occasion demands, bar or fabric reinforcement being laid in the trenches prior to deposition of the concrete. The principles of measurement of reinforced concrete, discussed in this chapter, may therefore be regarded as applying to all types of foundation, although in domestic construction reinforcement is usually confined to foundations of the raft or beam type.

The general order of taking-off and principles of measurement of raft foundations are basically the same as for strip foundations, as far as surface excavation and any necessary trench excavation, etc., are concerned; unreinforced edge beams normally being treated as ordinary concrete in foundations, as in the case of deep strip foundations. Raft foundations, however, sometimes give rise to special problems concerning working-space, especially where the raft is below ground level, so before discussing the principles of measurement of reinforced concrete as applied to raft slabs and beams, it will be desirable to first deal with the question of working-space in the excavation.

WORKING-SPACE

Figure 38 shows a section through a simple reinforced concrete raft foundation of the type set in the Quantities (1) paper of the 1964 RICS

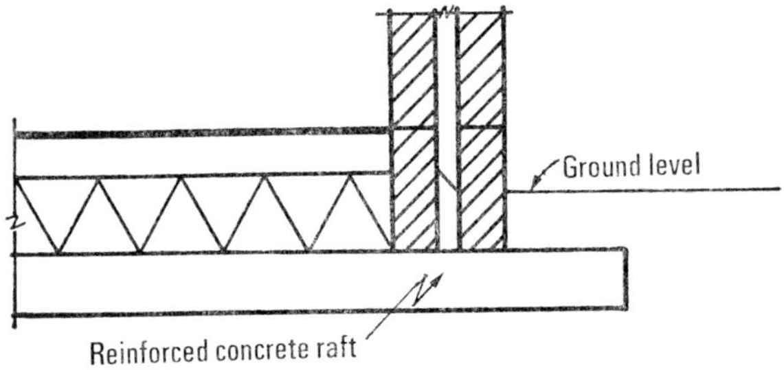

Fig. 38

intermediate examination. The measurement of surface excavation down to the underside of concrete raft (formation level) seems straightforward enough, but in fact a problem arises owing to the fact that formwork might be necessary to the edge of the concrete raft. Two questions must be asked here, (i) will formwork be necessary?, (ii) if so, how does this affect the excavation?

We shall see that if formwork is in fact required, working-space in the excavation will need to be provided for it. It is therefore necessary to decide at an early stage whether or not any formwork is required.

Formwork below ground level

'Formwork shall be measured as the actual surfaces of the finished structure which require to be supported during the deposition of the concrete. . . .' So states SM F20(a).

We have seen that in the case of normal unreinforced strip foundations the vertical sides of the concrete strip are supported by the sides of the excavation, and so do not normally require formwork, except transversally, or at a junction where no trench-side exists. The concrete is in fact merely tipped into the trench and levelled-off at the required level. Can this system of using earth as formwork be universally applied, in cases where concrete is required below ground level?

It is, in fact, sometimes necessary for the contractor to use formwork below ground level instead of relying on the earth to uphold the concrete, and where its use is deemed necessary by the designer, it must, of course, be measured. It is normally used in the following cases:

(a) Where the nature of the ground is such that its use as formwork would be unsuitable. Thus, topsoil, near the surface, may be too unfirm, or the substrata may consist of running sand, or other unsuitable material.
(b) Where reinforced concrete is required underground, proper formwork may be necessary in order to ensure that correct 'cover' to the

reinforcement is maintained and that the concrete is protected from deleterious substances in the soil during deposition and setting.

Formwork might also be used underground where the contractor has racked-back the side of the excavation for his own convenience, in order to avoid the use of planking and strutting, which is measured whether or not any is in fact required, under SM D20(b). In this case, it is not normally measurable, of course. In the case of (a) or (b) above, however, it should be measured at taking-off stage, and since formwork is not permanent work, and is therefore not shown on the drawings, the taker-off himself must decide whether or not such formwork will be required, and measure accordingly.

When in any doubt, the surveyor in practice will query this with the designer, whose responsibility it is to make the decision. For the guidance of students, however, the following suggestions are made:

(i) Where the concrete is reinforced, and the reinforcement extends close to the edge of the concrete, measure formwork.
(ii) Where concrete is at a very shallow depth, measure formwork if the drawing or specification notes suggest unfirm ground.
(iii) Formwork is not usually measured in the case of normal strip foundations, even if these are reinforced, unless they are shallow and in unfirm ground or the reinforcement is of a complex nature.
(iv) In all cases, read the specification carefully, and try to visualize the actual site operation, especially the pouring of the concrete. When in doubt measure formwork.

Working-space for formwork

SM D6(g)(ii) requires a standard allowance of 0·25 metres from the face of any work which requires formwork not exceeding 1 metre deep below starting level, and this allowance (twiced for both sides as necessary) is added-in on waste to the excavation measurements. Thus waste calculations for the area of surface strip for the raft shown in Fig. 38, assuming formwork to be required, would be in the following form:

Intl. dims.		7·50	3·75
walls	2/·27	0·54	0·54
conc. proj.	2/·30	0·60	0·60
w/s for f'wk.	2/·25	0·50	0·50
		9·14	5·39

Working-space for planking and strutting

This is not measurable, since SM D6(b) states that '. . . no allowance shall be made . . . for any extra space required to accommodate planking and strutting'. The provision of planking and strutting may need to be considered when deciding whether or not formwork below ground will be required, however. If the ground is firm enough to be used as formwork, planking and strutting will probably not be required, although it must, nevertheless, always be measured under SM D20(b). Where planking and strutting is actually required, however, the desirability of using formwork will be enhanced, since the planking and strutting will otherwise interfere with the surface of the concrete. In this connexion, the question arises as to how, with ordinary strip foundations, the planking and strutting can be recovered after the concrete has set. If it cannot, then it must be described as 'left in', in accordance with SM D21(f). Why it is never measured in this way under normal circumstances?

The answer is that in the case of normal strip foundations the planking and strutting is either stopped-off above concrete level, or extracted upwards before final set of the concrete has taken place. This would not be acceptable in the case of high quality reinforced concrete below ground level, however, when formwork (and consequent working-space) should be measured.

REINFORCED CONCRETE

The term *reinforced concrete* applies to concrete reinforced with steelwork in such a manner that both the concrete and the steelwork take a share in resisting structural stresses which may be set up.

Steel girders may be encased in concrete, and thus comprise a mixture of steel and concrete. In this case, however, the concrete usually serves merely to protect the steelwork, and is not intended to share any stresses. The resultant mode of construction in this case is not termed reinforced concrete, but 'structural steelwork', although steel sections are occasionally used as reinforcement in connexion with reinforced concrete; but this is unusual.

There are two main types of reinforced concrete:

(i) In-situ,
(ii) Precast.

In-situ reinforced concrete is the type normally used in foundations of the kind under consideration, and consists of concrete poured around reinforcement fixed in position on the site.

Reinforced concrete descriptions

The principles relating to the framing of descriptions for concrete, as discussed in Chapter 11, will apply equally to reinforced concrete, subject to the following considerations. SM F20(c) states:

> *Reinforced in-situ concrete and its reinforcement and associated formwork shall be given under an appropriate heading.*

This is achieved by inserting the word 'reinforced' into the concrete description. If this word is not present, the worker-up will assume plain concrete, and, to ensure that reinforced concrete is billed under the appropriate heading, strip foundations (for example) containing reinforcement should be described as:

> *Reinforced concrete* (1:2:4) *over* 150 *and not exceeding* 300 *mm thick in foundations in trenches.*

It should be noted in passing that not only should the concrete be described as reinforced, but the formwork also requires reparation under SM F20(c). 'Sawn formwork to vertical sides of *reinforced* foundations' is required to achieve this, although with manual abstracting, the worker-up can sometimes be relied on to separate any 'unreinforced' formwork by inspection of associated items in the dimensions.

Ground-beams

The concrete strip projecting downwards from a raft, in the form of an edge beam as shown in Fig. 39, will normally require a foundation

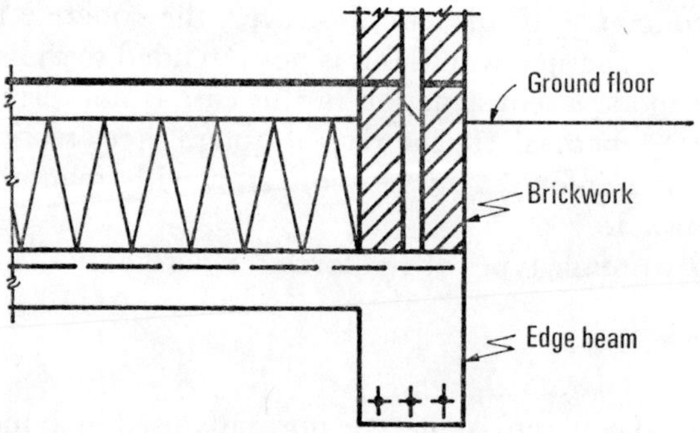

Fig. 39

trench to be dug, to receive the concrete. Like ordinary strip foundations therefore, this concrete, if unreinforced, is properly described as *concrete in foundations in trenches*. In cases where formation level is down to bottom of concrete, as in Fig. 40, no trench will be required, and *concrete in foundations* would be a more appropriate description.

In the case of splayed edge beams, as shown in Fig. 40, the concrete below bottom of slab level is cubed by taking the average width, and an item of *hand-packing hardcore to form battering face* is measured in accordance with SM D22(f) in lieu of formwork on the inside.

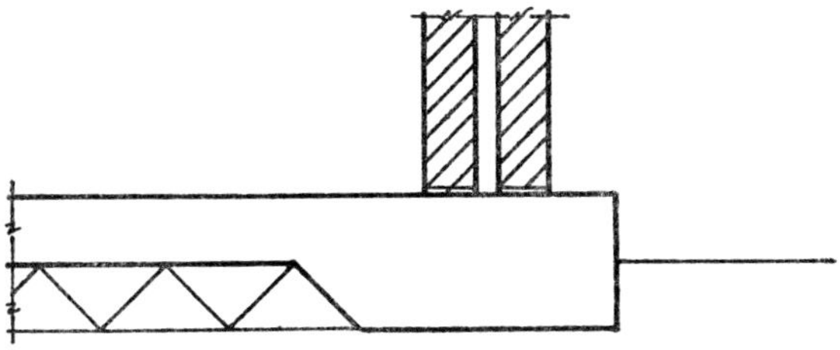

Fig. 40

At this point, however, it is necessary to examine SM F3(d), which states:

> *Ground-beams . . . shall . . . be given separately in cubic metres stating the sectional area . . .*

The question therefore arises, when does a concrete strip foundation become a ground-beam? What precisely is meant by the term ground-beam?

A beam is usually thought of as a horizontal member spanning between two supports, and such beams are sometimes used below ground to span between pile-caps or column bases. Is the term 'ground-beam' in SM F3(d) restricted to such cases? If so, edge beams of the type under consideration would be excluded from the definition. On the other hand, it may be argued that such a strip of concrete, *if reinforced*, acts as a kind of beam, spanning between irregularities in the bearing-capacity of the sub-soil.

Again, if the concrete is poured into formwork, and tamped around reinforcement—then from the point of view of the actual site operation, i.e., from the cost aspect, it is analogous to a reinforced concrete beam, regardless of actual function.

We come to the conclusion therefore that 'ground-beams' in SM F3(d) refers to foundations which comprise a beam-like structure, consisting of reinforced concrete poured into a confined space around reinforcement, and supported by formwork. Consequently if the edge beam is reinforced and requires formwork, it should be treated as a ground-beam under SM F3(d), not as foundations in trenches under SM F3(a). A suitable description for the reinforced edge beam shown in Fig. 39 would be as follows:

> *Reinforced concrete* (1:2:4) in
> *ground-beams over* 0·05 *but not*
> *exceeding* 0·10 *square metre in*
> *sectional area.*

It will be noted that the method of measurement (cube) is the same in either case, but that ground-beams require statement of the sectional area in accordance with SM F1(h), instead of thickness-range under SM F3(a).

Although the description of the concrete includes the word 'reinforced', this is merely a device to ensure that it appears in the bill under an appropriate heading, in accordance with SM F2(c), and does not refer to the actual reinforcement itself, which must be dealt with separately, under the sub-section of Reinforcement, SM F17–18. Reinforcement may be of two kinds (i) bar (or rod) reinforcement, or (ii) fabric (or mesh) reinforcement.

Bar reinforcement

It will be seen from SM F17(b) that bar reinforcement must be given in kilogrammes. To achieve this, each bar is measured linear by the taker-off, the diameter being stated in the description. Since each diameter of mild steel bar has a standard weight per metre lineal, this will enable the total weight for each size to be calculated at working-up stage. All the taker-off needs to concentrate on is measuring the exact length of each bar against a suitable description.

Measurement of bar reinforcement

In the case of complex reinforced concrete structures, a bar-bending schedule may be provided by the designer at taking-off stage, which will greatly facilitate measurement. In the case of simple reinforced foundations as under discussion, however, this will not be the case, and the taker-off must carefully examine the drawings and measure the precise length of each individual bar, *including the girth of any cranks, hooks, anchors, etc.*

Unless otherwise indicated, bars may be presumed hooked at each end, measurement being taken to the points indicated in Fig. 41, and the following allowances made[1]:

Each hook: 9 *Diameters (minimum* 100 *mm).*
,, *anchor:* 5 ,, *(* ,, 100 *mm).*

[1] See BS 4466:1969.

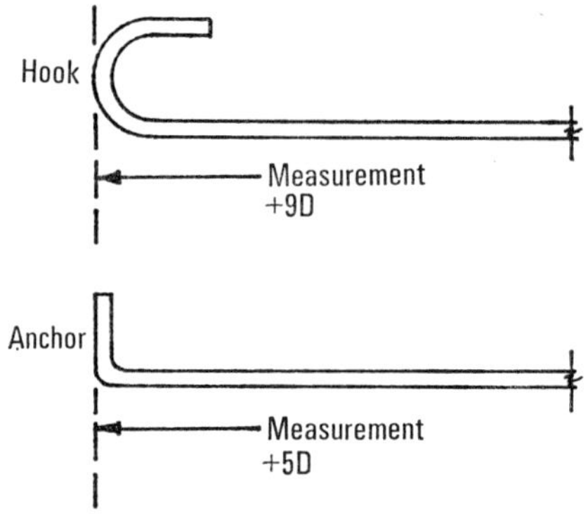

Fig. 41

The initial measurement is derived from the overall dimension of the concrete, from which twice the amount of stipulated concrete 'cover' is deducted on waste. This cover, if not specified, is usually taken as the diameter of the largest permissible piece of aggregate, e.g., 20 mm in the case of normal 1:2:4 concrete. Thus the waste calculations for a 12-mm longitudinal bar in a 3·5-metre slab, or section of ground-beam, would be as follows:

concrete		3·50
less cover	2/·025	·05
		3·45
hooks	2/9/·012	·216
		3·666

Descriptions of bar reinforcement

SM F17(a) requires particulars to be given of (i) kind and quality of steel, (ii) any tests of the bars, and, (iii) any restrictions as to hot or cold bending. With the exception of the *kind* of steel, these particulars are normally covered by preambles, and need not form part of the taking-off description. The *kind* of steel is usually stated, however, in order to differentiate between mild steel and high tensile steel, which could both occur in the same project.

Bearing this in mind, it will be apparent from SM F17 that a description of bar reinforcement should contain the following information:

(a) Bar diameter,
(b) Kind of steel,
(c) Type of bar,
(d) Location.

166 and the normal practice is to give this information in the order stated, thus:

12 *mm Diameter mild steel bars*
in stirrups in foundations.

It should be noted that the *location* must be classified under SM F17(b) (i)–(viii), any bars extending from (say) foundations to beds being described as *in foundations and beds.*

With regard to the *type of bar*, two main categories are implied by the SMM:

(i) Links, stirrups, etc., referred to in SM F17(g),
(ii) Ordinary reinforcing bars.

Links, stirrups, etc., do not normally occur in simple reinforced foundations of the type under discussion, and if no mention is made of the type of bar in the description, ordinary bars will be presumed, i.e., bars not classifiable as links, etc., under SM F17(g). Thus all bars in foundations may normally be simply described as:

20 *mm Diameter mild steel bars*
in foundations.

It should be especially noted that only bars classifiable under SM F17(g) as links, etc. (should these occur), merit a description of *type* under (c) above. Longitudinal bars, for example, may be so described *on the drawing*, which might designate some bars as 'anti-crack reinforcement', 'transverse rods', or by other functional names of this kind. There is no separate SMM classification of this sort; all bars which are not classifiable as links, etc., must simply be collected together under their diameters as ordinary bars, and given a *location* classification only, under SM F17(b)(i)–(viii).

It will be noticed that SM F17(a) requires bends, hooks, tying wire, distance blocks and ordinary spacers to be given in the description. It is standard practice, however, to relegate this information to a preamble or bill heading, and not to pad out the actual taking-off description with these details, which will apply to the reinforcement generally (but bars specially bent to curve must be so described under F17(d)).

Finally, an eye must be kept open for any bars over 10 metres long, which must be separated out under SM F17(c) and described in the following manner:

20 *mm Diameter mild steel bars*
over 10 *and not exceeding* 12 *metres*
long in foundations.

Bars of greater length are classified into their appropriate stages of 2 metres.

Fabric reinforcement

Fabric reinforcement comprises main steel bars with cross bars electrically welded to them, forming a grid or mesh. The size of the mesh (i.e., the nominal pitch or distance apart of the bars) is stipulated by the British Standard, which lays down a standard range of meshes, each with its corresponding nominal weight per square metre of fabric reinforcement.

The British Standard allocates a reference number to each mesh, and this number is all that is required to define any particular mesh and weight, fabric reinforcement usually being specified by this number, it being presumed to refer to the relevant British Standard unless otherwise stated.

SM F18(b) requires the mesh and the weight per square metre to be given, and if both of these are specified, they should both be given in the taking-off description. If the reference number only is specified (as is frequently the case) the student is justified in taking the view that this number (which refers to both mesh and weight) will satisfy the provisions of the SMM in this regard.

Measurement of fabric reinforcement

Under SM F18(b) fabric is measured super, laps being ignored in the measurement (but stated in the description).

A careful watch must be kept, however, for *linear* fabric. SM F18(c) demands that strips required to be in one width '*e.g., in foundations under walls . . .*' must be measured linear, and such strips often occur in raft foundations at points of contraflexure, in association with superficial fabric. Fabric in ordinary strip foundations would almost certainly come under the linear category, since it would probably be in the form of longitudinal strips, which would normally be in one width.

Having thus measured the fabric itself, the taker-off must then check for labour items measurable under SM F19. When linear items have been measured to any raking or curved edges of the fabric under SM F19(a), a search must be made for any notching of the fabric around obstructions, to be enumerated under SM F19(b).

Descriptions of fabric reinforcement

The main components of a fabric description, in order of drafting, are as follows:

(a) Reference number (or mesh and weight),
(b) Lap details,
(c) Location.

Bends, tying wire and distance blocks, referred to under F18(a) are usually relegated to a standard preamble or bill heading, as in the case of bar reinforcement. Location must be strictly confined to the bar reinforcement location classification, previously referred to; a typical description for superficial fabric being as follows:

> *Fabric reinforcement, reference*
> B503, *lapped* 150 *mm at joints and*
> *laying in foundations.*

In the event of the fabric being in narrow strips, as most likely in foundations of the type under consideration, a typical description for linear fabric would be:

> *Fabric reinforcement, reference*
> B503, *in* 1·5-*metre widths with* 150-*mm end*
> *laps, and laying in foundations*
> *under walls.*

Raft slabs

Having measured the concrete, formwork, and reinforcement up to the underside of the slab, the actual raft slab itself must next be measured. If this occurs at ground level, it will probably form the structural ground-floor slab as well, and we have seen that, with ordinary strip foundations, this slab (together with hardcore under) is usually measured in the 'Floors' section of the taking-off. If the raft is below ground level, and a separate floor slab provided, this might still be the case, but with raft foundations generally, it is usually convenient to measure all slabs, etc., below DPC level with the foundations, together with any necessary hardcore beds; the ground-floor *finishings*, together with any cement-and-sand screeded or floated beds being measured in either the 'Floors' or the 'Internal Finishings' section of the work. Any *timber* ground-floor construction is always measured in the 'Floors' section, but the student's attention is drawn to the necessity of always reading the question-paper carefully when taking examinations, to see whether any special instructions are given regarding sequence of measurement, or what is to be included in the candidate's taking-off. In the absence of such instructions all slabs below DPC and all hardcore beds should be taken with raft foundations.

Plain slabs

In-situ concrete slabs are divided by the SMM into two main groups:

(i) Beds, under SM F5,
(ii) Suspended floors, roofs, etc., under SM F7.

Plain or reinforced slabs supported by hardcore or earth are classed as *beds*, and this term will apply to raft slabs and ground-floor slabs. Thus the unreinforced raft in Fig. 40 would be measured under SM 5(a) and described as:

> 150 *mm Bed of concrete* (1:2:4)
> *with tamped surface.*

The normal practice is to put the thickness first, as shown, to facilitate correct bill sequence, and the surface treatment must be included in the description under SM F14(a). In the case of beds it is usual to assume a tamped finish unless otherwise specified, although an ordinary spade finish may be adequate for a bed laid to receive hardcore, as in Fig. 38 in which case no surface treatment need be mentioned.

Reinforced slabs

In the case of a raft or floor slab reinforced with bars or fabric the description would be:

> 150 *mm Bed of reinforced concrete*
> (1:2:4) *with tamped surface.*

Bars in beds

The reinforcement is measured as previously described for ground-beams. In the case of bar reinforcement, two factors will be required,

(i) The number of bars,
(ii) The length of the bars.

If the pitch of the bars is given on the drawing, it will be necessary to calculate the number of bars on waste by dividing the slab dimension by the pitch, having made allowance for the cover at each end. The result of such a calculation will not actually give the number of bars, however, but only the number of *spaces* between the bars, which will be one less than the number of bars. It will therefore be necessary to add one, as in the following example, which represents the waste calculations required to determine the number of bars specified as being at 150 mm pitch in a slab 3·50 metres wide:

slab dimension		3·50
less cover	2/·025	0·05
	150)	3·450(24 + 1

The timesing factor is therefore 24 + 1 equals twenty-five bars.

The distance between the two end bars must be divided by the spacing (in this case 150 mm). Note that when dividing in this way, the quotient should always be rounded to the nearest integer *above*, any fraction of a space being regarded as a whole space. This is because where a pitch is given, it will always be regarded as a maximum pitch, which would be exceeded if this fraction was ignored.

If each individual bar is shown on the plan, they will merely require counting, and the above waste calculation will not be required. The waste will then consist of calculations to arrive at the length of the bars only, and these will take the form as described for bars to the groundbeams, the extra length for hooks being added as necessary.

The description will take the form:

> 12 *mm Diameter mild steel*
> *bars in beds.*

thus complying with SM F17(b)(iii) as to location classification. Note that although the bars might be described on the drawing as '12-mm bars at 150-mm pitch', it would be wrong to describe them as '12-mm diameter bars *at* 150-*mm pitch*', for the pitch does not affect the price per kilogramme, and must not therefore be given in the bill description, which must comply strictly with the SMM in this regard.

Fabric in beds

This is measured as described for foundations, care being taken to watch for any strips required to be in one width, under walls, etc., which must be separated out and measured linear. Reference is made to *tension strips* in SM F18(c), but these are merely strips at points of contraflexure in the concrete, to take any tensional stresses which may be set up. In beds they usually occur under the walls, and need not be separately described as tension strips, but included as normal fabric under walls, the width being given. A suitable description for fabric in beds would be:

> *Fabric reinforcement, reference*
> B 503, *lapped* 150 *mm at joints and*
> *laying in beds.*

HARDCORE FILLING

Hardcore may be defined as broken brick, stone, or similar hard material, filled in and consolidated as a foundation for concrete beds, etc. It is used to make up levels under floors, ordinary earth backfilling being unsuitable for this purpose, owing to risk of subsidence. It is also used to isolate the concrete from the sub-soil, where this is thought desirable, and to form a dry and hard surface for its reception.

Other materials may be used for this last purpose, such as ashes for example, or even weak concrete, which is then known as a 'blinding-coat' since it blinds the surface of the excavation. Hardcore is most usual, however, and is dealt with under SM D22.

Except where over 300 mm thick, hardcore is measured super, and if it is the same area as the concrete bed, may be anded-on to this. It should be noted that hardcore should only be measured where shown or specified. If not so indicated, the presumption is that the concrete is merely required to be laid directly on to the surface of the earth, as is sometimes the case with light structures.

Descriptions of hardcore

SM D22(d) states that hardcore filling required to be deposited and compacted in layers shall be so described, stating the maximum thickness of the layers. This thickness is usually specified as 150 mm, and since all hardcore is normally required to be compacted in order to guard against subsidence, it is the usual practice to relegate this information to a standard preamble, which will also give details of the allowable composition of the hardcore. The description for cube hardcore will therefore be as follows:

Hardcore filling in making
up levels under floors.

Since this is cube, and therefore over 300 mm thick, the phrase 'in making up levels' as SM D22(c) is appropriate. A suitable description for superficial hardcore, however, would be:

150 *mm Bed of hardcore blinded*
with gravel, including
levelling and ramming ground
under.

As in the case of concrete beds, the thickness is put first, to facilitate sorting to bill sequence, and it will be observed that the above description for superficial hardcore falls into three main parts:

(i) Hardcore,
(ii) Surface treatment (blinding),
(iii) Level and ram.

Surface treatment of hardcore

This is dealt with under SM D22(e), which refers the reader back to SM D17(a) and (b). The two possibilities are:

(a) BLINDING. This consists of a layer of fine material (e.g., gravel, ashes), the purpose of which is to fill in the interstices in the surface of the hardcore, thus providing an even bed for the concrete. This treatment may be presumed necessary in the case of reinforced beds. It is measured super under SM D17(a), which specifically states that it may be given in the description of superficial items of hardcore, and this is the normal practice where hardcore is blinded.

(b) MECHANICAL COMPACTING. Thus surface treatment is dealt with under SM D17(b) where applicable, but is not normally specified to hardcore under the floors of the type of structures under consideration. Hardcore under roads, etc., usually requires this treatment, which normally takes the form of rolling.

It will be seen that blinding is the surface treatment likely to occur under floors. It should be noted that it must be included in the description of superficial hardcore only. In the case of cube hardcore (i.e., over 300 mm thick) *it must be measured separately* as:

Blind surface of hardcore
with gravel.

Specification details of the blinding will normally be given in the general hardcore preamble, and need not be repeated in the taking-off description.

Level and ram under hardcore

It will be recalled from Chapter 10 that in the case of normal strip foundations the item of *level and ram bottom of excavation* is measured to trench-bottoms only in the first instance, since an initial measurement of level and ram anded-on to surface excavation would fail to account for the overlap *p* shown in Fig. 15. This argument may not apply to cases of raft construction, where no overlap occurs, and if level and ram has been taken overall in the first instance, none remains to be measured with the hardcore. Unless this is the case, however, levelling and ramming of the earth to receive the hardcore bed (or concrete bed, if no hardcore) is measured at this later stage, and since the areas of the hardcore and this level and ram will be the same, they are included in the same description, under the last sentence of SM D17(a).

Note that in the case of cube hardcore the level and ram must be measured as a separate superficial item.

part 4

Superstructures

Brickwork

The taking-off sequence of superstructure brickwork for the conventional domestic building is as follows:

1. Areas of brickwork in the order
   (a) external walls,
   (b) internal walls,
   (c) projections,
   (d) chimney stacks.
2. Common brickwork labours.
3. Sundry DPC's and beams.

Some general principles relating to the measurement and description of substructure brickwork were discussed in Chapter 12, and these principles will hold good for superstructure brickwork. With this in mind, the first task will be to measure the areas of superstructure brickwork in the order given above, the object being to ensure that the whole of the brickwork in the structure is included. It will be appreciated that this brickwork is measured *gross*, no deductions being made for window or door openings, the adjustments for these being made at a later stage, when the windows and doors are measured. No labours are measured until the whole of the brickwork has been measured, so that the taker-off can concentrate entirely on setting down the dimensions of the actual areas of brickwork thus ensuring that no section of brickwork is inadvertently missed.

In short, the taker-off must aim at arriving at the exact cubic contents of all the brickwork in the building expressed in terms of brick-thickness and classified as walls, projections, or chimney-stacks.

In the case of a drawing of any complexity, it will help if the taker-off first colours the brickwork with a red pencil on the plans and elevations. The time spent in so doing will usually be found well worth while.

EXTERNAL WALLS

These will have already been measured up to general damp-proof course level, as described in the 'Substructures' section of this book. Since all external walls will normally extend down to foundations, the mean girth of the substructure external walls can usually be re-used for the measurement of the superstructure walls. The height for which this mean girth remains valid will vary according to the drawings, but for normal pitched-roof buildings it can usually be used for the brickwork up to main eaves level; the additional brickwork in respect of any gable ends, etc., being bracketed-in afterwards. For flat-roofed buildings on the other hand, the mean girth is usually valid up to the level where a change of wall thickness occurs, parapets walls, etc., being dealt with separately. If the general height of the building varies, the mean girth will only be valid for the lowest section, higher sections of brickwork being added-in afterwards. In any case, the first problem with which the taker-off will be faced is, to what height can the external walls be measured on the mean girth as calculated in substructures?

Pitched-roof construction

In considering the height of brickwork up to eaves level, regard must be had for *eaves filling* (SM G9), which is to be added to the general brickwork. Thus, while the drawings may show the brickwork extending to plate level, as shown in Fig. 42(a), the spaces *between* the rafters (on plan) will require to be bricked up solid to underside of roof coverings on completion of the roof construction, as shown in Fig. 42(b), in order to prevent

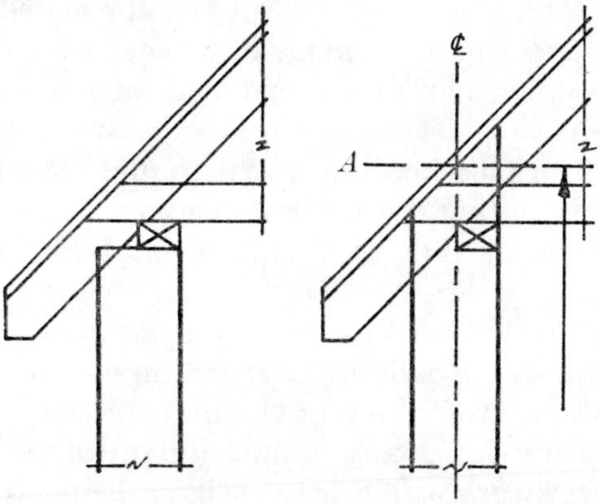

(a) (b)

Fig. 42

draughts in the roof-space. The additional brickwork involved is known as eaves filling, and since this is being added to the general brickwork, the height of the wall is in fact its height to *underside of roof coverings*; no deduction being made for the space taken up by the rafters, and any deduction for the plate being considered when the plate itself is measured in the 'Roofs' section of the taking-off.

Brickwork is measured '. . . the mean length by the average height' (SM G1(a)) and must therefore be measured up to where the centre-line cuts the underside of roof covering, see *A* in Fig. 42(b), thus giving the mean height.

It will probably be realized that this method of measurement will only give a mathematically correct result if applied to single lengths of wall, and will strictly be inaccurate if a mean girth is used. This is because the mean girth is itself an average, and to multiply an average by another average will not result in a mathematically correct area.

Thus it will be seen that measuring the average height of such a wall is supposedly equivalent to measuring the volume of brickwork up to line *A–C* (Fig. 43) and adding-in the triangle *ABC*. The volume of brickwork

Fig. 43

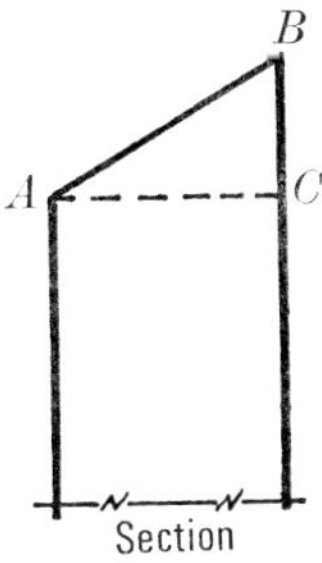

in this triangle, however, is not the base times half the height by the *centre-line*, but by the *centroid* of the triangle, and this clearly involves a shift of girth from the calculated mean girth.

In practice, however, the resultant error is a negligible over-measurement, and may be ignored.

Gable ends

Having measured the brickwork to the underside of roof coverings on the mean girth, any gable ends will now require to be added. The area of brickwork required for a gable end will not be a triangle with a base equal to the overall width of the building, if the walls have been measured to underside of roof coverings by the mean girth. Instead, the requisite triangle will be one with a base equal to the overall width *less two halves of the wall thickness*.

This can be seen from Fig. 44, where the external walls have been measured on the mean girth up to the average height *A–B* all round the

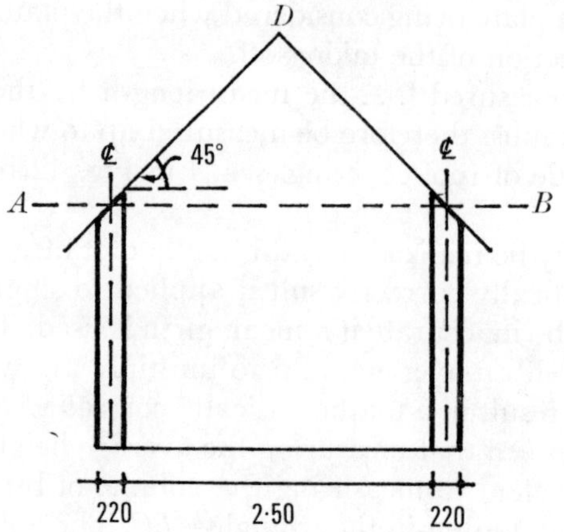

Fig. 44

building, including of course, the gable ends. Since the gable wall itself has been measured up to the line *A–B*, it is the triangle of brickwork above this height which is the extra involved.

Even this is not strictly correct, for an initial error due to multiplying the mean girth by an average height will still be present. If gable walls are measured in this way, however, the error can again be regarded as negligible.

Cavity walls

The external walls to habitable dwellings will usually be of cavity construction. Cavity walls were dealt with in Chapter 12 in connexion with substructures, but further points will arise when measuring cavity construction above ground level.

The cavity will probably be closed at eaves level by solid brickwork. Here the girth will require shifting to the centre-line of the solid construction above plate-level, and this section of solid wall measured right up to underside of roof coverings, taking the average height by the revised mean girth; care being taken not to measure this cavity-closing brickwork to any section of wall carried up as a gable end, since the cavity will not be closed at the gables.

As mentioned in Chapter 12, when the two skins of a hollow wall differ from each other, either in thickness or in brick-type, this will involve not only separate descriptions, but also separate *dimensions* for each skin. Above ground level this will almost certainly be the case in fact, because the outer skin will normally be in facing bricks, while the inner skin will be either common bricks or blocks.

Now it will be appreciated that facings generally are measured in the 'Facework' section of the taking-off, and are completely ignored when measuring the 'Brickwork and Blockwork' section. An exception to this rule is made in the case of walls being built entirely of facing bricks, however, and these are normally dealt with when measuring the brickwork; the outer skin of hollow walls being a case in point.

Thus the measurement of a hollow external wall will require three items. One for the cavity (using the mean girth); one for the inner skin (using the mean girth suitably adjusted on waste to the centre-line of the inner skin); and one (similarly adjusted again) for the outer skin in facings. The three items in question would be described on the following lines:

(1) *Form cavity* 50 *mm wide between skins of hollow walls, including galvanized twisted ties, four per square metre.*

(2) *Half brick skin of hollow wall in Flettons in cement mortar* (1:3). [*inner skin*

(3) *Ditto in Rustic Fletton facing bricks in stretcher bond, pointed with a neat weather-struck joint in cement mortar* (1:3). [*outer skin*

For a detailed discussion on the measurement and description of facings the reader is referred to Chapter 18. It will be sufficient to note at this stage that facings are measured to *one course below ground level.* If no facings have been measured in the substructures therefore, the outer skin must be measured down below DPC level, and a suitable deduction of common brickwork made. The inner skin and cavity will be measured from DPC level upwards of course, care being taken to note in the description of the inner skin any change in brickwork specification from that applying to substructures—Flettons instead of wirecuts, for example; or a change from cement mortar to gauged mortar.

Brickwork built entirely of facings

Such brickwork may be left until the 'Facework' section of the work is measured, but in order to prevent such brickwork being overlooked

180 altogether, some surveyors make a practice of measuring it at this stage as common brickwork, and afterwards adjusting. The necessity of making such an adjustment can be avoided, however, by looking ahead as it were, and measuring brickwork built entirely of facings as such, during the measurement of common brickwork.

We have already seen how the outer skins of hollow walls are dealt with in this way, and any other sections of wall built entirely of facings are best dealt with likewise. SM G25 applies here, and refers specifically to half-brick or one-brick walls built fair face both sides. These may be in either commons or facings, and if in facings will require to be built entirely of facings, as referred to in the clause.

It is unlikely that any one-brick walls will be built entirely of facings unless they are fair faced both sides. With regard to half-brick walls in facings; the outer skins of hollow walls have already been dealt with, but there remains the possibility of garage walls or garden walls which may come into this category.

A position in the building where walls may be expected to be one-brick fair faced both sides in the superstructure is at the parapets of flat-roofed buildings.

Parapet walls

Take the case shown in Fig. 45(a), for example. The parapet wall is a

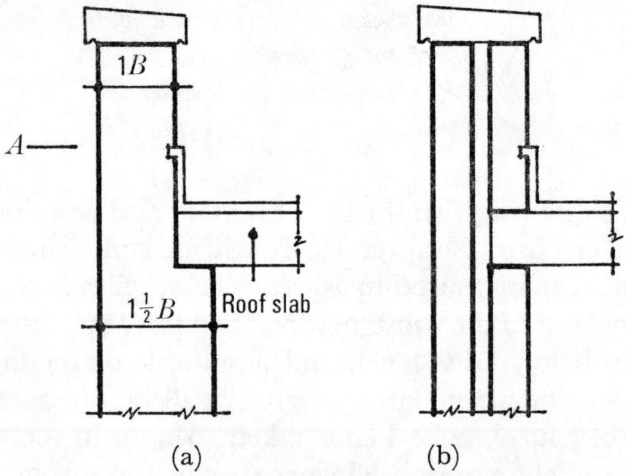

Fig. 45

one-brick wall, but note that the point at which it becomes a one-brick wall *fair faced both sides* is not until it extends above the level of the asphalt skirting in the inside. Thus the wall would require to be measured in three sections:

(1) to underside of roof slab as a $1\frac{1}{2}B$ *wall in commons*,

(2) from underside of roof slab to point A as a $1B$ *wall in commons,*
(3) from point A to coping level as a $1B$ *wall in facings, fair faced and pointed both sides.*

The girth would, of course, require adjustment from the centre-line of the one-and-a-half brick wall to the centre-line of the one-brick wall for items (2) and (3) above. Care must be taken here to ensure that the distance on section between the two centre-lines is calculated accurately on waste for this purpose.

It will be appreciated that the facings below level A on the outside face of the wall are measured as 'extra over' in the 'Facings' section of the taking-off, and not at this stage, with the general brickwork.

If the parapet wall is of hollow construction, as in Fig. 45(b), the inner skin will be in facings above the level of the asphalt skirting, but in commons below this level, and must be measured accordingly.

INTERNAL WALLS

This sub-section refers to internal *brick* walls, internal walls built of other materials being taken-off in the 'Partitions' section of the work. Internal brick walls are normally load-bearing walls which extend right down to foundations, but an eye must be kept open for any which are built up off the floor slab, and these measured accordingly. Internal party walls may be of cavity construction, and the specification for these walls (width of cavity, type of wall ties, etc.) may differ from that of the external walls, and descriptions must be drafted accordingly.

PROJECTIONS

Having measured all of the external and internal (brick) walls of the building, the taker-off must now turn his attention to measuring any chimney-breasts, piers, or similar projections from the walls, following the principles discussed in the chapter on substructure brickwork. The areas (on elevation) of all such projections must be set down (each in the order (a) length, (b) depth); the *width* of such projections being converted to brick thickness and written in the description, thus[1]:

2·00 3·00	$1B$ *in Flettons in cement mortar* (1:3) *in projections.*	
		[*chimney breast*
0·50 3·00	$\frac{1}{2}B$ *in ditto.*	(*piers*

[1] See also alternative descriptions overleaf.

It will be recalled (Chapter 12) that SM G4(i) applies here, and it is important to note that brickwork in projections is all *grouped together*. Thus the word 'chimney-breast' or 'pier' must on no account appear in the actual description itself, or this will have the effect of separating the chimney-breasts from the piers or other projections. The use of the same functional term *in projections* in each description will ensure that all the work is grouped together on the abstract, thus complying with the requirements of the SMM. The use of the terms 'chimney-breast', 'piers', etc., should thus be confined to the side-notes, as shown; and this should always be done, to give adequate annotation to the dimensions.

It should be noted that SM G4 calls for brickwork in projections to be '. . . *reduced to one-brick* and given separately in square metres'. This operation is carried out on the abstract. The taker-off must give the *actual* thickness of the projection in his description so that the abstractor knows the factor by which to multiply the dimensions in order to reduce them to one-brick in thickness. After this operation has been carried out on the abstract, all such items are then merged together, and the resultant bill description emerges as:

Reduced brickwork in
projections.

under the general heading of 'Flettons in cement mortar (1:3)'.

It is interesting to note that when a system of bill preparation which avoids the use of abstracting is used (a slip-sorting or computer system), the above-mentioned items would need to be measured as follows:

2·00	*Reduced brickwork*	
3·00	*in projections.*	
		$\times 1B =$
0·50	*Ditto.*	
3·00		$\times \frac{1}{2}B =$

This would ensure that the item was billed correctly; the reduction to one-brick thickness being carried out as part of the squaring operation in this case.

Plinths and band courses

In addition to vertical projections for piers and chimney-breasts, any horizontal projections for plinths and band courses must also be measured. The measurement of plinths and band courses often causes confusion, owing to the fact that they may be either in common brickwork or in facings. Since these projections more often than not occur on the outside of the building, they are usually in facings, unless the wall is rendered. If the wall was of common brickwork and *not* rendered, fair face would be measurable; which would bring the measurement of any horizontal projections under the rules for facework (see SM G14(b)). In this chapter only plinths and band courses *not* in facings will be discussed, the assumption being that the wall is rendered.

Plinths and band courses are measured under SM G4(i) in the same way as chimney-breasts and piers, viz.: from the wall-face as brickwork in projections. In the case of plinths and band courses, however, it will often be the case that the amount of projection is not a multiple of half-a-

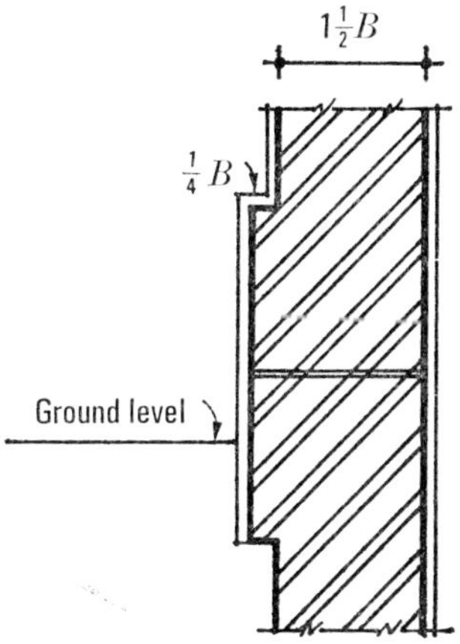

Fig. 46

brick, in which case rough cutting must be measured. Thus, the plinth in Fig. 46 would be measured (dimensions assumed):

6·00 0·75	*Reduced brickwork in projections.*
	& × $\frac{1}{4}B$ =
	Rough cutting
6·00 0·06	*Horizontal DPC a.b.*

band courses being similarly measured.

Note that in the above case a DPC is present in the wall where the plinth occurs, as will often be the case. The additional area of DPC is then taken with the plinth; and note especially that although this extra width of DPC is under one-brick wide, it is, nevertheless, measured *super*, since it is being added to a superficial item, the total width of which exceeds one-brick.

A problem sometimes arises regarding the measurement of garden walls having a plinth on either side of the wall. Take the case shown in Fig. 47, for example. Since there is a $\frac{1}{4}B$ plinth on either side of the wall,

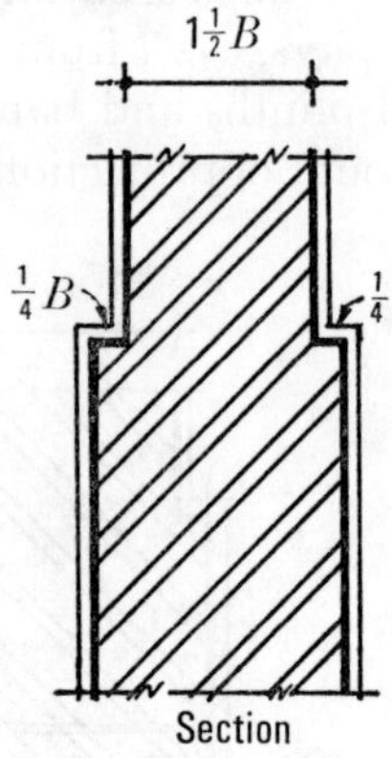

Fig. 47

the total thickness of the brickwork at plinth level is $1\frac{1}{2}B$ plus two $\frac{1}{4}B$ projections = $2B$. Since this is a multiple of half-a-brick, clearly no cutting is in fact involved. Therefore, in order to accord with the provisions of SM A1 and make the quantities accurately represent the work executed, the brickwork below the level of the offsets is best measured as a wall of two-brick thickness, and not treated as a $1\frac{1}{2}B$ wall with two plinths.

In general, if a plinth extends right down to foundations, it is reasonable in any case to measure the wall as a wall of the full thickness below plinth level, since below DPC level this will be its function. If the projection is not a multiple of half-a-brick, so that the wall below plinth-level is not a multiple of half-a-brick, rough cutting is given in the description under G3(b), which appears to make provision for just such a case.

Sunk band courses

If a band course in common brickwork is sunk instead of projecting, no extra brickwork is of course inolved. It then becomes a *rough chase*, and is measured as such under SM G11, stating the size and whether horizontal; no deduction of brickwork being made.

Oversailing courses

These may be regarded as footings which are inverted, and are measured under SM G4(i) in the manner described for footings in Chapter 11; the average thickness being taken. In the case of oversailing courses, however, as distinct from footings, rough cutting must be measured the same area as any courses which are not a multiple of half-a-brick in thickness, as required by the SMM.

CHIMNEY-STACKS

A chimney-stack may be defined as that part of a flue projecting above the roof-line (or parapet). Brickwork in chimney-stacks must be kept separate under SM G3(a)(v), together with isolated piers, if such exist. It is important to realize that above the roof-line no walling will have been measured, and the total thickness of brickwork to the stack must be taken. Thus the chimney-breast shown on plan in Fig. 48 must be

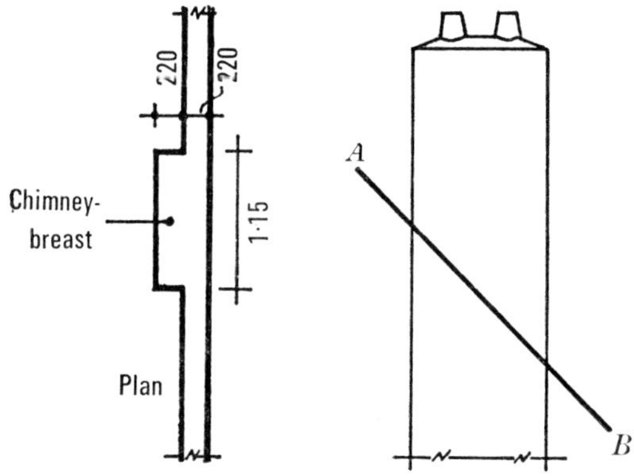

Fig. 48

measured up to the roof-line *A–B* on the section (taking the average height) for the width of the projection, viz.: 1*B* in this case, and described as

Reduced brickwork
in projections.
× 1*B* =

Assuming the stack to be the same thickness as the combined thickness of wall and breast, the item for brickwork above roof-level in this case would be:

Reduced brickwork
in chimney-stacks.
× 2*B* =

again taking the average height from the roof-line *A–B*, this time to the top of the brickwork in the stack.

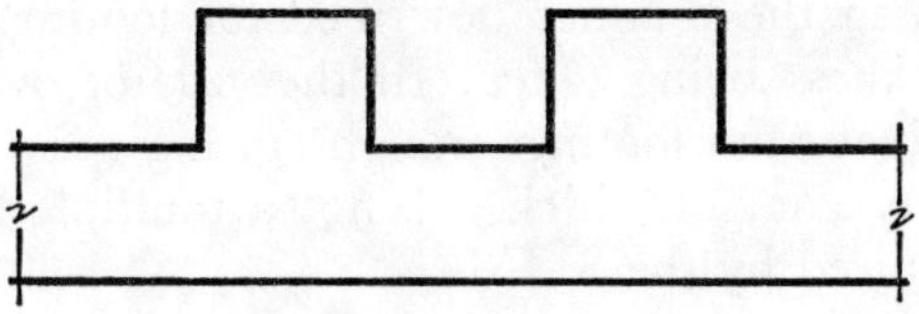

(a)

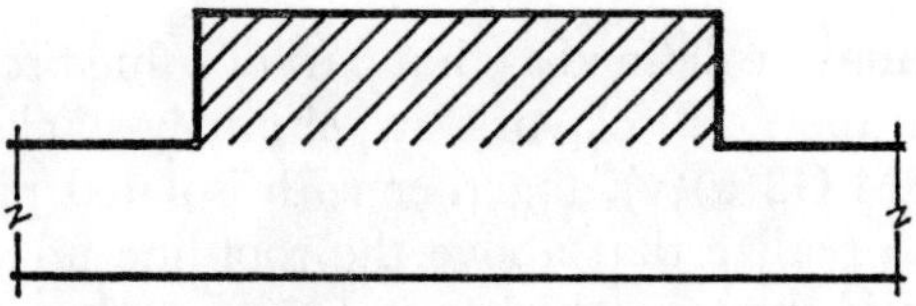

(b)

Fig. 49

It will be realized that no deductions are made for the actual flues at this stage, either in breasts or stacks. Neither are deductions made for fireplace openings; any such deductions for fires or flues being made in the 'fires' section of the taking-off. All breasts and stacks are therefore measured solid at this stage, and care must be taken to measure a chimney-breast as shown in Fig. 49(a) right across the fireplace, as in Fig. 49(b).

COMMON BRICKWORK LABOURS

Having measured the whole of the actual brickwork in the structure, the taker-off now applies his mind to measuring the labours involved in this brickwork. Up until now, all labours will have been ignored, and the labours now under consideration are confined to labours on *common* brickwork, since (apart from any brickwork built entirely of facing bricks) all work connected with facings is measured in the 'facework' section. Thus all facings are ignored at this stage, and the whole of the brickwork treated as common brickwork, so that the mind of the taker-off is not confused by the thought of *facings labours*, which are measured later.

Thinking of all the brickwork as common brickwork then, the taker-off now examines the drawings once again, looking this time for common brickwork labours, the two commonly met with at this stage being eaves filling, and rough cutting.

Eaves filling

As previously explained, eaves filling is added to the general brickwork under SM G9. The extra *labour* in eaves filling, however, has now to be measured as a linear item stating the thickness of the wall (second sentence in SM G9).

In other words, the brickwork itself has been measured, but the extra labour involved in cutting and fitting the bricks in between the feet of the rafters at eaves level has now to be included. The linear measurement will simply be the mean girth of the walls, less any sections of wall carried up as gables. This item will naturally only apply to pitched-roof buildings with eaves; the appropriate item for the building shown in Fig. 44, for example, being (dimensions assumed):

2/20·00	*Labour eaves filling to top of one-brick wall.*

Rough cutting

Rough cutting may be defined as any cutting of common bricks other than the cutting necessary for normal bonding purposes. Thus the cutting of closers for square quoins, etc., is excluded; and the mind of the taker-off must be directed to examining the drawings for situations where the bricklayer will be required to cut the common bricks for any purpose other than to achieve a correct bond. SM G10(a) lists the situations which fall into this category, and it may be desirable to now examine this list in detail.

ROUGH CUTTING AGAINST SOFFITS. The soffits of concrete suspended slabs will usually be supported by the brickwork itself, on which the slab is bearing. Since the concrete is not cast until after the brickwork is built in such cases, the brickwork cannot really be said to be cut up against the soffit although if a split course of bricks is involved, this item may be considered as measurable. In the case of a non-load-bearing partition wall, this item is measurable at the top, where the wall abuts a soffit. Non-load-bearing walls are rarely built of brick, however, so this form of cutting is seldom measurable.

ROUGH CUTTING AT SQUINT OR BIRDSMOUTH ANGLES. If squint quoins exist, rough cutting is measurable as described in Chapter 12.

ROUGH CUTTING AT REBATED REVEALS. This is measured when the reveal is dealt with in the 'Windows' or 'Doors' section of the taking-off—not at this present stage.

ROUGH CUTTING AROUND STEEL SECTIONS. Dealt with when the steelwork is measured—not at this present stage.

188 ROUGH CUTTING WITHIN THE THICKNESS OF WALLS OVER TWO-BRICK THICK. Such walls are very unlikely in the domestic construction with which the present-day intermediate syllabus is concerned.

ROUGH CUTTING TO PROJECTIONS NOT A MULTIPLE OF HALF-A-BRICK IN THICKNESS. This is anded-on to the brickwork in question as described earlier in this chapter—not measured with common brickwork labours.

ROUGH RAKING CUTTING. This will occur at the top of gable walls and party walls in pitched-roof construction.

ROUGH SPLAY CUTTING. Mostly occurring under external window sills which are set weathering. Measurable in this case in 'Windows' section—not at this present stage.

ROUGH CURVED CUTTING. To the tops of brick walls forming the ends of Nissen huts. Not usual otherwise.

It will be apparent from the foregoing that in a simple building of the type under consideration it is more than likely that any rough cutting measurable at this stage will be confined to rough raking cutting. Assuming, for example, that Fig. 44 shows a cross-section through a building with two one-brick gable ends, the rough cutting item in respect of the splay cutting at the top of the gable walls would be measured as follows:

		2·50
	walls 2/·22	0·44
	base of gable	2·94
	$\frac{1}{2}$/ =	1·47
	hypotenuse =	$\sqrt{2}$ × 1·47
	=	1·414 × 1·47
	=	2·08
2/2/2·08 0·22	*Rough cutting.*	

Note that the description is *rough cutting*, not rough *raking* cutting, since SM G10(a) states that rough cutting shall be deemed to comprise any or all of the categories mentioned above, including rough raking cutting. All categories must therefore be measured simply as *rough cutting*. It is not necessary to say more than this in the description if only one kind of common brickwork is specified, since the term rough cutting traditionally refers only to the cutting of common bricks not exposed to view. Should there be more than one type of common brickwork, however, in the same project, then one should write (for example):

Rough cutting to
Fletton brickwork

to comply with SM G1(b), which states that labours on different kinds of work shall be given separately.

Note that all rough cutting is measured super, and not linear. The linear dimension up the slope of the gable wall has been supered by the wall thickness before being timesed for two slopes and then two gables. It will be appreciated that it is the upper surface of the wall abutting the underside of roof coverings which has to be cut, and this is the area measured. It will also be realized that facings have been completely ignored at this stage, the whole thickness of the wall being measured for rough cutting, as if the wall was wholly built of common bricks which is how it will have been measured so far. Any adjustment for *fair* cutting in lieu of *rough* cutting is made after facings have been measured in the 'Facework' section of the taking-off.

DAMP-PROOF COURSES

The main damp-proof course will have been measured in substructures, but any DPC's occurring in the brickwork above this level (except those in connexion with window and door openings, which are measured with the windows and doors) are now dealt with.

The drawings are examined for DPC's to stacks, or in parapet walls, etc., and these are now measured. It should be noted that these DPC's are not kept separate from similar DPC's in the foundations; the only DPC's requiring to be separated under SM G44 being those with cavity gutters in hollow walls. In the case of stack DPC's, the flues are not deducted, since these will not exceed 230 × 230 mm in domestic work, and SM G44(a) allows no deduction for voids not exceeding 0·5 square metres. Thus a DPC to the stack shown in Fig. 48 would be measured as follows:

1·15 0·44	*Horizontal layer of lead-cored bituminous felt DPC, bedded in cement mortar* (1:3) (*no allowance for laps*) (*stack*

Note especially that the *location* of the DPC is given in a side-note, and *not* in the description, since this would have the effect of separating the item from other similar DPCs in other locations.

190 Any DPCs to parapet walls, etc., are similarly measured, those not exceeding one-brick wide being given linear.

STEEL BEAMS

Any beams occurring in the brickwork (other than in connexion with floors, or with window or door openings) are now measured.

The detailed measurement of structural steelwork does not form part of the intermediate syllabus, and will not be dealt with here. The measurement of an occasional steel beam in an unframed building may be said to come within this syllabus, however, and to this extent it will be necessary at this stage to touch briefly on the subject of structural steelwork. The following points should be noted:

(i) SM Q1(a) states that unframed steelwork shall be given under an appropriate heading.
(ii) Simple steel beams are classified as *plain girders* under SM Q4(b), and shall be so described and classified under SM Q1(g).
(iii) SM Q1(g) mentions joists, broad-flange joists and universal joists. However, the British Standard relating to structural steel sections (BS 4, Part 1:1962, amended 1963) divides girder sections into two types:

(a) *Universal beams* (having near-parallel flanges)
(b) *Joists* (having tapered flanges).

Universal beams are only made in the larger sizes, so that any girders met with in domestic construction are almost certain to be classifiable as 'joists'. The discrepancy in terminology between the SMM and the British Standard is therefore unlikely to worry the intermediate student.
(iv) The weight-group must be stated in accordance with SM Q1(h).
(v) The unit of billing for steelwork is by weight. Items are therefore measured linear, giving the exact weight per linear metre so that the total weight can be calculated at working-up stage.

It follows from the above that the measurement of a 4-metre length of 200 mm × 100 mm × 30 kg steel beam would appear as follows:

The following in unframed steelwork:

4·00	*Plain girder of joist section, over* 20 *but not exceeding* 50 *kg per linear metre.*
	× 30 *kg* =

It should be noted that the actual weight is not stated in the description only the weight-group as SM Q1(h). The actual weight per lineal metre appears as a timesing-factor to enable the total weight to be calculated at squaring stage. Note also that the *size* does not appear either! This may be stated in a side-note for reference purposes, but should not appear in the description.

The heading, which is written across the taking-off column, as shown, is underlined to indicate that it must be transferred to the bill at working-up stage. This is a 'special bill heading', referred to in Chapter 5 distinguished from a fingerpost heading by use of the phrase 'The following in . . .' at commencement of the wording.

The measurement of steel beams does not end here, for the ends must now be dealt with, followed by the paint or beam-casing.

Ends of beams

Each end of the steel section must now be dealt with, where it bears on the brickwork.

A concrete template (or padstone) as mentioned in SM F35 will be required, to support each end of a heavy girder, but a steel joist in domestic construction is most likely to rest directly on the brickwork requiring the measurement of either *building in* under SM G65(b) or *cutting and pinning* under SM G65(c).

It is usual to cover either eventuality in the description as follows:

2	*Cut and pin or build in end of small steel section to brickwork.*

Note especially that both SMM clauses require classification of size under SM A4(b), viz.: small, medium, large, or extra large; the description must be worded accordingly.

Painting of steelwork

If the steelwork is painted, this must be measured at this stage, taken with the beam. It is interesting to note that SM Q23 requires painting on steelwork *by the fabricators or erectors* to be measured by weight! In the case of domestic construction, however, painting would normally be carried out by the contractor, not the steelwork fabricator, or a specialist steelwork erector. It therefore becomes measurable under SM W6 instead. Under this clause painting on steelwork is all measured super, the beam being girthed around on waste as follows (200 mm × 100 mm section assumed):

web	2/·20	0·40
flanges	2/2/·10	0·40
		0·80

4·00 0·80	*Wire brush, clean down and apply two coats red lead oxide on surfaces of structural steelwork.* (*internally*)

The word 'internally' in the above is to ensure that this item is billed under the heading of internal work, as provided for in SM W1(a).

Casing of steelwork

If the steelwork is encased in concrete, this concrete casing must also be measured at this stage, under SM F9. The concrete is cubed-up (no deduction being made for the steel) and described as follows:

Concrete (1:2:4) *in horizontal beam-casing over* 0·05 *but not exceeding* 0·10 *square metres in sectional area.*

Note that 'horizontal' beam casings are required to be 'so described' by SM F9, which also requires the cross-sectional area of the concrete to be stated in accordance with SM F1(h).

Formwork to the sides and bottom of the beam-casing must now be measured under SM F21(a). The sides and bottom are supered in together as:

Sawn formwork to sides and soffits of horizontal beam casings.

Blockwork

Partitions

Having completed the brickwork section of the taking-off, the surveyor must now turn his attention to the internal partitions. These may be defined (for domestic work) as *internal walls other than those constructed of bricks.*

In simple domestic construction these may be of blocks, or timber studding covered with plasterboard. The measurement of timber stud partitions under SM N4 should present no special problems, and the general rules for carpentry work, as outlined later in this book, will apply here. The partitions dealt with in this chapter will therefore be confined to those built of blockwork.

Blockwork generally

The *Brickwork and Blockwork* section of the SMM devotes considerable space to the measurement of blockwork, the reason being that blockwork may be used in lieu of brickwork for the external load-bearing walls of the building, as well as for the partitions. In particular the inner skins of hollow walls are often blockwork instead of brick, and the difference in the rules of measurement between the two materials becomes apparent when dealing, for example, with attached piers and the like. Block partitioning is usually non-load-bearing, however, and does not normally involve piers or other projections. Load-bearing blockwork is dealt with on pages 198–9.

Block partitions

These will be built up off the floors, usually after the external walls and roof have been erected. The partitions will normally extend for one storey-height on each floor, and it follows from this that all the partitions to any one storey will be the same height. From this it will be apparent that the best approach to the measurement of the partitions (which are, of course, measured super) is to start by collecting together the girth of all partitions *on the same floor-plan* on waste. The total length can then be set down in the dimension column and timesed by the storey-height, thus achieving the measurement of all the partitions on each floor by means of only one superficial dimension. Each floor-plan is examined in turn, and the girths collected and totalled on waste, care being taken of course not to include partitions of different thicknesses in the same total, since the thickness must be given in the description.

The result, in the case of a two-storey building will look something like this:

living room	3·00
	0·75
	2·00
dining room	3·50
	2·75
study	2·50
	0·75
	1·50
ground floor	16·75
bedrooms	4·00
	0·75
	3·00
	2·50
	2·75
bathroom	3·00
	0·75
w.c.	1·75
	1·00
first floor	19·50

16·75 2·50	75 *mm Clinker block partition bedded and jointed in gauged mortar* (1:2:9) (*ground floor*)
19·50 2·25	(*first floor*

In this way even a very large five-storey building for example, would require only five dimensions for each thickness of partition (assuming no mezzanine floors). Thus economy of working is achieved.

Block partitions are measured under SM G33(a)(i), and it will be appreciated that no deductions are made for the door openings at this stage, which are adjusted in the 'Doors' section of the taking-off.

In this last connexion it is important to bear in mind that partitions extend over the tops of all doors, and this may not be apparent from the $\frac{1}{100}$ floor plans from which the taker-off will be working. This is because the floor plans are assumed to be taken at 1 metre above floor level when drawn. Because, therefore, the floor plan is taken well below the tops of doors, it will appear as if the partition ceases altogether when a door is encountered. This effect is particularly marked when doors are close together, when a section of partition may easily be missed by the taker-off. Figure 50 shows a case in point, where the landing appears on the

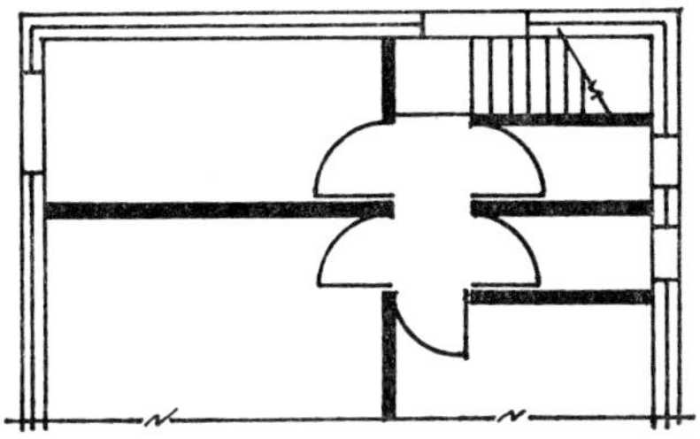

Fig. 50

plan to have no walls at all, and the careless taker-off may easily fall into the trap of measuring the partitions as shown.

The best plan is to carefully prepare the drawing before taking-off the partitions by colouring-in all block partitions with a (say) blue pencil (red being for the brickwork), colouring right through all door openings. The true lengths of the partitions will then stand out on the plan, and the likelihood of any section being missed is avoided. This procedure is standard taking-off practice in many offices, and the extra time taken is well worth while. This system should also be adopted in examinations, except in the case of very simple plan-shapes.

Linear items to block partitions

Having measured the whole of the area of the partitions in the manner described above, the taker-off must now turn his attention to the measurement of labour items, etc. These will normally fall into the following categories:

(a) Rough cutting (SM G37)
(b) Damp-proof courses (SM G39)
(c) Bonding to brickwork (SM G44).

Rough cutting to blockwork

Unlike rough cutting to brickwork, this item is measured *linear*, and

196 SM G37 demands that the various classifications be kept separate, and not all included in the same item. Classification is as follows:

(i) ROUGH CUTTING AGAINST SOFFITS. Non-load-bearing partitions of the kind under consideration are built up off the floors, and it is usually convenient to construct the brickwork, floors, and roof before building the partitions, which will then be required to be cut up against the soffits of the upper floor and roof joists. This item is therefore measured to the tops of all block partitions under SM G37(a)(i).

(ii) ROUGH CUTTING AT IRREGULAR ANGLES AND INTERSECTIONS. This will only occur at any junction on the plan which does not form a right-angle. In the example shown in Fig. 51 for instance, twice

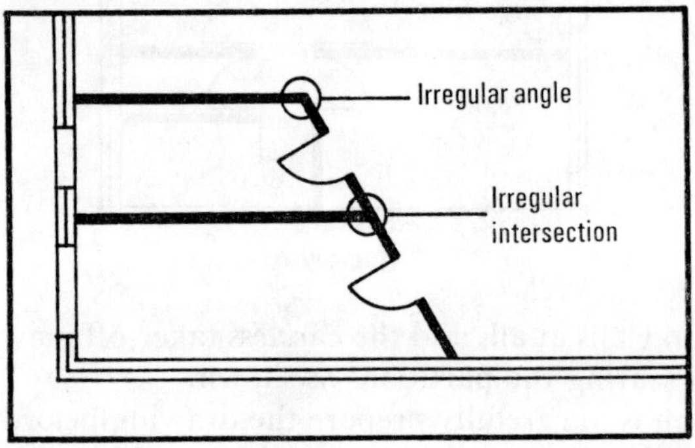

Fig. 51

the height of the partition would be taken as '*rough cutting on 75 mm partition at irregular angles and intersections*' under SM G37(a)(ii).

(iii) ROUGH RAKING AND SPLAY CUTTING. Rough raking cutting would be measurable where a partition met the underside of a sloping ceiling. This should be described as '*Rough raking cutting against soffits*' since it contains an element of SM G37(a)(i). The same thing applies where a spandrill partition meets the soffit of a staircase, although this is usually measured in the 'Stairs' section of the taking-off—not with partitions.

If a partition met a sloping soffit at right-angles to the line of slope, then this would be *splay* instead of *raking* cutting. Both are grouped together under SM G37(a)(ii).

(iv) ROUGH CURVED CUTTING. An arched opening in a block partition would entail this item, but this would be measured when the opening is adjusted—not with partitions.

From the foregoing it will be apparent that unless irregular junctions or sloping soffits, etc., exist, the only rough cutting item to be measured at this stage will be *rough cutting against soffits*. This will normally be measurable to the whole of the partitions, and the total girths already calculated can therefore be used for this purpose, thus:

16·75 19·50	*Rough cutting on* 75 *mm clinker block partition against soffit of ceiling joists.*

When this item is measured against joists, as opposed to continuous soffits (e.g., concrete soffits), it is usual to mention this in the description, as shown, in case any difference in price is thought to apply. It is not the usual practice to carry this policy to the extent of differentiating between partitions which run *across* the joists, and those which run *with* the joists, however.

It should be noted that no item is measurable in connexion with right-angle junctions or intersections of the partitions (see SM G37(b)).

Damp-proof courses in partitions

Ground-floor partitions will probably require DPC's if built off a solid floor, unless a membrane runs underneath. It is important not to miss this item, and the drawings and specification must be examined at this stage to establish whether such DPC's are intended. If so, it may be convenient to have them anded-on to the item for rough cutting against soffits, unless this includes upper-floor partitions as well, in which case it will have to be a separate item.

Bonding to brickwork

This item now remains to be measured, the full height of the partition being taken for every junction with brickwork shown on the plan. These junctions will stand out if the brickwork has previously been coloured-in with a red pencil. SM G39 states that the extra material for bonding must be given in the description. This is because the partition will have been measured up to the face of the brickwork only, the extra blockwork actually inside the brickwork not having been included. Forming pockets or chases in the brickwork to receive the blockwork must also be included, and the description will appear thus:

Extra labour and material cutting and bonding 75 *mm partition to brickwork, including cutting chase* 50 *mm deep every alternate course.*

Note that the thickness of the blockwork must be stated under SM G39, which clause incidentally makes no mention of ends bonded at irregular intersections with other types of construction. Where a partition meets brickwork other than at right-angles, the description should be worded as follows:

Ditto all as last described
but at irregular intersection
with brickwork.

Load-bearing blockwork

As previously indicated, the use of blockwork may not be confined to partitions, but may occur in the external or internal walls, as a load-bearing material used in lieu of brickwork. Any such blockwork in conjunction with brickwork is usually measured in the 'Brickwork' section, as girths, etc., are then readily to hand. The rules for the measurement of such work are the same as those described for partitions, above; the SMM does not differentiate between load-bearing and non-load-bearing blockwork. These rules are very similar to the rules for measuring brickwork, except for the labour items, as we have seen. There is a basic difference in the measurement of piers, however, and since these may well occur in load-bearing blockwork some discussion is merited here.

Blockwork piers and chimney-stacks

These are grouped together under SM G33(a)(v), whether the piers are isolated or not.

Since chimney-stacks and isolated piers are usually (though not necessarily) built of brickwork, however, and since no special problems are usually presented by these, we shall confine ourselves at present to the measurement of attached piers.

These may well occur on the block inner skin of a hollow wall, or on a solid block load-bearing wall, and are dealt with under SM G32(c) which states, *inter alia*:

The thickness of attached piers
shall be taken as the combined
thickness of the wall and the
pier.

In other words, instead of being measured as in Fig. 52(a) as a brick pier would be measured, they are measured as in Fig. 52(b), *the wall being deducted out for the length of the pier.*

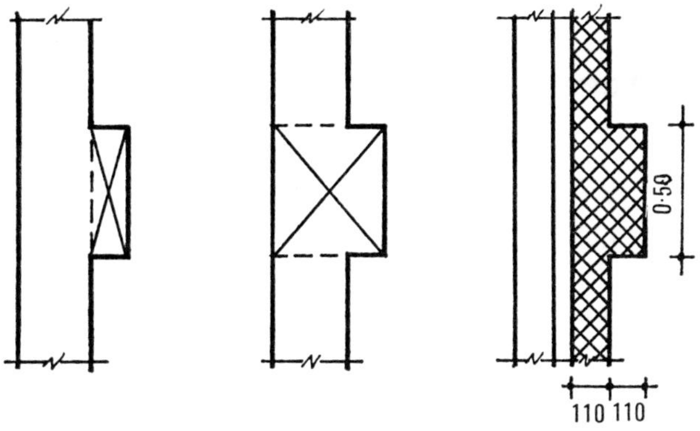

(a) (b) (c)

Fig. 52

Thus, the pier on the hollow wall in Fig. 52(c) would be measured as follows (height assumed):

	wall	0·11
	pier	0·11
		0·22

0·50 2·50	220 *mm thick piers in concrete blocks to BS 2028 Type B, bedded and jointed in gauged mortar* (1:2:9). & *Deduct* 110 *mm Skin of hollow wall in ditto a.b.*

Note that under SM G32(c) a pier having a length of over four times its thickness is classified as a wall. This means four times the *total* thickness, not four times the projection, so that the pier in Fig. 52(c) would be a wall only if it were over 880 mm wide.

Chapter Eighteen

Brick Facework

The face of a wall built of common bricks or blocks will require special treatment where it is exposed to view, in the interests of appearance and utility. The alternative treatments usually met with are as follows:

1. *Internally:*
   (a) plastering
   (b) fair face.
2. *Externally:*
   (a) block facework
   (b) cement rendering
   (c) vertical tiling
   (d) timber cladding
   (e) brick facework.

Internal plastering and *fair face* work are measured in the 'Internal Finishings' section of the taking-off; not with Facings. *Block facework* to imitate stone is used in some districts. It is best dealt with as 'cast stonework', and measured under the rules for natural dressed stone in accordance with SM K38. Since dressed stone is specifically excluded from the RICS intermediate syllabus, this form of facing will not be dealt with in this chapter.

Cement rendering is measured under the rules for plasterwork, under 'Internal Finishings'. *Vertical tiling* is measured under Section M of the SMM, which deals with roof tiling, discussed later in this book; while *timber cladding* is measured under the rules for carpentry work, also dealt with in a later chapter.

We are thus left with only *brick facework* to be dealt with in this chapter, and this is the type of facing usually met with in buildings of the kind under consideration. It must be emphasized, however, that the 'Facework' section of the taking-off includes the measurement of the treatment of the whole of the external face of the building which is exposed to view. Thus if a building is partly in brick facework, partly tile hung, and partly rendered externally, these different treatments are all measured in the 'Facework' section. This is subject to the general rule that any rendered panels, etc., forming part of the treatment of a window or door opening are usually measured under the 'Windows' or 'Doors' section of the taking-off.

The term *facings* may therefore be defined as a general term referring to any external finish to the face of a building; the term *facing bricks* being used for bricks which are employed for this purpose. The term *brick facework* applies to the face of a wall finished in facing bricks and pointed to give a finished surface.

Although the facings section of the taking-off may include any type of finish to the building, however, the term *facings* is often used to refer specifically to brick facework, or more specifically still to the actual facing bricks themselves, depending on the context.

The order of taking-off this section of the work, is as follows:

1. Areas of facework
2. Facework labours
3. Architectural features (plinths, copings, etc.).

AREAS OF FACEWORK

It will be recalled from the chapter on brickwork that the first task in measuring the common brickwork is to concentrate solely on setting down the whole of the areas of brickwork in the building, ignoring all openings and labour items.

The same principle now applies to the measurement of the facework. The taker-off must carefully go over the drawings, starting at the bottom, and setting down the whole of the area of brickwork *exposed to view*, from just below ground level to the top of the highermost chimney-stacks. For reasons previously given, the outer skins of any hollow walls will have already been measured for facings in the 'brickwork' section. Likewise, any other walls built entirely of facing bricks. It is with this in mind, therefore, that the taker-off must now concentrate solely upon the external face of work previously measured as common brickwork, care being taken not to include any facings previously measured.

Extra over common brickwork

SM G14(d) requires facework to be given as '. . . extra over the brickwork on which it occurs'. The term *extra over* means that the estimator must price the item for facework as the extra cost over and above the cost of building the wall in common bricks. This is because only the outer face of the brickwork will be built of facing bricks; the remainder being common bricks, which are more economical. The facing bricks will, of course, need to be bonded-in to the common bricks, the extent of bonding depending on the type of bond specified for the brickwork generally.

202 Because this will affect the price of the extra-over item, the bond is required to be given in the description of the facework under SM G14 (c)(iii), to cnable the estimator to make allowance for this.

Facework is therefore measured *on the exposed face* of the brickwork (SM G1(a)) as extra over the brickwork on which it occurs, and it is this exposed face which the taker-off must now measure.

Height of facings

As previously explained in connexion with the outer skins of hollow walls, facework is measured to *one course below ground level*, to allow for any irregularities in the ground. The distance from this point to the point where the eaves soffit conceals the wall must be calculated on waste, and the height of the brickwork exposed to view up to eaves level will then have been obtained (see *H* in Fig. 53).

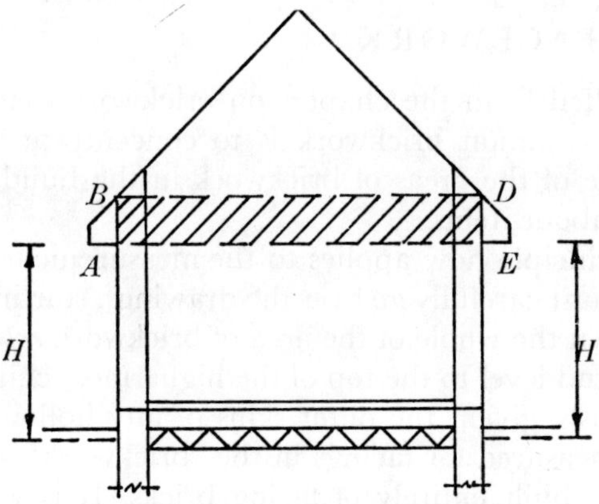

Fig. 53

The mean girth of brickwork is now set down on waste, and shifted to the outside face of the walls by adding eight times half the wall thickness. This will now give the general area of facings up to eaves level.

Facework to gables

The question now arises; what is the correct amount to add in respect of any gable walls?

The additional area will not be a simple triangle only, but a quadrilateral plus a triangle, as may be seen from Fig. 53. There will be a strip of facings *ABDE* on the gable wall, above eaves-soffit level on the side walls, in addition to the triangle above the line *BD*.

Facework to parapets

As indicated on page 180, solid parapet walls are likely to be one-brick thick, faced both sides, and as such will have been measured down to a point where the brickwork disappears behind the skirting of the roof-coverings on the inside of the parapet wall (at point *A* in Fig. 45(a)). Extra-over for facings to walls carried up as parapets in this way therefore must be measured for height only up to point *A*, where the one-brick wall commences. In the case of solid parapet walls of over one-brick thick, the facings will, of course, be measured right up to underside of parapet. The girth must then be shifted to the inside of the wall, and the facings on the inside face of the parapet measured down to point *A*, and added to the general facings.

When the parapet is a hollow wall, both skins above roof level will have been taken in facings when the brickwork was measured, as described in Chapter 16, leaving no further facework to be accounted for at this stage.

Descriptions of facework

Facing bricks are normally the same size as those in the body of the work and do not require to be purpose-made. The composition and mix of mortar for pointing may or may not be the same as that in the body of the work, but if not it must be given in the description. It is the usual practice, in fact, to give the mortar mix in the description, in any case.

If the foregoing is borne in mind, an examination of SM G14(c) and (d) will indicate that a description for facings will normally contain the following four ingredients:

(1) Extra over common brickwork
(2) Kind and quality of bricks
(3) Bond
(4) Method of pointing and mix.

It is important that the student should remember to include all four of these factors in his description, and should not make the common error of missing out the bond, for instance. In each of these factors is dealt with in the order given, a description similar to the following will emerge:

Extra over common brickwork
for facing with red sand-
faced facing bricks P.C. £20.00
per thousand, in Flemish bond,
and pointing with a neat weather-
struck joint in cement mortar
(1:3).

204 The SMM states that facework shall be given as extra over the brickwork *on which it occurs* (SM G14(d)). Facework to walls, piers, chimney-stacks, etc., is to be all grouped together, however, (SM G15(a)), so that the term 'common brickwork' may be used unless more than one type of common brick is used in the building. When a different type of common brick is specified below DPC level, the term 'common brickwork' will not be valid in a facings description. In this case the wording would be 'Extra over wirecuts . . .' below DPC level, for example, and (say) 'Extra over Fletton brickwork . . .' above this level. Thus a change of common brickwork at DPC level would require two separate descriptions of facings, and care must be taken to ensure that the facings are separated in this way.

P.C. facing bricks

It will be noticed that after the mention of the type of facing brick in the above description, a P.C. amount per thousand has been inserted, thus making the item a 'P.C. measured item', as referred to in Chapter 3. This is fairly common practice in the case of facing bricks, which vary in quality and price, and this device enables each tendering contractor to base his price on the P.C. amount given, knowing that this amount will be adjusted to the actual invoice price of the bricks at settlement of accounts.

The actual decision as to whether a P.C. price for the bricks is to be included or not is the responsibility of the Architect, since it is he who will have to subsequently approve the source of supply. This decision should be incorporated in the specification notes from which the bill of quantities is prepared, and in an examination therefore, the student must carefully peruse the specification notes to see what P.C. price (if any) is to be allowed for the facings. Should a P.C. price be specified, it must be inserted in the description, and *underlined,* as shown, to ensure that its appropriate adjustment is not overlooked when the time comes.

Facework to projections

Any vertical projections for piers, chimney-breasts, etc., occurring on the outside face of the building must now be added-in for facings. The facework on the *front* of a pier will, of course, compensate for the facework to the wall not required because of the pier, if the pier has so far been ignored. It is only the facework to the *sides* of piers and chimney-breasts which will be an extra therefore, and must now be added-in, unless this was done on waste when arriving at the original girth of the facings.

A word of warning is necessary here, in respect of the sides of any projecting piers or chimney-breasts *not exceeding half-brick wide.* These are classed as 'returns' under SM G15(b) and are to be measured *linear* and described as facework to *margins*. This applies also to any external vertical breaks in the wall (including those on the inside face of parapet walls) which are not exceeding half-brick wide, as well. If previously included in the girth of general facework, these must now be measured linear, and an adjustment made by deducting out the area of general facework previously measured. All returns not exceeding half-brick wide are thus measured linear and described as

Extra over common brickwork
for facings as before described,
but to margins.

Note that the width is not given in the description, since margins are to be grouped together irrespective of actual width, the maximum width being deemed to be half-brick, as stated by SM G15(b).

Projections on walls faced both sides
Consider the projection illustrated in Fig. 54.

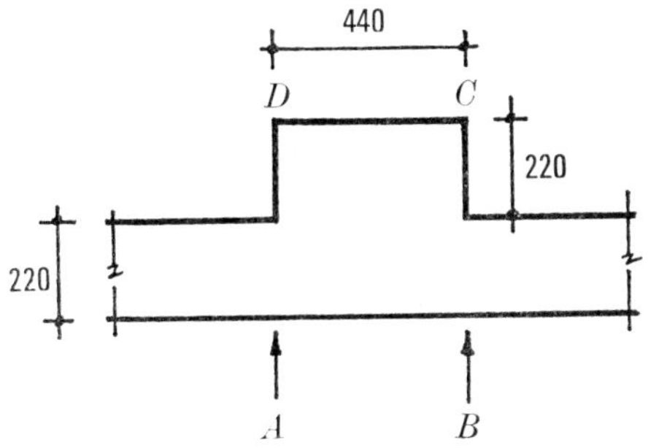

Fig. 54

This shows the plan of a 440 × 220 mm pier projecting from a one-brick thick wall. There would be no special problems if the inside face of the wall were plastered, or if the wall was over one-brick thick. Supposing, however, the one-brick wall as shown is *faced both sides*. It will then have been measured as such in accordance with SM G25(a) in the general brickwork section. Where the pier occurs, however (from point *A* to point *B*), the one-brick wall faced both sides ceases to exist. It will therefore have been deducted out and measured as common brickwork, i.e., a one-brick wall in commons; the projecting pier itself having been measured as 'brickwork in projections' under SM G4(i). It might be objected that no one-brick wall exists at all at this point; but since the

206 SMM demands that the projections of attached piers be measured *beyond the face of the wall* (SM G4(i)), there is no alternative method of measurement in such a case.

Thus the whole area *ABCD* in the Figure will have been measured as common brickwork, and an area of general facings will require to be measured to the projecting pier, and to the wall between points *A* and *B*, thus (height assumed):

pier		0·44
returns	2/·22	0·44
		0·88
wall		0·44
		1·32

1·32	*Extra over common brickwork*
3·00	*for facings as before described.*

It should be noted that if the two returns were not exceeding half-brick wide, they would require to be measured linear under SM G15(b) and described as margins.

The above arguments apply also to projections for piers, etc., on hollow walls.

Piers entirely in facings

Consider the case of a half-brick wall *faced both sides*, with an attached pier projecting half-brick from one side of the wall.

In this case the total thickness of the wall plus the pier is one-brick; and both wall and pier would therefore be built entirely of facing bricks. Projections for piers are to be measured from the wall-face under G4(i), but does this apply to brickwork built entirely of facing bricks under SM G24–28?

SM G25(a) states that the classification of walls shall be as Clause G3(a). There is no mention of projections here, however.

The purpose of keeping one-brick walls faced both sides separate is to enable the estimator to allow for the extra cost of selecting headers of optimum length. This would also apply to the pier under discussion, and the only method of displaying this factor to the estimator would seem to be to measure the pier super, and to describe it (on the analogy of SM G25(a)) as:

One-brick pier in facings
as described, faced and
pointed both sides.

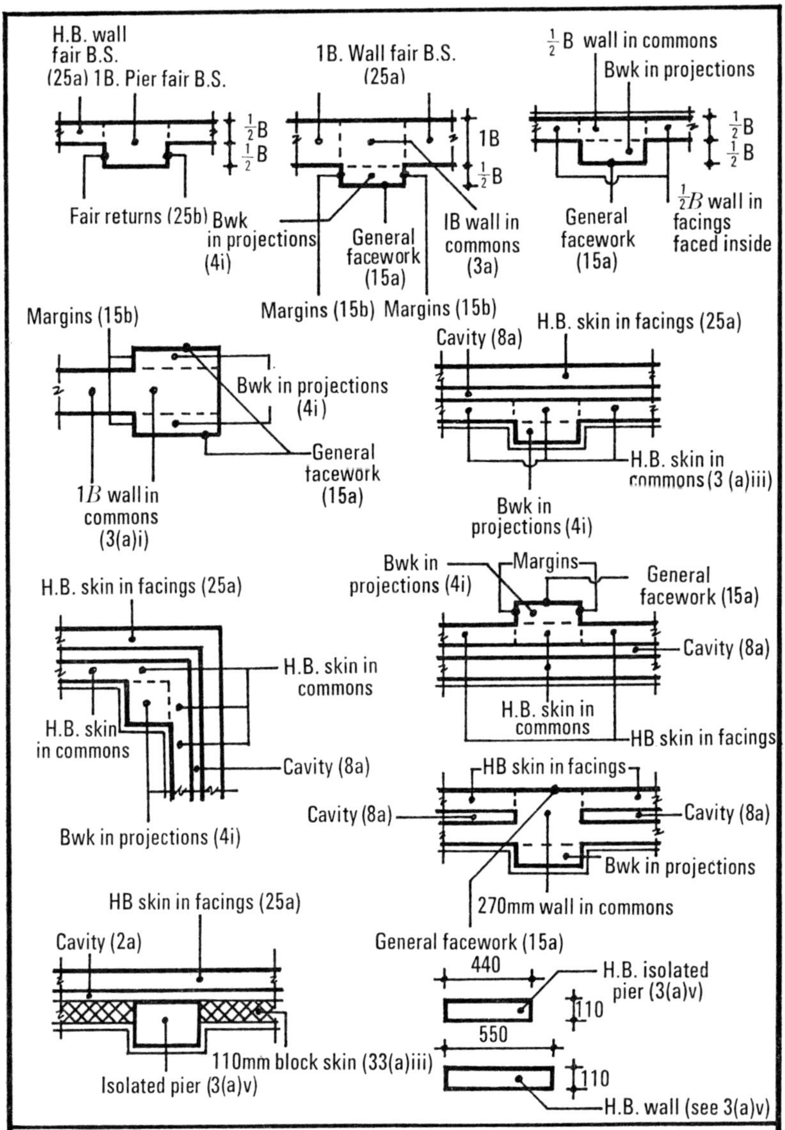

S.M.M. TREATMENT OF VERTICAL PROJECTIONS ETC. IN BRICKWORK

Note: Figures in brackets refer to section 'G' clause numbers.

Fig. 55

208 Note that the G3(a) classification called for in G25(a) includes *isolated* piers (G3(a)(v)) in any case and these would, of course, be measured in the same way, but described as 'isolated'.

Fair returns

An important point to bear in mind in connexion with brickwork entirely of facings is that the measurement of 'margins' is an item which is extra over common brickwork; and common brickwork does not exist in brickwork built entirely of facings. In lieu of margins therefore, *fair returns* are provided for in SM G25(b); and these are measured linear in half-brick stages, regardless of width, and described as (for example):

Fair returns in facings
as described, not exceeding half-brick
wide.

[*sides of*
piers

The above case is illustrated in Fig. 55, which deals with the measurement of vertical projections generally.

Note that where a pier projects half-a-brick from a half-brick wall in facings, it is measured only as above described if the wall is *faced both sides*. In the case of an outer skin of a hollow wall for instance, the section of wall behind the projection could (in theory) be built of commons; and in any case selection of headers is not involved. This case is also illustrated in Fig. 55, which the student is recommended to study with care.

Battered facework

This is referred to in SM G15(f), and denotes facework which is sloping instead of vertical. This may occur in the case of an external chimney-breast which diminishes in size at gathering-over of the flue. Figure 56

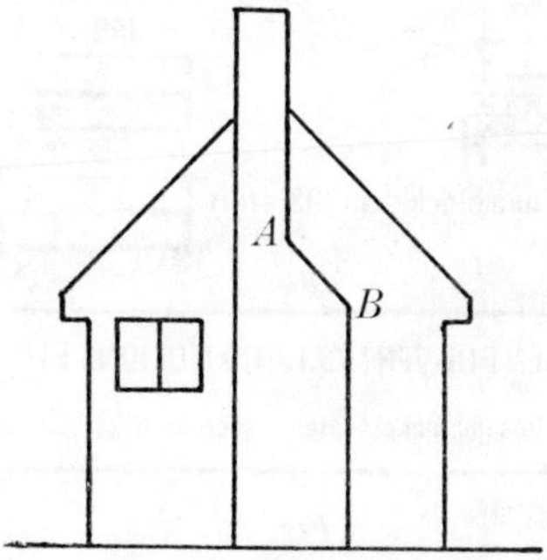

Fig. 56

shows a case in point. Here, the battering return *AB* would require to be measured super under the above clause as:

Extra over common brickwork
for facings as before, but
battering at a rate of 300 *mm*
every four courses
in height.

unless the returns were not exceeding half-brick wide, in which case a linear measurement of:

Ditto but to margins, battering
at a rate of 300 *mm every*
four courses in height.

would be necessary, to comply with SM G15(b). Note that the rate of batter must be given, as required by SM G15(f).

Curved facework

Here the radius *on face* must be given in accordance with SM G1(d), and cutting included in the description, thus:

Extra over common brickwork
for facings as before described,
but curved on plan to 3·5 *metres*
radius, including cutting.

Chimney-stacks

These will simply be girthed around on waste, and measured by the average height the area being added to the general facings. It should be remembered that only the actual area exposed to view is being measured. The height of stacks projecting through the roof coverings should therefore be calculated from top of flashing level, rather than roof-covering level.

FACEWORK LABOURS

Having measured the whole of the exposed face of the brickwork for facings, in the manner described above (and having checked to make sure that no more facings require to be measured), the taker-off can now turn his attention to measuring the facework labours. By facework labours are meant the items dealt with in SM G16 and G17; and apart from *fair chases* (not often met with in the class of work under discussion) it will be found that these facework labours are confined to (a) *fair cutting*, and (b) *fair angles*.

Fair cutting

Rough cutting was defined in Chapter 16 as any cutting of *common* bricks other than the cutting necessary for normal bonding purposes.

Fair cutting may now be defined as any cutting of *facing* bricks other than the cutting necessary for normal bonding purposes.[1]

There is a difference of measurement between the two however, owing to the fact that the facing bricks only occur on the outer face of the wall, making a *linear* measurement of cutting more logical than a superficial one, as in the case of rough cutting. Hence all fair cutting is measured linear.

If an item is to be measured linear, instead of super, it will, of course, be necessary to give its width. But the actual width of the fair cutting will depend on the width of the facings; in other words, on the depth to which they penetrate into the wall. This will vary according to the bond of the brickwork, as we have seen, and the depth will therefore require to be assumed.

The assumed depth is half-a-brick. This is stated in SM G16(a) which rules that fair cutting shall be deemed to penetrate half-brick into the wall. It will not be necessary to state this in the description of course (since the SMM deems it to be the case), but we shall need to bear this depth in mind when deducting any superficial areas of rough cutting previously measured which are now to be replaced by linear items of fair cutting. These are classified under SM G16 as follows:

FAIR CUTTING AT VERTICAL ABUTMENTS. This is measurable when the facings abut a vertical face of some other material, e.g., a concrete column. Not very probable in elementary construction.

FAIR CUTTING AGAINST SOFFITS. This implies a soffit which is existing when the facings are executed. Not very probable in elementary construction.

FAIR CURVED CUTTING. Measurable to the extrados of segmental arches, etc. (taken with 'openings'), but not often encountered in the 'facework' section of the taking-off.

FAIR CUTTING ON BRICK VAULTING. Hardly applicable to elementary construction.

FAIR RAKING AND SPLAY CUTTING. Fair *splay* cutting may be encountered underneath tile sills set weathering, but this would be

[1] Strictly speaking, *fair cutting* could also apply to the cutting of common bricks forming a fair face to the wall; but this would normally be measured with 'internal finishings'.

measured in the 'windows' section of the taking-off, not with facings. Fair *raking* cutting is often met with in the 'Facework' section, however.

Fair raking cutting

As will be seen from the above, fair cutting is quite likely to be confined to fair *raking* cutting at this stage of the taking-off, in the case of a simple building of the kind under consideration.

The drawings must be examined for instances where the facing bricks require to be cut to a raking angle, an obvious instance being to the top of gable walls.

It will be recalled that an item of rough cutting was measured to the top of the gable wall shown in Fig. 44, during measurement of common brickwork labours. The slope height of the gable wall was supered by the width of the wall, giving, for two such gables (see page 188):

2/2/2·08 0·22	*Rough cutting.*

Assuming the gable to be faced externally, however, the outer 110 mm will not actually require any cutting to the common bricks (rough cutting), but to the facing bricks (fair cutting). The adjustment therefore now has to be made as follows:

2/2/2·08	*Fair raking cutting in facings.* & *Deduct* *Rough cutting.* super × 0·11 =

Fair raking cutting is described as 'in facings', since it could conceivably apply to cutting on fair-faced brickwork in commons.

The result of this adjustment is illustrated in Fig. 57, overleaf.

Fair angles

Having measured all the fair cutting, the taker-off must now search the drawings for any cases of fair angles requiring measurement under SM G17.

Fair vertical angles are only measurable in the case of glazed brick facework; a very unlikely external finish to buildings nowadays. *Fair chamfered, rounded, or moulded angles* are also unlikely to be encountered very frequently.

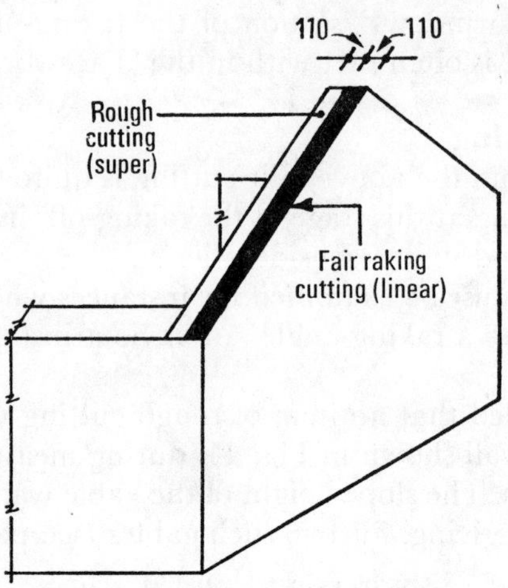

Fig. 57

Fair battered angles on the other hand, would in fact require to be measured to the example shown in Fig. 56, in connexion with the battered facework to the return *AB*, described earlier in this chapter. The length of the line *AB* would be measured linear as:

Fair battered external angle in facings

&

Fair battered internal ditto.

the external angle being where the battering return to the chimney-breast meets the face of the breast; the internal angle being where it meets the wall face.

Fair squint and birdsmouth angles

These now remain to be dealt with, and if any squint quoins exist, the taker-off will recall having measured rough cutting across the angle in the general brickwork section, as previously discussed in Chapter 12 (see Fig. 23, page 122).

The adjustment of squint quoins for fair cutting is best illustrated by way of an example, but first it will be necessary to attempt some definitions.

SQUINT QUOIN. This refers to any sharp corner of the brickwork which forms an angle other than a right-angle.

SQUINT-ANGLED INTERSECTION. The intersection of two walls other than at right-angles.

SQUINT ANGLE. This term, as used in SM G17, refers to the junction of two brick faces meeting at an irregular angle *exceeding* 180°.

BIRDSMOUTH ANGLE. This is used in the above clause to mean the junction of two brick faces meeting at an irregular angle of *less than* 180°.

It is interesting to note that SM G10(a) refers to 'rough cutting at squint or birdsmouth angles' when 'rough cutting at squint quoins or intersections' would seem to be more appropriate especially since squints are not kept separate from birdsmouths in the case of rough cutting.

However this may be, squints require separation from birdsmouths under SM G17 (fair angles), as the following example will show.

Let us take the case of the squint quoin shown in Fig. 58(a). Let us further suppose that the wall shown is *faced both sides*, and that in consequence of the squint quoin, rough cutting has been measured across the angle in the general brickwork section, as previously described. The adjustment now required to be made in the facings section is as follows:

FAIR SQUINT ANGLE. A vertical linear measurement is required of the squint shown on the plan.

FAIR BIRDSMOUTH ANGLE. A separate item is required (similarly measured) for the birdsmouth.

Rough cutting has been measured across the angle, however, and some of this will now require to be deducted. To what extent will this deduction be necessary?

We have seen that fair cutting is deemed to penetrate half-brick into the wall (SM G16(a)), and this allowance is presumed to apply also to fair angles. The adjustment in the case of Fig. 58(a) would be made as follows therefore (height assumed):

3·00	*Fair cut birdsmouth angle in facings* & *Fair cut and rubbed squint angle in ditto.*
2/3·00 0·11	*Deduct* *Rough cutting.*

It will be noticed that the birdsmouth has been described as 'cut', and the squint as 'cut and rubbed'. This is because the method of forming

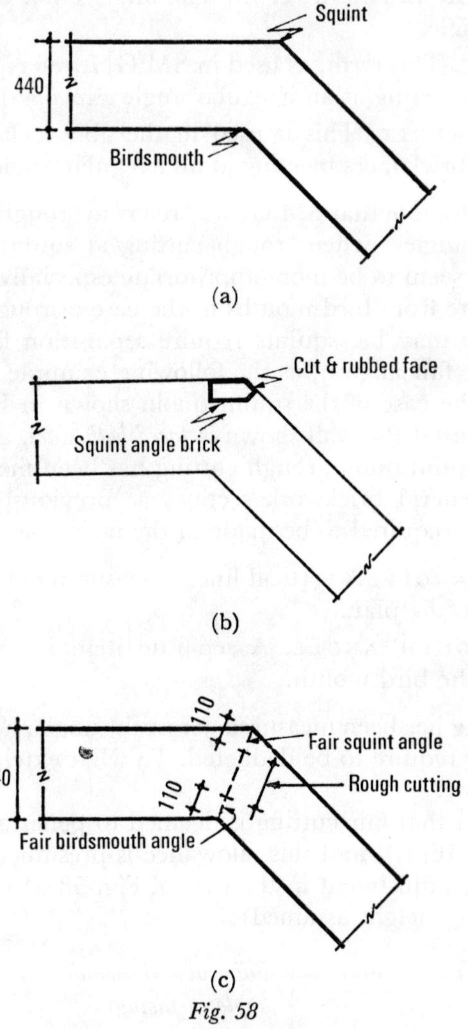

Fig. 58

the angle is required to be stated, in accordance with SM G17(c), which offers three alternatives, (1) cut, (2) cut and rubbed, and (3) purpose-made. This refers to the treatment of the brick forming the angle, which in the case of a birdsmouth will simply require to be cut at the back to form a bond. In the case of a squint, however, the exposed face of the squint angle brick will require to be cut, and the cut surface will need to be rubbed fair to match the rest of the facings, as shown in Fig. 58(b).

Alternatively, the squint angle bricks may be purpose-made to the required shape; and when this is specified, the description must be worded accordingly, thus:

Fair purpose-made squint
angle in facings

which would be necessary if the facing bricks were of a type unsuitable for cutting and rubbing. In this case the squint angle bricks are often given a P.C. price per hundred, thus:

Fair squint angle in purpose-
made facing bricks P.C. £3.00
per hundred.

The work is now measured as shown in Fig. 58(c), and the adjustment is complete.

It should be noted that a fair squint and birdsmouth angle exist in Fig. 56, also, not on plan in this case, but in elevation. Thus, the projection of the chimney-breast would require to be measured linear, and taken as a fair horizontal squint angle at point *B*, and a fair horizontal birdsmouth angle at point *A*.

Chapter Nineteen

Architectural Features

Having satisfied himself that all fair cutting and fair angles in respect of brick facework have been measured, the taker-off can now turn his attention to the measurement of any architectural features on the face of the building; starting at the bottom with any plinths, and working his way up the elevations measuring any band courses, etc., finishing with the measurement of any copings.

Chimney-stacks must be examined for any band or corbel courses; in fact the whole of the facings area previously measured must now be inspected, it being borne in mind, however, that any brickwork or other feature in connexion with window or door openings is normally left to be dealt with at a later stage, when the openings are adjusted in the 'windows' and 'doors' section of the taking-off.

In practice, the architectural features dealt with at this stage in connexion with elementary construction are likely to be confined to plinths, band courses, and copings.

Plinths in facework

It will be recalled from Chapter 16 that plinths in common brickwork, like band courses and other projections in commons, are measured from the wall-face as 'brickwork in projections' under SM G4(i). The assumption was made in Chapter 16 that no brick facework was present; the brickwork being rendered (see page 183).

In the case of faced work, however, band courses are separately provided for under SM G18 and G19 as discussed later in this chapter. *Plinths* are not separately provided for, however, in the Brick Facework section of the SMM, and this tends to lead to the conclusion that plinths in facings should be measured with the general brickwork as 'brickwork in projections', and the facework measured with the general facework as extra over common brickwork.

On this reasoning, the only remaining item to be measured in the 'architectural features' section of the taking-off in the case of a square-top plinth is an item of 'facework to margins' to the top edge.

There is, however, the matter of SM G15(g), which provides for facework sunk or projecting less than half-brick from the general face of the wall to be so described, stating the depth, and including the extra material. If this clause envisages a plinth, then plinths in facework (provided they do not project more than half-a-brick) would be ignored altogether in the general brickwork section, and measured in the facings section as

Extra labour and material
setting facings forward 50 *mm*
to form plinth,
including rough cutting.

No deduction of general facework is made in this case either, and a margin is measured to the top edge as before.

At least one well-known authority measures facework plinths in this way, and the method has much to commend it.

If the plinth extends below facings level, however, this method of measurement would involve splitting the plinth up; the section below facings being measured as brickwork in projections. This might be an argument in favour of the former method of measurement.

Alternative methods of plinth measurement

We have seen in the case of rendered plinths (Chapter 16) that where these project on either side of a wall in such a manner that the wall below plinth-level is a multiple of half-a-brick in thickness, there is a strong case for measuring the wall below plinth-level as a wall of full thickness. This avoids the measurement of non-existent rough cutting, measurable if the plinths were taken as brickwork in projections under SM G4(i).

This argument also applies if the wall is faced. Even more forceably in fact if the wall below plinth-level happens to be a one-brick wall faced both sides, required to be measured as such under SM G25.

We must therefore come to the conclusion that there are in fact no less than three legitimate methods of dealing with a plinth in facings. These may be summarized as follows:

(1) THE PROJECTION METHOD
- (a) The extra brickwork is measured as common brickwork in projections under SM G4(i).
- (b) Facework is added to the general facings as extra over commons under G14(d).
- (c) A margin is measured to the top edge under G15(b).

(2) THE EXTRA MATERIAL METHOD
- (a) No projection is measured with common brickwork.
- (b) Facework is added to the general facings as above.
- (c) An item of extra labour and material setting facings forward is measured under G15(g).
- (d) A margin is measured as above.

(3) THE FULL THICKNESS METHOD
- (a) The wall below plinth-level is measured as a wall of the full thickness; facings being measured as appropriate.
- (b) A margin is measured as above.

The taker-off must select whichever of the above methods is most appropriate to the case under consideration.

Plinths on hollow walls

A plinth may be formed by projecting forward a few courses of brickwork in the outer skin of a hollow wall, as shown in Fig. 59(a). Is there any extra involved at all in this case, apart from the margin to the top edge?

There may be an extra labour involved in forming the wider cavity, and to cover this eventuality it is usual to measure the area of the plinth as:

Extra labour setting forward
half-brick skin of hollow
wall in facings 25 mm
to form plinth.

If the 'plinth' goes down to concrete strip, the wall below plinth level is best treated as a wall of the full thickness, especially since longer wall ties may be required in this case.

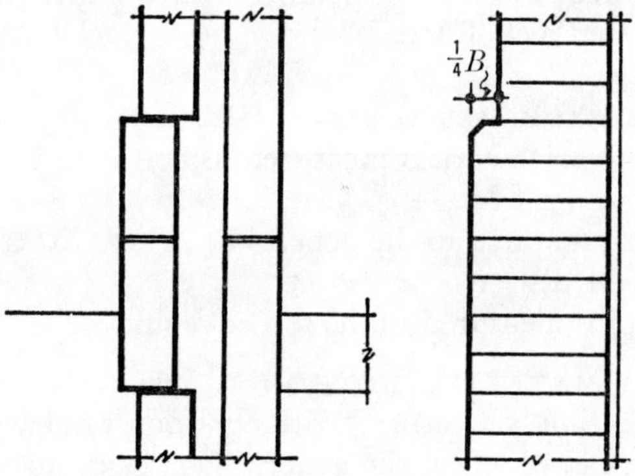

(a) (b)

Fig. 59

Note that any such plinths in hollow walls extending below *ground level* will require an adjustment in the thickness of the cavity filling.

Plinth cappings

These are measured linear under SM G19(a), with angles enumerated under 19(b). Since they are regarded as items of facework in the SMM clause, they are measured as extra over common brickwork, which will have to be measured as 'in projections' right to the top of the plinth capping. If general facework has already been measured over the capping, this will require to be deducted when the capping is measured.

Thus the plinth with the splayed capping shown in Fig. 59(b) would require the following items of measurement:

In the Brickwork section:

super	*Reduced brickwork in commons in projections*
	$\times \frac{1}{4}B =$ *(plinth*
	&
	Rough cutting

In the Facework section:

(general facings assumed measured)

linear	*Extra over common brickwork for splayed plinth capping in facings, one course high and* 55 *mm projection, including pointing as before described.*
	&
	Deduct *Extra over common brickwork for facing as before*
	Super $\times 0{\cdot}06 =$
number	*Add* *External angles to last plinth capping.*

If this plinth had a square top, on the other hand, the only item measurable, in addition to general facework, would be:

linear	*Extra over common brickwork for facework as described, but for margins.*
	[*top of plinth*

The phrase 'facework as described' refers the estimator back to the general facings description for the brick-type, bond, pointing and mortar-mix.

Facework to band courses

Band courses in facings are of two main types:

(1) *Plain bands*, dealt with under SM G18.

(2) *Ornamental bands*, measured under SM G19.

220 A plain band would be in facing bricks which differ in kind or size from the general facings; but a brick-on-edge or a brick-on-end band is classed as an ornamental band, not a plain band. The difference in measurement between the two kinds of bands is a question of size. Thus *plain* bands are effectively defined for this purpose in SM G18 as those not exceeding 300 mm wide; bands over 300 mm wide being dealt with as facework to walls; and if sunk or projecting would be measured under SM G15(g).

In the case of *ornamental* bands, the width is immaterial however, and ornamental bands are all measured as such, whether over 300 mm wide or not.

Either of these two main types of band may be:

(a) Flush
(b) Projecting
(c) Sunk.

Flush bands

An example of a flush brick-on-end band course is shown in Fig. 60. This would be measured linear, and the angles numbered under SM G19(a) as follows:

linear | *Extra over common brickwork for facework as described to horizontal brick-on-end flush band 220 mm high.*

&

Deduct Extra over common brickwork for facing as before.

Super × 0·22 =

number | *Add external angles to last flush band.*

Projecting bands

SM G4(i) expressly includes projections for bands to be classified as brickwork in projections. Under the heading of 'Brick facework generally', SM G14(d) states that facework shall be given as extra over the brickwork on which it occurs. Clauses G18 and 19 refer not to *bands*, but to *facework to bands*. It might appear therefore that projecting bands in

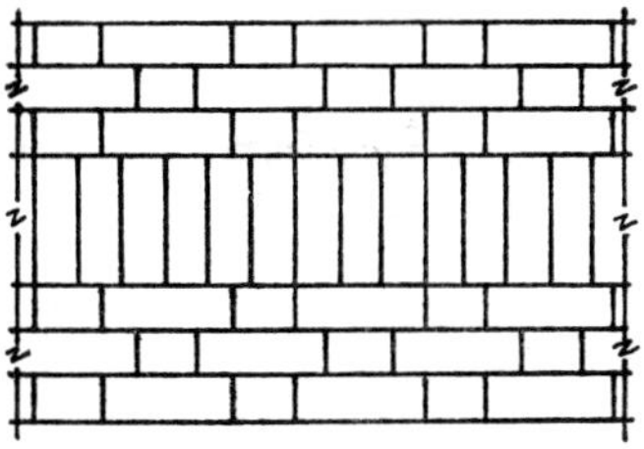

Fig. 60

facework should be measured for common brickwork under SM G4(i), and again for facework under SM G18 or 19, in the same manner in which a plinth may be measured.

However, G18(b) states that '. . . *rough cutting within the thickness and extra material shall be given in the description of the bands*'. This is stated again in G19(a) in respect of ornamental bands. To measure projecting bands in the manner described above therefore would mean that the rough cutting and extra material would be included twice over, if G18(b) and G19(a) are to be complied with.

This cannot be the intention of the SMM; and it is therefore the practice not to measure projecting faced band courses for common brickwork, but for facings only, as in the case of plinths which are measured under SM G15(g).

Thus the projecting faced band course in Fig. 61(a) for example would be measured in the facings section only as (length assumed):

50·00	*Extra over common brickwork for facework as described to horizontal plain band set projecting* 55 *mm for three courses in height, including extra material and cutting, and facework to two margins.*
	&
	Deduct Extra over common brickwork for facing as before.
	super × 0·22 =
(say) 4	*Add External angles to last plain band.*

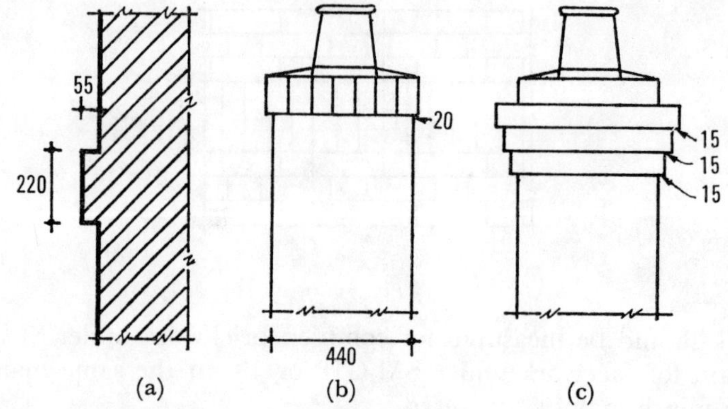

Fig. 61

Similarly, the band course to the chimney-stack shown in Fig. 61(b) would be measured as follows:

4/·44	1·76
8/·02	0·16
	1·92

1·92	*Extra over common brickwork for facework as described to horizontal brick-on-edge band set projecting 20 mm for one course in height, including extra material and cutting, and facework to one margin.*
	&
	Deduct extra over common brickwork for facing as before.
	super × 0·11 =
4	*Add external angles to last brick-on-edge band.*

Consider now the band course to the stack shown in Fig. 61(c).

Is this in fact a band course, or are these courses oversailing courses, as mentioned in SM G4(i)? If so, they should be measured as common brickwork in projections (with rough cutting taken); general facework not deducted; and margins measured separately under G15(b), no angles being measurable in this case.

The brickwork above the projecting courses in question does not oversail however, and these courses are therefore best regarded as an ornamental band under G19(a), which includes '. . . brick-on-edge bands . . . dentilled bands . . . moulded string courses . . . *and the like*' (author's italics).

Would this include band courses in which the ornamentation consisted of corbel courses? There seems to be no reason why not, and since this involves a method of measurement which presents the estimator with a better picture of the work than the one suggested above, such features as the one in Fig. 61(a) are best described as follows:

> *Extra over common brickwork for facework as described to horizontal projecting band three courses in height; lowest course projecting* 15 *mm, middle course projecting* 30 *mm and upper course projecting* 45 *mm from face, including extra material and cutting, and facework to four margins.*

General facework is now deducted out, and the angles measured as usual.

Facework to sunk bands

The principle here is the same as that for projecting bands, except that no extra material is involved; a typical description for a sunk plain band being:

> *Extra over common brickwork for facework as described to horizontal plain band, sunk* 30 *mm deep for three courses in height, including cutting, and facework to two margins.*

General facework is deducted, and angles measured, as for projecting bands.

Moulded string courses, etc.

Ornamental string courses and cornices are provided for in SM G19, and

224 facework to ornamental quoins is covered by SM G21; but apart from the simple types of feature discussed above, ornamental brickwork is unlikely to be encountered nowadays in the type of construction under consideration; and is in any case not considered applicable to the intermediate syllabus.

Figure 62 illustrates some of the horizontal features described in this chapter, and gives an indication of the appropriate method of measurement.

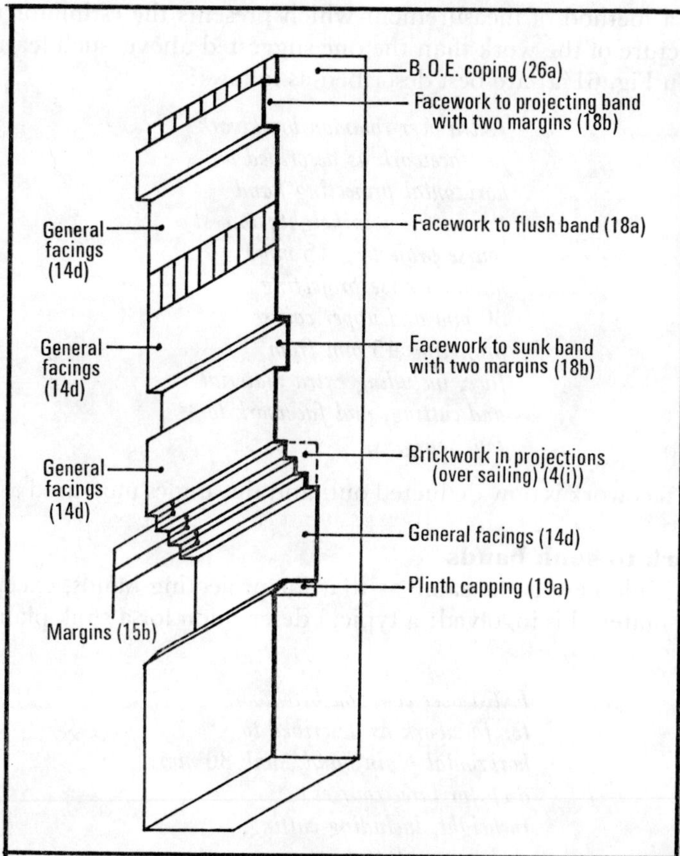

S.M.M. TREATMENT OF HORIZONTAL PROJECTIONS IN FACEWORK

Note: Figures in brackets refer to section 'G' clause numbers

Fig. 62

Brick copings

These are dealt with under SM G26(a), and will usually be of the brick-on-edge variety. A course of creasing tiles may occur beneath the coping, and this must be measured as *extra over brickwork* in accordance with SM G20.

It should be noted that all copings are measured for full value of labour and materials, and are not extra over brickwork. It therefore follows that:

(a) brickwork and facework must be measured up to *underside* of coping only, and
(b) any creasing course must be measured separately from the coping.

Typical coping descriptions are as follows:

linear	*Horizontal brick-on-edge coping size* 220 × 110 *mm in Rustic Fletton facing bricks, and pointing as described to top and both sides.*
number	*Extra for external angles including cutting.*
number	*Extra for internal ditto including ditto.*
number	*Extra for fair ends.*

Copings to piers and projections

These are best dealt with as *cappings*, and measured under SM G27. The method of measurement is best illustrated by an example and Fig. 63 shows a suitable section of wall with piers, finished at the top with a brick-on-edge coping.

SM G27 states that cappings to pilasters and cappings to isolated piers shall each be measured separately. A pilaster is in fact an attached pier, and the coping in the figure would therefore be measured right through to where the end pier becomes isolated, above coping level; i.e., from *A* to *B* as shown. The description would be as above.

The portion of coping over the projections *C*, *D*, and *E* would now be measured as:

Ditto, but as capping to pilaster size 440 × 110 *mm projection, including pointing to top and two returns.*

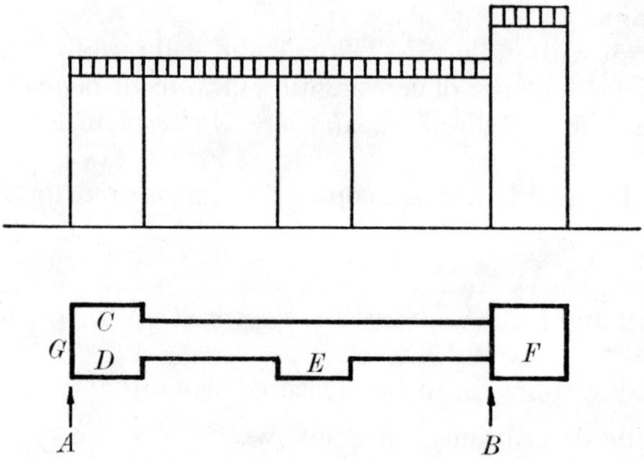

Fig. 63

Note that pointing to the *face* is offset by the coping already measured through. The cap at *F* would be described as:

Ditto, but as capping to
isolated pier size 440 × 440 *mm*
including pointing to top
and four faces.

A fair end to the coping would now be measured at point *G*; but is anything measurable at point *H*, where the coping abuts the pier?

It can be argued that the laying of the end brick against the pier is no different from laying any of the others, since any cutting involved would probably occur elsewhere along the length of the coping. However, the estimator is the best judge of this matter, and he need not price a fitted end if he thinks no extra is involved. SM G26(b) refers to 'ends', not necessarily fair ends only; and the measurement of fitted ends is on the whole advisable where appropriate, in order to be on the safe side.

Stone copings

As previously explained, the measurement of dressed stonework is excluded from the intermediate syllabus; but copings may well be in dressed stone, or in reconstructed stone, or in stone-faced precast concrete, or some similar material. If so, they are still measured in the 'Facework' section of the work, as are brick copings, and would require to be dealt with at this stage.

Stone copings are measured under SM K20, and concrete ones under SM F34, and it will be seen that both are measured linear, with angles and ends numbered.

In the case of stone or concrete copings, caps over attached piers would be integral with the coping at that point, to avoid a longitudinal joint; so that the main coping should be deducted where any piers or pilasters occur, and a pier cap taken for the total width.

Thus if we assume the coping in Fig. 63 to be of reconstructed stone, instead of brick, it would require to be measured *between* the piers as (say):

Reconstructed stone coping size 280 × 100 *mm*
twice weathered
and twice throated, and bedded
and jointed in cement mortar
(1:3).

The pier caps would then be numbered as

Reconstructed stone pier cap
size 500 × 390 × 100 *mm thick,*
twice weathered and throated,
and bedded and jointed to 280 × 100 *mm*
coping (measured
separately) on both sides,
in cement mortar (1:3).

Ditto but size 500 × 500 *mm*
three times weathered and
throated and bedded and jointed
to coping on one side.

Ditto but four times weathered
and throated, to isolated pier.

As in the case of the brick coping, it is usual to take a fitted end where the coping abuts brickwork, although as previously explained, it is doubtful whether an extra actually exists at this point.

Stone Walling

Natural stone may be used as a building material instead of bricks or blocks. In domestic work it may be used for walls, for the outer skins of hollow walls, for chimney-breasts, stacks, etc., or boundary walls.

Stonework is divided into two categories:

(1) DRESSED STONEWORK. This term is used to mean '. . . dressed blocks of natural stone accurately worked to given dimensions and laid in mortar in courses with fine joints' (SM K2(b)).

Dressed stonework, cast stonework, and clayware work (terra-cotta, etc.) are grouped together as *Masonry* under Section K of the SMM.

(2) RUBBLE WORK. This is defined by SMJ2(b) as natural stones which are

(i) irregular in shape or roughly dressed, and
(ii) laid dry or with comparatively thick joints.

Dressed stonework is sometimes called 'ashlar work'. It is excluded from the RICS intermediate syllabus, and will not be further dealt with in this chapter.

Rubble work is referred to as 'stone walling' in the above syllabus, and 'rubble walling' in the heading to Section J of the SMM, which lays down the rules of measurement for rubble work. An examination of Section J will show that the rules of measurement for rubble work are somewhat similar to those for brickwork; and the method of measurement is therefore best comprehended by comparing it with that for brickwork and brick facework, and noting the differences in treatment.

Stonework heading

The cost of the work will be greatly influenced by the type of stone used, and its availability within the area. Also by the availability of local craftsmen. For this reason it is customary to put stonework under a special heading in the bill and this heading is written in the dimensions by the taker-off before proceeding with detailed measurement of the work. As in all such cases of headings requiring transfer to the bill, every word is underlined, and the wording commences with 'The following in . . .':

The following in random stones
of approved local sandstone
bedded in lime mortar (1:3).

As indicated in this sample heading, rubble work is often carried out in local stone, by a traditional craftsman called a 'stone waller' as opposed to a 'stone mason', who works dressed stonework. In some parts of the country these local craftsmen are used to build 'hedges' of dry stone walling which may be seen in the countryside in 'stone' districts, and it is in these localities where rubble work is more likely to be specified in the construction of buildings.

Rubble facework

In considering the difference between the measurement of rubble work and brickwork, one of the chief factors to bear in mind is the treatment of facework, which effects measurement in the following way:

BRICKWORK. General facework is measured 'extra over' common brickwork. Common brickwork may therefore be measured and described first, without regard to the facework, except in certain instances (e.g., walls built entirely of facings).

RUBBLE WORK. General facework is not extra over stonework. Unless bounded to or built against a backing of other material (SM J7(a) and (b)) it is *included in the description of the walling* (see SM J2(c)(iv), etc.).

This means that faced rubble work must be kept entirely separate from unfaced work.

This difference in treatment is a result of the fact that in rubble work the facing stones are often of the same kind as those used in the body of the work.

Work below ground level

In view of the above, rubble walls below ground level are measured not up to DPC, but up to 100 mm below ground level, where they become faced, and therefore described separately. A description of the following kind would apply below ground, or where a wall was not faced on either side:

super	*Rubble work in walls* 300 *mm thick.*

Walls are described as such, and the thickness given as for brickwork (SM J(3)); but in millimetres instead of brick thickness. All work is under the heading, so that details of mortar mix, etc., are already taken care of.

230 **Stone piers and chimney-breasts**

It is here that another important difference between the treatment of brickwork and rubble work occurs:

BRICKWORK. Attached piers are measured from the wall-face, the wall itself being measured straight through.

RUBBLE WORK. Attached piers are taken as '. . . *the combined thickness of the wall and the pier*'. The wall is measured *between* the piers (see SM J2(d)).

The initial procedure for measuring a rubble wall is as for a brick wall, i.e., ignoring all projections in the first instance and measuring straight through on the mean girth. When any projecting piers or chimney-breasts are measured, however, an adjustment must be made, and the wall deducted out for the width of the pier or chimney-breast.

Take the case of Fig. 64(a) for example, which shows a plan below ground level of the brick wall illustrated in Fig. 63.

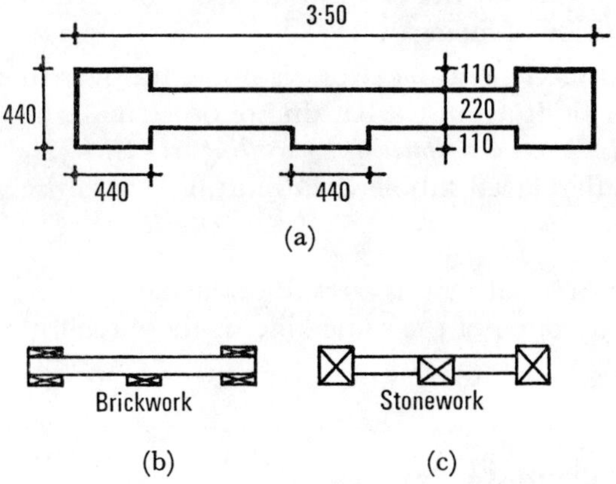

Fig. 64

If the wall were of *brick* it would be measured as shown in Fig. 64(b), as follows (height assumed):

3·50 1·00	*One-brick wall in commons* *in cement mortar* (1:3).
0·44 1·00	$\frac{1}{2}$*B in projections.* *(piers*

If of *rubble stonework*, as Fig. 64(c), as follows:

3·50 1·00	*Rubble work in walls* 220 *mm thick.*

0·44 1·00	*Ditto in piers* 440 *mm thick.*
0·44 1·00	*Ditto in piers* 330 *mm thick.*
0·44 1·00	*Deduct Rubble work in walls* 220 *mm thick.*

The same height is taken here as a comparison, but as previously explained, brickwork would be taken to DPC, and stonework to 100 mm below ground level. The stonework would appear under its heading as usual.

Stone above ground level

It will be recalled that the order of taking-off for superstructure brickwork was (a) external walls, (b) internal walls, (c) chimney-breasts, (d) stacks, followed by the facework. We have seen that in the case of rubble work, however, faced work is not extra over, and faced rubble work must be kept separate from unfaced work. Apart from any internal walls (and internal walls will quite probably be in some other material in any case) all the rubble work above ground level is likely to be faced; the face being included in the description of the walling. For this reason the order of taking-off will be different from that of for brickwork.

A suggested order of taking-off for superstructure rubble work is as follows:

(1) Walls
(2) Piers, chimney-breasts, and stacks
(3) Fair returns
(4) Copings and similar features
(5) Labours.

Because facework is not extra over, sides of piers and chimney-stacks are dealt with as *fair returns* under SM J10, and this will therefore form a subsection of the taking-off.

Superstructure walls

Measurement will be similar to that for walls below ground level, viz., as for brickwork: with walls measured through on the mean girth; piers, etc., being ignored and adjusted later. A sample description for walls exposed to view on the outside face is as follows:

Rubble work in walls 300 *mm*
thick as before, but faced one
side with picked natural faced
stones and flush pointed in
cement mortar (1:4).

Note that pointing and mortar-mix is stated, in accordance with SM J2(c)(v) and (vi), mortar for bedding and jointing being given in the heading.

Any *curved* walls are, of course, separated as for brickwork (see SM J1(e)) with the mean radius stated. *Tapered* walls (i.e., walls with battering faces) are also kept separate, under J4; a typical description for a tapered wall being:

Rubble work in tapered walls average
380 *mm thick,*
battered one face at a rate of 25 *mm*
to every 300 *mm in*
height, and faced and pointed
both sides as described.

'As described' may be used if this item follows work in which facing and pointing has in fact been described. Note that (i) the average thickness, (ii) whether one or both faces are battered, and (iii) the rate of batter have each been described in accordance with SM J4. The calculation required to arrive at the average thickness should be shown on waste.

Faced piers, chimney-breasts, etc.

The method of measurement is the same as for substructure piers, illustrated in Fig. 64(c), except that the description will include the facing. For example if the work shown in Fig. 64(a) was faced all round above ground level, it would be measured as follows (height assumed):

3·50 2·00	*Rubble work in walls* 220 *mm thick, faced with picked natural faced stones, and flush pointed in cement mortar* (1:4) *both sides.*
2/0·44 2·00	*Ditto in piers* 440 *mm thick and ditto all as last.*
0·44 2·00	*Ditto in piers* 330 *mm thick and ditto.*

3/0·44 2·00	*Deduct* *Ditto in walls* 220 *mm thick faced both sides as before.*

It should be noted that piers and stacks are grouped together in the case of stonework (see SM J3(iii)), and that attached and isolated piers having a length of over four times the thickness are classed as walls (SM J2 (d)).

Note also that *attached* piers are not kept separate from *isolated* piers, as they are in the case of brickwork.

Chimney-stacks

A chimney-stack would be measured in the same way as for brickwork, but described in a manner such as the following:

super	*Rubble work in chimney-stack* 440 *mm thick, faced and pointed as described both sides.*

If similar work in piers 440 mm thick is also measured, this would be grouped with the above stack, and the description altered to 'Rubble work in piers and chimney-stacks 440 mm thick . . . etc.', in accordance with SM J3(iii).

Fair returns on rubble work

Having measured all walls, piers, breasts and stacks in the above manner, a sub-heading of 'fair returns' is inserted in the dimensions, and these are now measured. By way of illustration, the work shown in Fig. 64(a) and measured above as for faced stonework, would be dealt with for fair returns as follows:

3/2/2·00	*Fair returns on rubble work* 110 *mm wide.* *(returns*
2/2·00	*Ditto* 440 *mm wide.* *(at ends*

Again, the sides of the chimney-stack described above, assuming the stack to be (say) 440 mm wide, would be taken as:

linear	*Fair returns on rubble work* 440 *mm wide* *[sides of stack*

234 Note that *all* fair returns are measured *linear*, regardless of width, under SM J10.

Rubble copings

These are measured under SM J13, which envisages the kind shown in Fig. 65. As will be seen, the height of the coping stones will vary, and J13

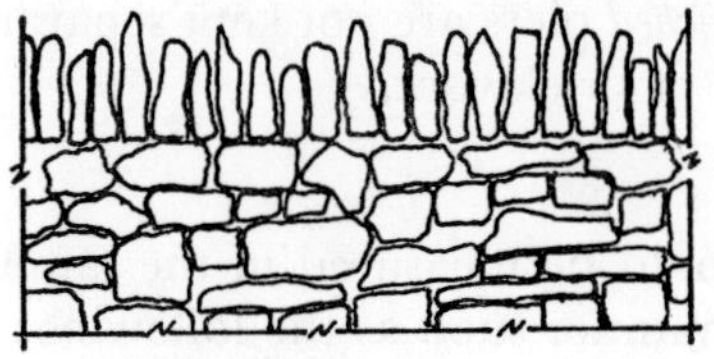

Fig. 65

provides, therefore, for the *average* height to be stated. A typical description would be as follows:

Coping 300 *mm wide and*
average 220 *mm high of*
random stones and pointing
to match facework, including
levelling random walling
under.

The reason for the last phrase in this description is on account of SM J8. Levelling uncoursed rubble work for copings and the like is to be measured linear. It is the usual practice to include this item with rubble coping descriptions, as allowed for by SM A3(d).

Note that angles and ends are not measurable to rubble copings under J13. Any pier-caps in rubble coping work would be dealt with in a similar manner to those in brick-on-edge coping work, described in a previous chapter.

Dressed stone or concrete copings

If the rubble wall is topped with a coping other than one formed of rough stones, the rules of measurement applying to the particular material concerned must be adhered to. Rubble work is sometimes finished with a dressed stone or concrete coping for example, and where this occurs, it is measured as described for the stone coping discussed in the chapter on architectural features.

An important point to note in this connexion is with regard to the item for levelling uncoursed walling. It can only be included in the coping description under SM A3(d) if the coping is measured under Section J. It is only valid to include items together in this way if both items occur in the same work-section of the SMM; otherwise different trades will be merged together in the bill, which might upset pricing arrangements at tendering stage.

In the case of a dressed stone coping therefore, levelling will require to be measured separately under the coping as (for example):

linear	*Extra for levelling uncoursed rubble work* 300 *mm wide.*
	[*under coping*

Labours, etc., on rubble work

It will be observed from the SMM that *rough cutting* is not measurable in the case of rubble work (see J9(a)). *Fair cutting* is measured linear (J9(b)) stating the thickness of the work, and would be measured to tops of any gables, etc., in the manner described for brickwork; except that there would be no rough cutting to deduct, and the wall thickness would be stated. Note that no distinction is made between square, raking, and curved fair cutting in rubble work however; it all goes in simply as 'fair cutting', and would be measurable at abutments for example, if rubble work abutted other material such as brickwork.

Eaves filling is measurable under J5 as in the case of brickwork, and so are *fair squint angles* and *fair birdsmouth angles* under J11(b). No deduction of rough cutting will be necessary in the case of these last, and there will, of course, be no cutting and rubbing of the fair squints either.

Chapter Twenty-one

Fires and Vents

We have seen that in measuring the brickwork the taker-off will have treated all chimney-breasts as being solid, openings for fireplaces having been ignored. These now require to be adjusted, and all work in connexion with the fireplace itself measured in detail.

The surveyor must now examine the drawings for any such fireplace openings, and deal with each one in turn. In present-day construction some form of central heating may well be specified, and in this case it may be that no fireplaces are to be provided. A brick stack may be required for the boiler, and this will have been measured in the brickwork section; but the boiler itself, together with the flue connexion, will be measured in the 'Heating' section of the taking-off; not in the 'fires' section. Even so, a traditional fireplace may well be specified for the living-room, to supplement the central heating and provide a source of radiant heat in very cold weather.

When all fireplaces have been accounted for, *air vents* required by the byelaws in rooms without fireplaces are measured; together with any vents which may be required for other purposes. Thus the work in this section of the taking-off will usually be divided up as follows:

(1) Ground-floor fires
(2) Upper-floor fires
(3) Vents.

Fireplaces in upper floors are usually dealt with separately owing to the slightly different treatment required in respect of the hearth construction. The general order of taking-off for fires will be the same in all cases however, and is as follows:

(a) Fireplace
(b) Opening
(c) Surround
(d) Hearth construction
(e) Flue.

Flues may be provided for built-in gas fires fixed flush with the wall. The work involved in this case, however, is probably best measured in 'work in connexion with gas services', rather than in the 'fires' section of the taking-off.

The fireplace itself will normally consist of a firebrick lining, or *fireback* as the British Standard terms it, as shown in Fig. 66; together with a *grate*

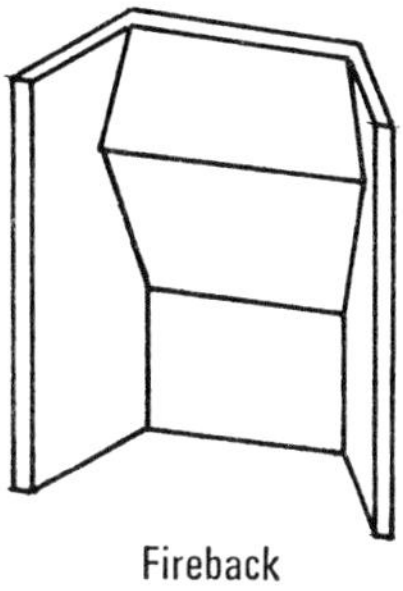

Fig. 66

of some kind or other. The grate may be a simple 'basket and fret' forming an open fire of the traditional kind or, more likely now-a-days, an all-night-burning or *continuous-burning* grate, the grate itself often being called a 'continuous-burning fire'.

These two items, the fireback and grate, may be purchased separately, but are included in the same description under SM G64(a), which refers to firebacks as 'concrete and brick backings'. This is because concrete filling is placed between the back of the fireback and the surrounding brickwork of the fireplace opening, as shown in Fig. 67. Since the cost of

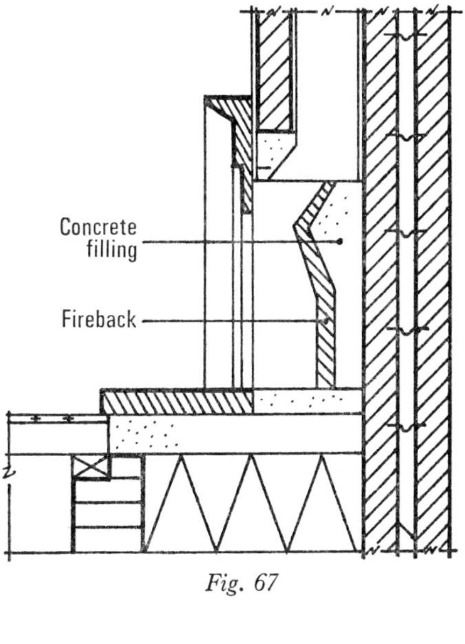

Fig. 67

the grate can vary, this item is often made the subject of a P.C., the P.C. price including the fireback. A typical description is as follows:

1 *Continuous-burning fire, complete with fireback lining P.C.* £10·00 *and setting to* 570 *mm opening including weak concrete filling at back, and all necessary fireclay cement.*

Note that SM G64(a) requires the *width* of the fireplace opening to be given. This is because although the height of the fireplace opening (and hence the lining) is fairly constant, the British Standard gives alternative widths; which will affect the amount of concrete filling required. These are nominal widths of the opening in the surround, which, as Fig. 68 shows, will normally be less than that of the brickwork opening into which the fireback is inserted.

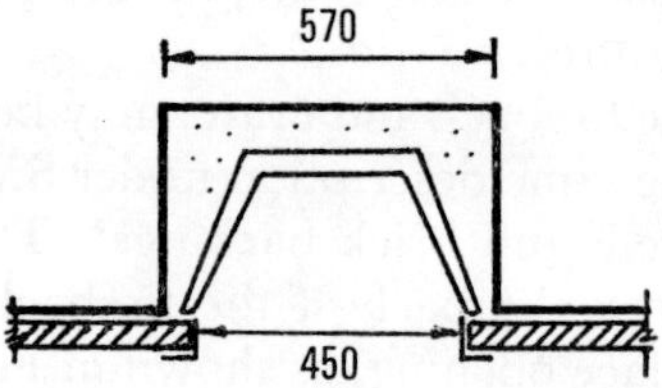

Fig. 68

Adjustment of opening

Brickwork having previously been measured solid, it will now be necessary to deduct the brickwork for the actual fireplace opening, and insert a concrete lintel to support the brickwork above. Only the actual fireplace opening is deducted, not the flue for flues are not deducted unless exceeding 0·25 square metres in sectional area (see SM G1(a)(ii)). This area might be exceeded in the case of an industrial chimney-shaft, but hardly in the case of a domestic flue of the type under consideration.

For purposes of this deduction, the gathering-over above fireback height is regarded as part of the flue, and the deduction is taken from underside of lintel down to the point where the chimney-breast maintains full thickness, thus (height assumed):

0·57 *Deduct Reduced brickwork in projections.*
0·75 × $1\frac{1}{2}B$ = *(opening*

The precast lintel is measured under SM F34, and it will be appreciated that such lintels include the reinforcement under F27(c)(i); formwork being *deemed* included (see F27(d)). A suitable bearing on the brickwork must be allowed for, and the lintel measured as follows:

		0·57
	bearing 2/·11	·22
		0·79
0·79	120 × 150 *mm Precast concrete* (1:2:4) *splayed lintel with one* 12 *mm diameter steel reinforcing rod.*	
2	*Extra for stooled ends*	
0·57 0·15	*Deduct* *Reduced brickwork in projections.*	
		$\times \frac{1}{2}B$ –

Note 1. The splayed lintel, as shown in Fig. 67, is taken 10 mm thicker than the brickwork over, to allow for thickness of cement rendering on the inside.

Note 2. Stooled ends (square ends) are required, to enable the splayed lintel to bear properly on the brickwork. These are numbered under SM F34.

Note 3. No deduction of brickwork is made for ends of lintels built in as the work proceeds (see SM G56(b)). The deduction for the lintel is therefore taken as the net size of the opening.

Surround

Fire surrounds are provided for under SM G64(b), which also mentions hearths. Present-day surrounds and hearths are usually pre-slabbed, i.e., made up in the workshop of concrete slabbing faced with glazed tiles. The surround and hearth are made to match, bought as a complete unit, and fixed in position as shown in Fig. 70. The hearth may be regarded as part of the surround in cases such as this, and should not be confused with the *structural* hearth, taken in the next sub-section.

It is usual to P.C. fire surrounds, as these vary considerably in price. A typical description would be as follows:

1	*Pre-slabbed fire surround size* 1400 × 875 *mm high of glazed tiles with hearth to match, P.C. £*30.00 *and setting and pointing in cement mortar and cleaning-off on completion.*

240 | 1·40 0·88 | *Deduct* Render and set in Sirapite to walls

&

Deduct Apply two coats plastic emulsion paint to walls.

Although the wall plaster and decorations have not yet been measured, it is, nevertheless, usual to make the deduction at this stage, since the size of the surround is readily to hand. As will be seen from Fig. 70, the surround is fixed on top of the plaster, since to make good the plaster up to the surround would result in an unsatisfactory joint between the two. No labour item on the plaster is therefore measurable at this point.

Strictly speaking, the plaster deduction is slightly smaller than the surround, but this difference may be regarded as negligible, the practice being to and-on to the decorations as shown; particularly since the height of the hearth has not been taken into consideration—although sometimes the surround goes down to floor level instead of sitting on the hearth, depending on the design.

Hearth construction

Ground-floor hearth construction comprises *fender walls* as shown in Fig. 69, filled in with *hardcore*, as shown in Fig. 70. The *structural hearth* rests on the hardcore, the top surface being at floor level, as shown.

The brick fender walls are one-brick wide, supporting both the timber plate and also the edge of the concrete hearth. Both fender walls and plate are usually measured in the 'floors' section of the taking-off, when the floor timbers will require to be adjusted round the opening. They have much the same function as sleeper walls, which are also measured with their plates in the 'floors' section, not with 'fires'.

Only the hardcore and the concrete to the hearth require to be measured at this juncture therefore. Typical items are as follows:

1·50 0·45 | *Concrete* (1:2:4) 130 *mm thick in hearths.*

(in No. 1.)

1·50 2/0·45		*Sawn formwork to edge of concrete hearth* 130 *mm high.*

	1·50	0·45
2/½/·22	·22	·11
	1·28	0·34

1·28 0·34	*Hardcore filling* 300 *mm thick under hearths, well rammed and blinded to receive concrete.*

Note 1. Hardcore filling would be measured cube if over 300 mm thick, in accordance with SM D22(c). Blinding is included in the description as permitted by SM D17(a).

Note 2. Provision is made for isolated *suspended* hearths in SM F7(d), and these are to be enumerated. This refers to upper-floor hearths, which are suspended where projecting from the chimney-breast. This would in any case hardly be an appropriate method of measurement for

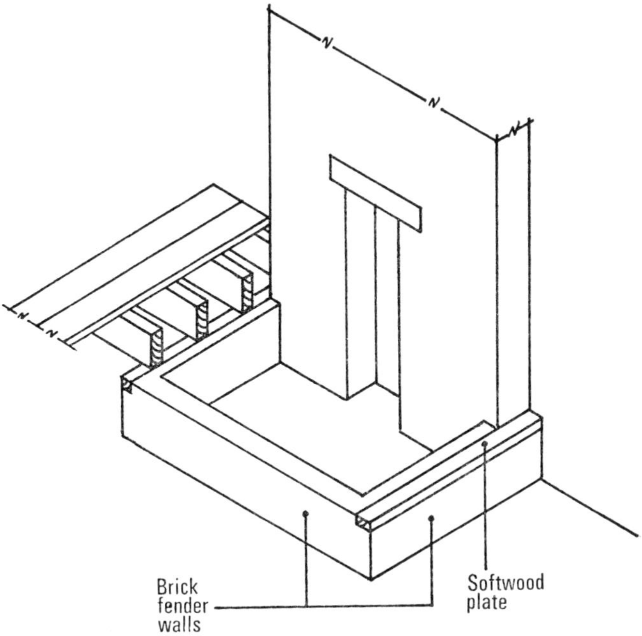

Fig. 69

242 hearths such as the above, since the portion within the fireplace opening is a different width from the remainder hence the two dimensions above. It is recommended practice therefore, failing any SMM guidance, to measure ground-floor hearths super as shown; with the number stated, to give the estimator an indication of approximate size.

Note 3. Formwork is measured linear under SM F22. It may not be actually used where a joist runs parallel with the edge of the hearth, but strictly speaking the concrete should not abut the structural timbers, and formwork should therefore be measured where no brickwork occurs at the edges of the hearth.

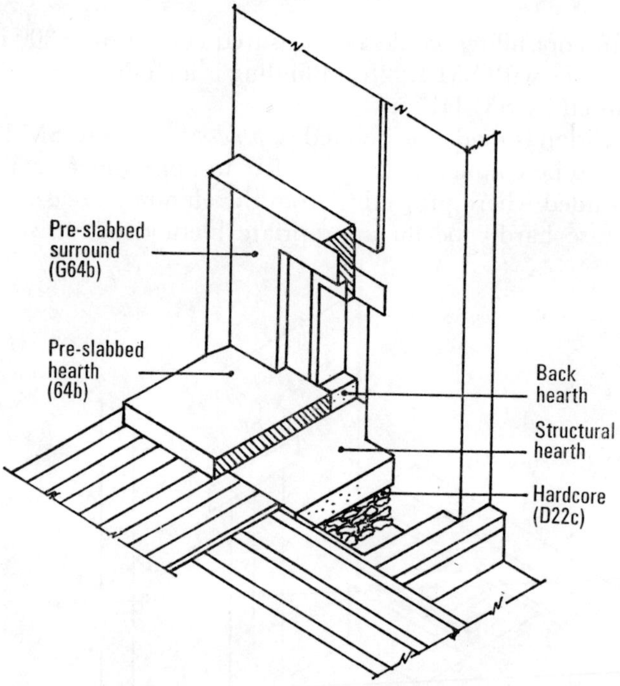

Fig. 70

It will be noticed from Fig. 70 that a concrete back hearth is necessary where a pre-slabbed front hearth is specified, in order to bring the base of the opening up to the level of the pre-slabbed hearth. This back hearth will need to be trowelled smooth to receive the grate, and is measured as follows:

0·57 0·33	*Fine concrete* 100 *mm thick in hearths, trowelled smooth.* *(in No. 1)* [*back hearth*

This again is measured super and the number stated, as for the lower hearth. A finer aggregate than normal would be desirable in this situation, to facilitate placing and trowelling. No formwork would be required, since this work would need to be carried out after the pre-slabbed hearth was in position.

Flue

This sub-section comprises work to the actual flue itself, which will normally include a lining inside, and a chimney pot at the top. These items are not measured with the stack in the 'brickwork' section, but at the present stage, in the 'fires' section of the taking-off.

SM G61(a) refers to parging and coring of flues, i.e., rendering the inside of the chimney and passing a ball or other object down the flue on completion, to ensure a free passage. Under the new building regulations, however, flues are required to be lined instead, the linings being measured under SM G61(b). This clause refers to fireclay and precast concrete linings. Refractory brick linings are mentioned in G61(c), but these are used for industrial chimney-shafts rather than domestic flues. Linings are measured linear, typical items being:

linear	220 × 220 *mm Square section fireclay flue lining to BS 1181, with rebated joints, bedded and jointed in cement mortar* (1:3).
number	*Extra for bend.*
number	*Extra for cutting to form easing.*

The last item is measurable to slight bends, where it is thought that a curved section of lining would be inappropriate.

If the lining is measured from the top of the fireplace opening at lintel level, this will offset against the rendering which will probably be done in lieu of the lining at this point. Alternatively a numbered item of 'Rough render to chimney opening above fireback' can be taken for this.

It is important to remember at this point that rendering will also be

244 required to the outside face of any chimney-breast passing through a timber floor or roof, to comply with fire regulations. This will probably not be shown on the drawings, and is measured *at this stage*, with the flues. It is measured super and described as

> *Rough render face of chimney-*
> *breasts passing through*
> *floors or roofs.*

Finally, the chimney pot is measured, together with its 'flaunching' (cement fillet) under SM G63, a typical description being:

> 200 × 200 *mm Square based natural*
> *clay chimney pot* 300 *mm*
> *long, set to show*
> *about* 150 *mm, and bedding*
> *and flaunching in cement*
> *mortar* (1:3).

UPPER-FLOOR FIREPLACES

The warming of bedrooms today is best achieved by either central heating or electric radiators or convector heaters. Fireplaces in bedrooms are therefore a thing of the past.

Living-rooms on the upper floors of flats may require fireplaces however. In this case the structural floors will probably be of concrete construction, no isolated hearths being required. Most modern blocks of flats will, however, incorporate some other form of heating.

There remains the possibility of upper-floor fireplaces in two-storey maisonette construction however, and this may involve fireplaces in the living-rooms on upper floors which are constructed of timber.

It will be appreciated that the only difference in principle between ground-floor and upper-floor fireplaces is due to the different method of hearth construction. If the floor is of concrete, the upper-floor hearth construction will probably involve little more than a back hearth similar to the one taken above for ground-floor fireplaces. This is because the pre-slabbed hearth will sit directly on the structural floor, which being of concrete will take the place of the structural hearth.

In the case of an upper floor in timber, however, a concrete suspended hearth will need to be provided. This will be similar to the concrete ground-floor hearth shown in Fig. 70, except that there will be no hardcore to support it. Instead, the front will be supported on a fillet spiked to the joist, and the rear on the brickwork of the chimney-breast, as shown in Fig. 71.

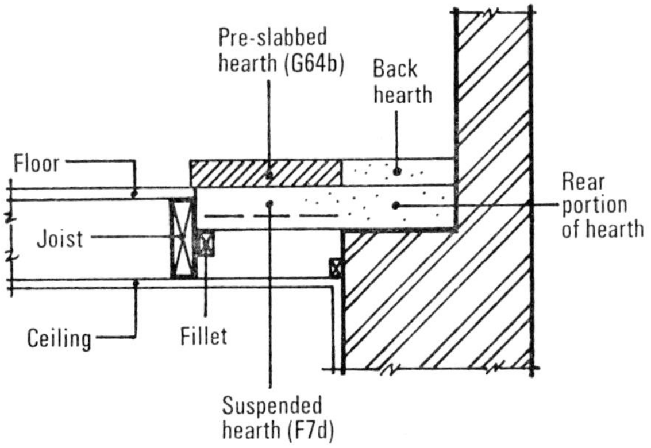

Fig. 71

At the sides of the suspended portion, the hearth will rest on a similar fillet spiked to the sides of the short lengths of joist called *cradling-pieces* shown in Fig. 72, which support the edges of the floor-boards next to the sides of the hearth.

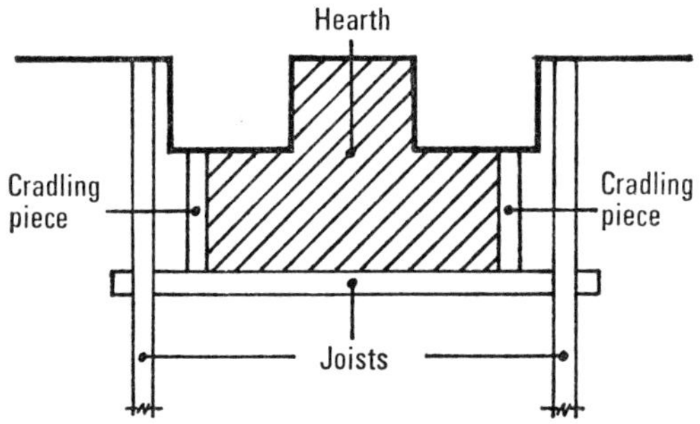

Fig. 72

If we now assume this hearth to be of the same size as the example given for a ground-floor hearth, the items required for the hearth construction in the case of this upper-floor fireplace would be as follows:

0·57 0·33	*Concrete* (1:2:4) 130 *mm thick in hearths* *(in No. 1)*
1	*Isolated suspended hearth size* 1400 × 500 × 130 *mm thick, filled in on and including sheet metal formwork (left*

(description continued overleaf)

*in) and fillets spiked to joists, including layer of fabric reinforcement reference B*503.

Note 1. The rear portion of the hearth is measured super, as in the case of the ground-floor hearth.

Note 2. The front portion is numbered in accordance with SM F7(d); formwork and reinforcement being included.

Note 3. The cradling-pieces are measured as joists in the 'floors' section of the taking-off—not with 'fires'.

Note 4. The back hearth required to make the level up to top of pre-slabbed hearth as shown in Fig. 71 would require to be measured in the same way as for ground-floor fires.

AIR VENTS

On completion of the 'fires' section of the taking-off, the drawings are now examined for vents. Rooms without fireplaces will be expected to require air vents, which may or may not be shown on the drawing. The specification notes should be examined in this connexion, and bathroom and kitchen vents should not be overlooked. If a larder is provided, this will require a permanent means of ventilation, which should be measured accordingly.

Air-bricks and vents to hollow floors are not taken at this stage, but are measured in the 'floors' section. Similarly, air-bricks which ventilate the cavity of hollow walls are usually measured in the 'brickwork' section of the taking-off, should any such air-bricks be required.

The provision of a vent in a wall will normally call for the following operations to be considered:

(1) an opening to be formed in the wall,
(2) a duct to be formed across the cavity of a hollow wall,
(3) an air-brick or grating to be provided externally,
(4) a vent or grating to be provided internally,
(5) plaster to be made good around the internal vent.

Forming the opening

SM G60 provides for the forming of small openings in walls to be numbered, stating the size and wall thickness (and material) and including any necessary arch or lintel in the description. Openings for small vents of the kind under consideration do not usually require any such support over the top, although splay cutting of a brick or two to form a 'Welsh arch' over might be carried out on the inner skin perhaps, where the brickwork is not supported by an air-brick. The labour involved in any such rough cutting would be considered covered by the formation of the opening however, and would hardly warrant the description of an arch, unless the opening was larger than 225 × 225 mm.

It should be noted that brickwork is not deducted for such openings, unless they exceed 0·10 square metre in area (see SM G1(a)(i)).

Bridging the cavity

The best method of achieving this is to line the sides of the opening with slates, bedded in cement mortar; it being essential to use a material which is impervious, in order to prevent the moisture from reaching the inner skin. Alternatively, a short length of drain pipe may be used for this purpose; but this is rather in the nature of a round peg in a square hole, and slates have the advantage of being thin enough not to obstruct the vent in any way.

In some cases the cavity is not bridged at all, but this allows stale air from the cavity to enter the room, which may be undesirable.

External air-bricks

Terra-cotta air-bricks are normally used on the outside face, since these look well, and do not require painting, as do cast iron gratings. They are numbered under SM G60. Since they are built-in and pointed with the facing bricks, an item of making-good to facings is not necessary.

Internal vents

Fibrous plaster air vents are quite usual nowadays, as these are simply fixed by bedding in position on top of the wall plaster, using gypsum plaster as an adhesive. These vents look neat and unobtrusive, and are decorated to match the wall. The decoration of the vent is not measured separately in this case, but taken in with the wall decoration in the 'internal finishings' section. If a metal grating is specified, however, this will require painting as a separate operation from the general wall decoration, and the item for painting the grating must be measured separately at this present stage under SM W12(b). This item must be taken to the external air-brick as well, should this be of iron.

Making-good of plaster

The plaster will require making-good around the air vent, or around the opening before the vent is fixed if a fibrous plaster vent is specified. This making-good is numbered under SM U6(j), stating the size of the vent *in stages of* 0·025 *square metres*. It will be seen that the above clause groups ventilators with brackets, pipes, and other sundry items for the purpose

248 of making-good, hence the idea of the stages; and these items may be conveniently termed 'sundries' for the purpose of merging them together at abstracting stage. If the making-good around vents is also termed 'sundries' in the description, this will ensure its correct grouping on the abstract.

Measurement of vents

It will be seen from the above that all items in connexion with an air vent are enumerated. The only problem concerning measurement therefore is to decide which items can be included in the same description, and which must be measured separately.

The answer to this problem is that items carried out by different tradesmen must be measured separately; so must items which will occur in different sub-sections of the same work-section in the bill.

Thus we may include the air-brick, the forming of the opening, and the bridging of the cavity together, since they are all executed by the bricklayer and billed in the same sub-section. The plaster air vent must be measured separately. The making-good cannot be included with the air vent on the grounds that it is also executed by the plasterer however, for the item of making-good is to be grouped with sundries in the bill, as explained above. It should also be noted that if external and internal vents both require painting to be measured under SM W12(b), the paint on both items cannot be included in the same description on the grounds

1	225 × 225 *mm Square hole pattern terra-cotta air-brick and building in, including forming opening in* 270 *mm brick cavity wall and lining with slates bedded in cement mortar* (1:3).
	&
	225 × 225 *mm Fibrous plaster louvred air vent bedded in gypsum plaster.*
	&
	Make good wall plaster around sundries, 0·050 *to* 0·075 *square metre in size.*

that both lots of paint are executed by the same tradesman, since SM W1(a) requires external painting work to be separated from internal work in the bill.

Bearing the above in mind, we shall see that the items required for a typical vent in a 270-mm cavity wall will be as shown on page 248.

Fly-screens

In the case of larder vents, a fly-screen may be specified internally, in lieu of the louvred vent. These are best numbered, as in the case of vents, a typical description being as follows:

1	*Fly-screen size* 225 × 225 *mm of fine copper gauze framed with* 40 × 15 *mm splayed softwood cover fillet, mitred at angles, and plugged.*
4/0·23	*Knot prime stop and apply two coats oil colour to wood frame not exceeding* 100 *mm girth.* *(internally)*

Here the paint on the cover fillet requires to be measured linear under SM W4(a). The items for the external air-brick and opening, and making-good plaster internally, would of course, need to be taken with this fly-screen, as for the plaster vent measured above.

Chapter Twenty-two

Carpentry Work

Before proceeding with the measurement of floors and roofs, which in domestic construction may well be constructed of timber, it is desirable that the surveyor should have a clear understanding of the principles of measurement and description of carpentry work generally. Once these are firmly grasped no difficulty should be found in ensuring that the

250 taking-off proceeds smoothly, with properly framed descriptions, as far as the measurement of timber is concerned.

THE MEASUREMENT OF TIMBER

It is the practice to measure each individual piece of timber its extreme length. Items are therefore not measured on the *mean* girth, but on the *extreme* girth. The purpose of this is to allow for any lap joints or tenons; and no deduction is made if the tenon is only a stub-tenon, or the timber is butt-jointed. Thus if two pieces of timber meet at an angle they are measured as shown in Fig. 73(a); if they intersect as Fig. 73(b). In cal-

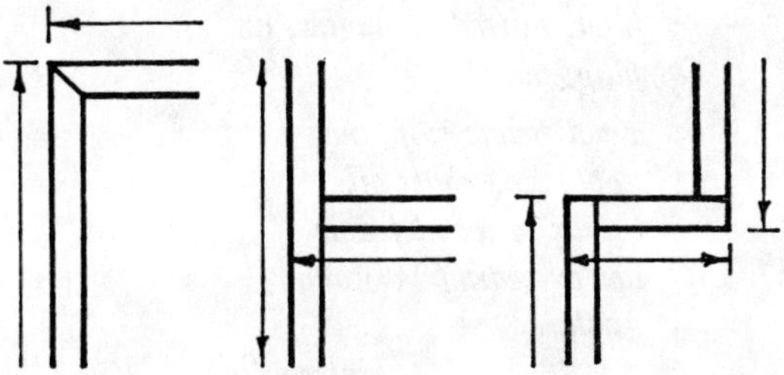

(a) (b) (c)

Fig. 73

culating the extreme girth of a wall-plate the extra timber at *internal* angles requires to be added also, in order that the girth will be that which would result if each piece of timber was measured its extreme length, as shown in Fig. 73(c).

The custom of measuring timber through the joints in this way is due to SM N3(b), which provides for timbers being measured '. . . the length as fixed in the work', which is not quite the same thing as the mean length.

Measurement of carpentry joints

As will be seen from SM N6(a), all joints are deemed to be included, except (a) *dovetailed* joints, and (b) *scarfed* joints. Dovetailed joints are rare in carpentry (as opposed to joinery) work, and apart from the very occasional dovetailed angle to a lantern-light curb, are unlikely to be met with.

Scarfed joints on the other hand are used to join two similar structural timbers together lengthwise; unless the timber is in compression only, in which case a halved joint is used instead. Thus timbers which might possibly contain a scarf joint would be:

(1) Ridges,
(2) Hip of valley rafters,
(3) Joists.

in descending order of probability.

It will be appreciated that such joints will probably not be shown on the drawings, and the surveyor himself will have to decide where to measure them. The answer to this problem is given in SM N3(b), which states that 'unless otherwise shown or specified, joints in continuous lengths shall be taken every 6 metres . . .' Thus it would appear that a scarf joint should be measured every 6 metres to any individual member performing one of the three above-mentioned functions.

This is not strictly so however, for it may sometimes be undesirable for a timber to be jointed in the middle, either for structural reasons, or on the grounds of economy or convenience. Among such cases are the following:

(a) BRIDGING JOISTS. If floor or ceiling joists were to span over 6 metres, each joist would be in one length, to avoid the labour of scarfing every individual joist as well as on structural grounds. Neither would it be a good policy to scarf trimming timbers, even if space allowed.

Bridging joists over 6 metres long supported in the middle by a beam or load-bearing wall would be lapped at the point of support—not scarfed.

(b) RAFTERS. As with bridging joists, common rafters would not be scarfed, neither would rafter-trimming members.

It is possible that hip or valley rafters could be scarfed, but far more likely that a long single length would be used if these were over 6 metres long, for structural reasons.

(c) PURLINS. These will probably be halved at the point of support where the strut or roof principal occurs. Such supports would occur at less frequent intervals than 6 metres, so avoiding scarfing.

(d) STRUTS. A scarf joint would be structurally unsuitable for a member acting as a strut. Where a strut exceeds 6 metres in length therefore (as in the case of raking shores perhaps), it should be taken in one length.

It will be apparent from the foregoing that a scarf joint is only likely to occur in the case of ridges, where one should always be taken every 6 metres of straight length.

Where the taker-off decides that a structural timber is (a) over 6 metres long, and (b) would be executed in one length to avoid scarfing,

252 the length of the timber must be given in the description in stages of 1·5 metres, in accordance with SM N3(c), as for example:

6·75	200 × 38 *mm Softwood valley over* 6 *metres and not exceeding* 7·5 *metres long, including cutting to rafters both sides.*

On the other hand, in the case of timbers requiring a scarf joint, the timber is measured as usual, and the item for the scarf joint numbered in accordance with SM N6(a) as:

1	*Extra labour and wedges scarfing* 175 × 25 *mm ridge.*

Allowances for joints

We have seen that apart from dovetailed joints, scarf joints are the only joints requiring a separate individual item of measurement in carpentry work.

There are two types of joints which require an *allowance* to be made in respect of extra measurement of the timber where the joint occurs however, as follows:

SCARF JOINTS. Twice the depth of the timber must be added to the measured length of the member in respect of every scarf joint, under SM N3(b)(ii). This is *in addition* to measurement of the actual scarf joint itself.

HALVED JOINTS. An allowance of 150 mm must be added to the measured length of the member in respect of every halved joint, under SM N3(b)(i). Note that the halved joint itself is not actually measured as such.

This allowance must be taken once for every 6 metres of continuous plate or other member not requiring a scarf joint, in accordance with SM N3(b).

The same allowance is provided for dovetailed joints, should these be encountered.

The above allowances are added to the length of the timber on waste, suitably annotated accordingly.

Trimming

There is another kind of joint-allowance provided for in the SMM, and this is in respect of trimming timbers.

These will require (in theory) rather complicated joints by way of tusk-tenons, dovetailed-notches, and the like. These joints are not themselves measurable items, but SM N6(e)(vi) provides instead for numbering the complete opening for which the trimming timbers are required, in addition to the normal measurement of the timbers themselves. A typical numbered item covering the trimming of, say, the ceiling joists around a trap-door would be as follows:

1	*Extra labour trimming* 100 × 50 *mm ceiling joists* *around opening size* 1·00 × 0·75 *metres.*

This enables the extra cost of forming the tusk-tenons and dovetailed-notches to be allowed for over and above the normal labour costs of preparing and fixing the ceiling joists.

Summary of carpentry joint requirements

The above requirements regarding jointing may now be summarized:

(1) *Timbers over* 6 *metres in length*

(a) *in direct compression only (plates, etc.):*
allow 150 mm every 6 metres on waste.

(b) *other timbers:*
either (i) number scarf joint
and (ii) allow twice depth of timber on waste
or (iii) give length of timber in 1·5-metre stages.

(2) *Trimming timbers*

(i) measure timbers as such
and (ii) number opening for labour trimming.

(3) *Dovetailed joints*

(i) number dovetailed joint
and (ii) allow 150 mm on waste if in continuous length.

THE DESCRIPTION OF TIMBER

It might be as well at this stage to define the difference between carpentry and joinery; since the rules of measurement and description vary slightly between the two, and any discussion of carpentry descriptions must take account of the possibility of associated joinery work.

Carpentry timber is deemed to be *sawn* on all faces unless otherwise described (see SM N1(b)). A sawn face is usually sufficient in the case of timbers not exposed to view. Carpentry timber may therefore be loosely defined as timber *not exposed to view*.

Joinery timber is deemed to be *wrought* on all faces (see SM P1(b)). A wrought finish is necessary in the case of timber exposed to view. Joinery timber may therefore be loosely defined as timber *exposed to view*.

The above criterion will not quite hold good in all cases. Soffit-bearers, for example, are not exposed to view, but are nevertheless joinery work (P9(b)). Similarly, the feet of rafters may be exposed to view, but would be wrought ends of carpentry not joinery timbers (N6 (e)(iii)).

If, however, we accept that unexposed timbers in close association with joinery work are joinery, and vice-versa, the above criterion will stand us in good stead when deciding whether to measure an item as carpentry work under Section N, or joinery under Section P of the SMM.

We can expect that most items of timber in connexion with floors and roofs of elementary construction will be found in the SMM. Not all will occur in Section N however, for some items which we might think of as carpentry work are in fact joinery and occur in Section P. As we might expect from the foregoing definition, these are items exposed to view, and include:

(a) floor-boarding (SM P3)
(b) eaves and verges (SM P9)
(c) fascias and barge-boards (SM P11).

Since these items are in Section P they are deemed to be wrought, and it will not therefore be necessary to state this in the description.

Apart from these special instances, the timber for floors and roofs will be not joinery, but carpentry timbers; the describing of which will now be examined.

Kind and quality of timber

Particulars of this are to be given, in accordance with SM P1(a)(i). There are two mains kinds of timber:

(a) softwood
(b) hardwood

and only the former is likely to be found in general carpentry work and should be described as such.

The *quality* of the timber is a matter for general specification, and this will therefore form the subject of a preamble—not usually written at taking-off stage. Thus the quality will not appear in the description.

Nominal size

The SMM states in Clause N1(b) that 3·2 millimetres shall be allotted from the nominal size for each wrought face.

From this we may conclude that the nominal size may be regarded as the cross-sectional size (or 'scantling') of a piece of timber after it leaves the saw. If the timber is subsequently planed and finished smooth, it is then said to be wrought. This process will reduce its size to the 'finished' size.

Since carpentry timbers are deemed to be sawn on all faces, the nominal size will be the actual size, unless the timber is described as wrought. Joinery timbers, however (floor-boarding, etc., as mentioned above) will have a nominal size (before planing) and a finished size (after planing), and the estimator will require to know which of these is intended by the size stated in the surveyor's description.

This problem is best overcome by inserting a preamble clause in both the carpentry and joinery work-sections of the bill stating that all sizes are nominal unless otherwise described. This ensures consistency, and avoids the necessity of mentioning this factor in the descriptions.

A word of warning is necessary here regarding full-size and half-full-size details. These must be drawn to *finished* sizes in order to make the parts fit together on the drawing. In the event therefore of a taker-off scaling from (say) a half-full-size section through eaves, it would be necessary for him to realize that he was scaling a finished size, and to add twice 3·2 mm to every scaled dimension of the soffit and fascia, in order to arrive at the nominal size.

The likelihood of this sort of thing being necessary in an intermediate examination is small in the case of floor or roof construction, and the candidate can assume that sizes given on the drawings and in the specification notes are nominal unless otherwise stated.

It is reasonable also for a candidate to assume that a 'nominal size' preamble applies to his dimensions; any Section P (joinery) work described as to a finished size (flooring for instance) in the question-paper therefore requiring to be so stated in the candidate's description, as for example:

25 *mm* (*Finished*) *softwood*
flooring . . . etc.

All Section N (carpentry) work (unless it happens to be wrought) will have the same nominal as finished size in any case—none of the above considerations being involved.

Fixing of timber

Carpentry timber may be fixed in any one of the following ways:

(a) nailed
(b) wedged
(c) plugged
(d) bolted
(e) bedded in mortar.

The alternatives for joinery timber would also include gluing, screwing, dowelling and pinning; but carpentry timbers and joinery work (flooring, etc.) in floor and roof construction are usually confined to the first five methods of fixing, dealt with as follows:

(a) NAILING. Timber is deemed to be fixed with nails (SM N1(d)). This term would include spikes, i.e., large carpentry nails. If timber is so fixed, no mention of fixing is required in the description.

(b) WEDGING. This may be considered equivalent to nailing, since nails may in fact be used to supplement the wedging.

(c) PLUGGING. This refers to inserting plugs in the wall prior to nailing, and is necessary when fixing a timber longitudinally to brickwork, concrete, or stonework.

Plugging is dealt with under SM N24, and may be given in the description of the timber concerned, since it is executed by the carpenter. The wording of the SMM clause suggests that plugging of brickwork, concrete or stone may be included together, the timber being merely described as 'plugged' in each case. Few carpentry timbers will require plugging.

(d) BOLTING. A plate fixed to the side of a steel beam or the side of a wall will require bolting, plugging being inappropriate in such cases. This implies four operations:

(i) providing and fixing the bolts (SM N31)
(ii) drilling the holes in the timber (SM N26)
(iii) drilling the holes in the steelwork (Q17) or cutting and pinning the bolts to brickwork (G56(c)) or forming mortices in concrete (F66) if ragbolts are used
(iv) providing and fixing the timber.

The timber itself is measured in the usual way, and described as '. . bolted on (bolts and holes measured separately)'.

Items (i) to (iii) above require to *each be measured separately*, each individual bolt being numbered, together with its holes, etc.

(e) BEDDING IN MORTAR. Since this operation is carried out by the *bricklayer*, it must be measured separately under G25(a). Plates will normally require bedding, but if buried in the wall (with brickwork not deducted) the item of bedding is not measurable (see the above clause).

The actual plate itself requires no fixing description; plates being assumed bedded unless described as '... spiked to joists' for example.

So far as the framing of timber descriptions is concerned, we may conclude from the above that no mention of fixing need be made in the description of a carpentry timber unless it is plugged or bolted.

Short lengths

Structural timbers not exceeding 1·25 *metres long* (except rafters at hips and valleys and timbers in trimmers and partitions) shall be so described (SM N3(d)).

Does 'trimmers' mean what it says, or could it include any trimming member, e.g., a cradling-piece for example?

Leaving aside this somewhat hairsplitting question ('when in doubt, measure short lengths'); it will be found that owing to the exceptions mentioned in the above clause, short lengths of carpentry timber rarely need to be described as such. An eye must be kept open, however, and where necessary the words '... in lengths not exceeding 1·25 metres' added to the description.

It should be noted that this type of short-length provision does not apply to joinery work. Fascias and eaves soffits, etc., do not come under this rule therefore. Instead, the following short-length rules apply:

FASCIAS AND BARGE-BOARDS. Clause P11 states that short lengths shall be given as SM A5(b). This defines short lengths as being *not exceeding* 300 *mm* and requires these to be enumerated as an *extra over*, irrespective of actual length.

EAVES AND VERGES. There is no provision of any kind for short lengths in the case of these items.

Framing of timber descriptions

It will be found helpful to adopt the following order of presentation when framing descriptions of timber:

(1) cross-sectional size
(2) type of timber
(3) special labours

258 (4) functional name
(5) fixing details
(6) other requirements.

Thus a moulded fascia plugged to the edge of a concrete flat would emerge as:

(1) 150 × 25 mm
(2) softwood
(3) moulded
(4) fascia
(5) plugged
(6) —

which is the exact bill description require, i.e.,

150 × 25 *mm Moulded fascia plugged.*

Or, in the case of a ridge, for example:

(1) 175 × 25 mm
(2) softwood
(3) — (no special labour)
(4) ridge
(5) — (nailing deemed included)
(6) including cutting and fitting rafters both sides (see N5)

which again produces the correct description:

175 × 25 *mm Softwood ridge, including cutting and fitting rafters both sides.*

Similarly for a gutter-board:

(1) 250 × 32 mm
(2) softwood
(3) cross-tongued
(4) gutter-board
(5) — (nailing deemed included)
(6) and bearers (see N14(b))

resulting in:

250 × 32 *mm Softwood cross-tongued gutter-board and bearers.*

Again, in the case of rafters:

(1) 100 × 50 mm
(2) softwood
(3) — (no special labour)
(4) rafters
(5) — (nailing deemed included)
(6) —

the resultant description being:

100 × 50 *mm Softwood rafters.*

Chapter Twenty-three

Floors

The types of structural floor likely to be encountered in connexion with simple buildings relevant to the RICS intermediate syllabus are as follows:

(a) concrete ground floors
(b) timber ground floors
(c) timber upper floors.

Concrete upper floors are used in multi-storey construction, the problems of measurement involved being similar to those in respect of reinforced concrete flat roofs, dealt with in a later chapter. Concrete is not normally used for upper floors in simple two-storey domestic work however, for which timber is generally considered more economical.

Floor finishings

A timber ground or upper floor will consist basically of floor-boarding supported on joists. Since the floor-boarding spans between the joists, it may be regarded as both a floor-covering and a structural part of the floor, since it must be designed to carry the load between the joists. For taking-off purposes, floor-boarding spanning joists in this way is regarded as part of the structural floor, and measured in the 'floors' section of the work—not with 'finishings'.

A concrete floor, on the other hand, will require a finish other than of concrete, and in this case the floor covering or finish will not carry any structural load, even if it happens to be some form of timber floor-boarding. In the case of concrete floors therefore, the floor finishing is measured in the 'finishings' section of the work—not with 'floors'.

CONCRETE GROUND FLOORS

These are often referred to as 'solid' ground floors, to distinguish them from floors comprising timber joists supported on sleeper walls resting on a concrete bed, the usual construction in the case of timber ground floors, or 'hollow' floors, described later in this chapter.

It will be recalled from Chapter 15 that in the case of raft foundations the raft slab itself may well form the structural ground floor, and should this be the case nothing will remain to be measured in the 'floors' section, since we have seen that both the hardcore and the concrete slab are normally dealt with in 'substructures' in this instance. The floor finishings will in any case be dealt with in the 'finishings' section of the taking-off.

In the case of substructures other than raft foundations however, the excavation only will have been dealt with over the area of the building, and the structural ground floor will now require to be measured.

The general principles of measurement of concrete ground floors will follow those described for raft slabs in Chapter 15, since both will consist basically of a concrete bed laid on hardcore. The considerations applicable to the descriptions of plain and reinforced concrete beds and hardcore filling discussed in Chapter 15 will not be repeated here; but the following additional notes may be of assistance in connexion with the measurement of concrete ground floors generally.

Hardcore

A minimum thickness will be specified, and this will probably be not exceeding 300 mm. On a perfectly flat site, this minimum thickness can be taken as the actual thickness, and measured accordingly under SM D22, including both the blinding and the levelling and ramming ground under in the description, as indicated on page 171. Should the thickness exceed 300 mm, the hardcore will be measured cube, and the two associated items (blinding and ramming) will then require to be measured as separate areas, as previously described in Chapter 15.

On a sloping site however, it will be necessary to find the *average* thickness of the hardcore, and care must be taken to ensure that if any part exceeds 300 mm in depth, that part is dealt with separately as a cube item, and not included in the depth-calculation of superficial hardcore.

Figure 14 shows a case in point, where the hardcore varies from 100 mm to 450 mm in thickness. An average would give $\frac{1}{2}(100 + 450) = 275$ mm average thickness. It is not the *average* thickness which is referred to in SM D22 as a criterion of superficial measurement however, and although the average thickness is not exceeding 300 mm, the *actual* thickness of part of the hardcore does in fact exceed this, and must therefore be cubed.

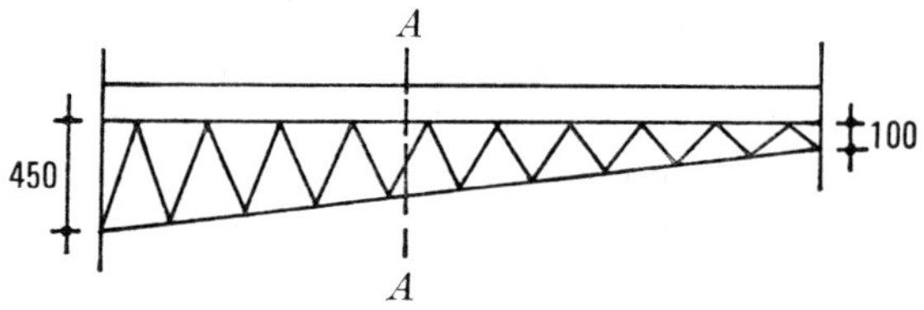

Fig. 74

The correct procedure is for the taker-off to move his scale vertically along the section until 300 mm of thickness can be read off, and a pencil-line drawn (*A–A*) marking the division between cubic and superficial hardcore. Hardcore to the right of this line can then be averaged thus:

$$\begin{array}{r} \cdot 100 \\ \cdot 300 \\ \hline 2)\cdot 400 \\ \hline \textit{av.} = \quad \cdot 200 \\ \hline \end{array}$$

<u>*super*</u> 200 *mm (av.) Bed of hardcore blinded with gravel, including levelling and ramming ground under.*

and the volume of hardcore to the left of the line cubed as:

$$\begin{array}{r} \cdot 300 \\ \cdot 450 \\ \hline 2)\cdot 750 \\ \hline \textit{av.} = \quad \cdot 375 \\ \hline \end{array}$$

<u>*cube*</u> *Hardcore filling in making up levels under floors.*

 followed by its associated items:

super	*Blind surface of hardcore with gravel* & *Level and ram bottom of excavation.*

Should the ground level at the upper end of the site rise above floor level, it may be necessary to divide the hardcore into (a) minimum thickness, (b) average superficial thickness, and (c) cubic thickness. Should the site slope in more than one direction, it will be necessary to plot the position of the line *A–A* on the plan, in the form of a contour, and measure the areas on either side of the contour-line.

Building paper

A layer of building paper is sometimes specified between the hardcore and concrete. This is measured super, and described in the following manner:

Layer of building paper to
BS 1521 *Class B, lapped at*
joints and laid on hardcore
bed to receive concrete.

The British Standard covers two grades of building paper, Class A and Class B. Unless otherwise specified, Class B is probably sufficient for this purpose, but one or other should be mentioned; a mere reference to the British Standard number is ambiguous, unless clarified by a preamble.

Concrete beds

As noted in Chapter 15, concrete slabs supported by hardcore are classed as beds under SM F5, and, like the building paper above, can probably be anded-on to the hardcore, provided the latter is all superficial. It will be recalled that the surface treatment must be included in the description of superficial beds under SM F14(a).

We saw in Chapter 11, when dealing with foundation concrete, that this surface treatment may be expected to comprise one of the following forms (see page 108):

(1) spade face
(2) floated finish
(3) trowelled finish
(4) tamped finish.

Mention was made of the fact that *spade face* is not considered a 'treatment' under SM F14(a), and is therefore not regarded as a measurable item. The description of a superficial concrete bed with no mention of surface finish therefore, will be taken by the estimator to refer to a bed with a spade face.

The surface treatment required in any individual case will depend upon the construction immediately above the concrete bed.

BEDS TO RECEIVE SCREEDS. Typical construction is shown in Fig. 75(a). The screed itself, and everything above, is measured in the 'internal finishings' section of the work. Only the hardcore and concrete

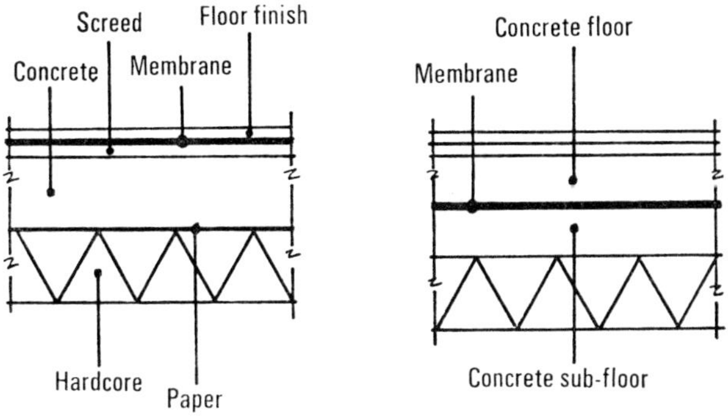

Fig. 75

bed would be measured in the 'floors' section. Beds receiving screeds are usually finished with a tamped surface. (Note: by 'screed' is meant a thin bed of cement and sand laid as a separate operation, to receive floor finishings.)

BEDS RECEIVING MEMBRANES. If the membrane is below screed level it is measured in the 'floors' section. Figure 75(b) shows a typical 'sand-which' floor construction, with a membrane between two concrete beds, a form of construction sometimes used in damp conditions.

The surface finish to a slab receiving a membrane will depend on the type of membrane involved:

(a) *Tamped surface.* This may be required for the application of liquid membranes such as Tretrol, Synthaprufe, and similar materials.

(b) *Floated surface.* Recommended by BS CP102:1963 for asphalt, hot pitch, and hot bitumen; also for felt sheeting. Polythene sheeting would also probably require a floated surface.

264 BEDS RECEIVING FLOOR FINISHINGS. It is possible that floor finishings may be laid directly on the concrete, with no screed. Should this be the case, the surface treatment will depend upon the type of floor finishings. In the event of this situation being encountered, the following notes may prove useful as a guide:

(a) *Tamped surface.* This may be suitable for inflexible floor finishings laid in mortar, e.g., quarry tiles; or for certain in-situ floor finishings, e.g. granolithic paving.
(b) *Floated surface.* This would probably be necessary if wood-block flooring was laid directly on the bed, with no screed.
(c) *Trowelled surface.* This would probably be required for flexible floor finishings, e.g., thermoplastic tiles, linoleum, rubber flooring, etc.

Owing to the possible differences in thickness of adjacent floor finishings however, it is usual to specify a screed, in most cases involving a tamped surface to the bed, as described above. It is emphasized that in any case the floor finishings themselves, *together with any screeds to receive them*, are measured in the 'finishing' section; not with 'floors'.

Membranes

Some form of damp-proof membrane is usually specified in solid floor construction. This may occur above screed level when it will usually consist of the adhesive material used for bedding the floor finishings, in which case it is measured with the screed and floor-coverings in the 'finishings' section of the taking-off.

Where it occurs below screed level, however, it must be measured with the ground-floor construction, and can usually be anded-on to the concrete slab. Care must be taken here however, because the membrane may continue across walls going down to foundations, in order to form an unbroken impervious layer; in which case its area will be greater than that of the slab.

It is usual to state in the description the type of surface upon which the membrane is to be applied, e.g.,

super	*Two coats Tretol or other approved bituminous waterproofing compound on concrete beds.*

The area of any such membrane overlapping brick walls will be relatively small, and in view of the fact that it will be executed in one continuous operation, is usually included all in the same description, not kept separate as 'on brickwork'.

TIMBER GROUND FLOORS

The order of taking-off for timber ground floors is much the same as the order in which the work will be executed, viz.:

(1) hardcore and concrete beds
(2) sleeper and fender walls (including DPC's)
(3) plates
(4) joists
(5) floor-boarding
(6) air-bricks

Hardcore and concrete beds

A bed of over-site concrete will be necessary in order to support the sleeper walls and comply with the building regulations and there will probably be a bed of hardcore under. The measurement of these beds will be the same as for those described above in connexion with solid floors. There will be no membrane to deal with however; and the top surface of the concrete will be spade face instead of tamped, requiring no surface finish to be mentioned in the description.

Concrete beds beneath timber floors will probably be only 100 mm thick, and of 1:3:6 concrete. If the hardcore is not exceeding 300 mm thick it can be anded-on to the concrete bed, since the areas will be the same. Blinding the hardcore and levelling and ramming the ground under must not be forgotten when framing the hardcore description, thus:

super	150 *mm Bed of hardcore blinded with ashes, including levelling and ramming ground under.*
	&
	100 *mm Bed of concrete* (1:3:6).

If the site is sloping, the hardcore thickness must be averaged, as in the case of solid floors, if the concrete bed is horizontal.

Cases have been known to exist however, where the over-site concrete below a timber floor has been laid to follow the slope of the ground, in

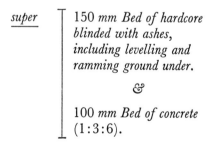

266 order to avoid additional excavation and hardcore. Where the ground floor is not supporting the weight of any partitions, this may be acceptable, and the beds must then be measured accordingly, on the following basis:

SLOPING HARDCORE. No provision is made for this in the SMM, and it may be assumed that in the case under consideration no extra is involved over laying a horizontal bed. If the hardcore bed was laid horizontally, and the top surface *graded to falls*, this would be different, and would require stating in the description under SM D17(a). Similarly with the levelling and ramming ground under; if the ground is level with respect to the original ground level, as in this case, it would hardly merit the description of 'grading to falls', and should be measured as for horizontal ground.

SLOPING CONCRETE BEDS. Beds laid to slopes not exceeding 15 degrees from horizontal require describing as such, under SM F5(b). It is interesting to note that if over 15 degrees, formwork would require to be measured to the top surface under SM F20(a); although this is hardly likely to arise in the case under consideration.

In the event of a sloping over-site bed, rough cutting would, of course, require to be measured at the bottom of all sleeper and fender walls.

Sleeper walls

Construction will be along the lines shown in Fig. 76, with sleeper walls at 2-metre centres, supporting the plates and joists. In order to achieve the shortest span for the joists (thus avoiding overlaps in the timber) the sleeper walls should be set out parallel to the *longest* wall of the room.

It will be appreciated that the position of the sleeper walls will probably not be shown on the plan, and the surveyor must therefore set them out

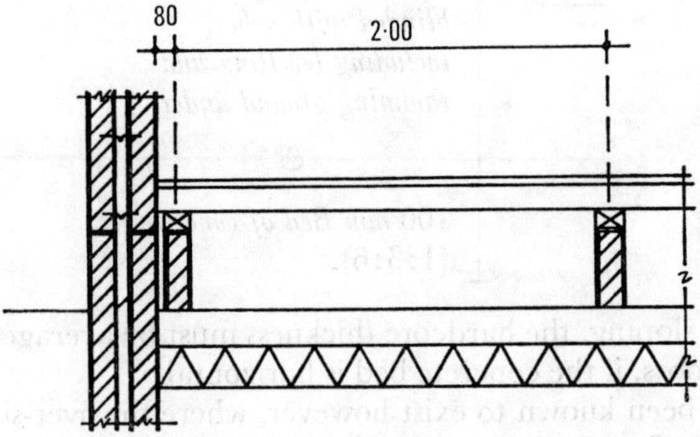

Fig. 76

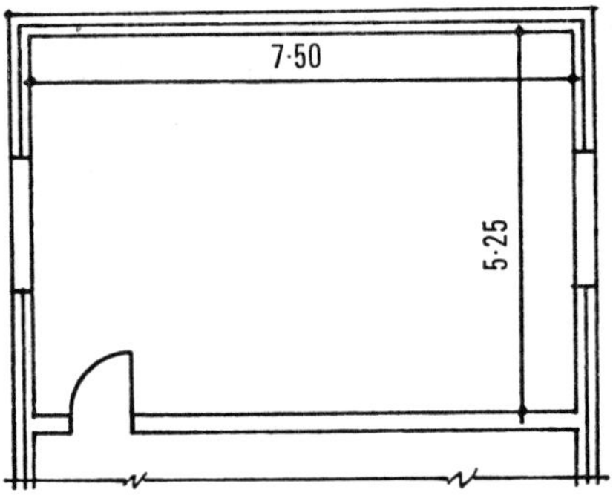

Fig. 77

himself, in order to arrive at the correct total of the lengths. The procedure in the case of the room shown in Fig. 77 for example, would be as follows:

Example

PROBLEM

Find the total length of sleeper walls for the floor of the room shown in Fig. 77.

SOLUTION

Since the sleeper walls will run parallel to the longest wall, the *number* of sleeper walls required will be found by dividing the *width* of the room by spacing of the sleeper walls, thus: (on waste):

		width	5·25
	less 2/·025	·05	
	$\frac{1}{2}$/2/·110	·11	0·16
			2·00)5·09(3 + 1
4/7·50 0·30	*Half-brick sleeper walls in wirecuts in cement mortar* (1:3) *built honeycombe.*		
			[*height assumed*

Note 1. An examination of the above waste calculation will show that before dividing by the spacing, the distance from the external wall to the *centre-line* of each end sleeper wall has been deducted from the room dimension of 5·25 metres. This is made up of 25 mm air-space plus

half the thickness of the sleeper wall (55 mm) = 80 mm total (see Fig. 76) on each side of the room, making 160 mm in all.

The resultant 5·09 is the distance from centre-line to centre-line of the end sleeper walls. This is divided by the spacing (2 metres), the result being the number of *spaces* between the sleeper walls.

Since there is a sleeper wall at either end, the number of sleeper walls will always be *one greater* than the number of spaces between them. Hence the 'plus one' shown above, at the end of the waste calculation.

Note 2. In dividing the distance by the spacing in this way, the answer should always be taken to the nearest integer *above*, and never to the one below, since the specified spacing of the sleeper walls is a maximum, not a minimum spacing.

Note 3. There is no SMM classification for sleeper walls, but since they will be built honeycombe, a separate description will be necessary to differentiate them from solid half-brick walls, and the one given above is the usual form.

Fender walls

Since these help to support the floor (see Fig. 69), they are usually measured with the floor, rather than with the fireplace.

The mean girth of any fender walls must be calculated on waste in the usual manner, and supered by the height, as follows:

		1·50
	2/·45	0·90
		2·40
2·40 0·30	*One-brick fender walls in wirecuts in cement mortar* (1:3).	

Note 1. If the concrete hearth laps to the centre-line of the fender wall, as shown in Fig. 67, the dimensions of the front hearth will give the mean girth of brickwork without further adjustment.

Note 2. Fender walls are usually described separately from ordinary one-brick walls, on the analogy of 'dwarf supports' as a separate classification given in SM G3(a).

Damp-proof courses

These will occur at the tops of all sleeper and fender walls, and must be measured as described for general DPCs in the substructures section of this book. The material specified will probably be the same as that used in the main walls, and the same description will therefore apply; the total lengths of the sleeper walls being taken from the previous waste calculation, and set down as (for example):

4/7·50	*Horizontal layer of felt DPC all as before described, but* 110 *mm wide*

the DPC to any fender walls being described as 220 mm wide, still measured linear under SM G44(a).

Timber plates

For measurement of carpentry timbers generally see Chapter 22.

It will be recalled that plates require an allowance for laps of 150 mm in every 6 metres of length. Plates to the above sleeper walls would therefore require to be measured thus:

	7·50
lap	·15
	7·65

4/7·65	110 × 50 *mm Softwood plate.* & *Bed plate in cement mortar* (1:3).

Note. The bedding of the plate, measurable under SM G52(a) is not thought worthwhile adjusting for deduction of the laps, but is anded-on to the plate itself, as usual.

Plates to fender walls must be measured the extreme girth, not the mean girth, to allow for laps at the corners. Thus, the plate to the fender wall measured above would require adjusting from the *mean* girth of the wall, to its *extreme* girth, as follows:

	2·40
corners $\frac{1}{2}$/8/·11	·44
	2·84

2·84	110 × 50 *mm Softwood plate* & *Bed plate as before* [*fender walls*

270 Here, the girth is required to be shifted from the centre-line of the one-brick wall to the outside edge, a distance of 110 mm. The girth adjustment formula is $A = M \pm 8D$ (see substructures section) and D in this case is 0·11. The parameter 8 assumes four external angles however, and in this case there are only two; hence $\frac{1}{2}$/8/·11 in the above waste calculation.

Timber joists

We have seen that the sleeper walls will span the *length* of the floor, and the joists the *width*. The problem of finding the *number* of joists remains, and this is calculated on waste in the same manner as the number of sleeper walls, viz., by dividing the total width by the spacing.

The position of the joists will not normally be shown on the drawing. The specification notes must be examined to determine the spacing, and the following sequence of operations carried out on waste:

(1) calculate the distance between the centre-lines of the two end joists (take the room dimension and deduct from it (a) the air-space and (b) half the joist-thickness—twiced for both ends)
(2) divide by the spacing, taking the answer to the nearest integer above.
(3) add one, in order to convert from spaces to number of joists.

If we assume that:

(a) the end joists are placed 25 mm from the wall-face, and
(b) the joists themselves are 50 mm thick,

the above calculations can be expressed as follows:

$$N = \frac{d - 100 + 1}{S},$$

where N = number of joists,
d = internal dimension of room in millimetres,
S = spacing of joists.

Failing any specification notes to the contrary, the above assumptions may be considered reasonable, since a 25-mm clearance will allow air-space without undue overlap of the boarding, while 50 mm for width of joist may be regarded as normal, allowing sufficient width to prevent splitting when the boards are nailed. The above formula will only give correct results if it is remembered that rounding-up in the case of any remainder is necessary, in order to maintain maximum spacing, in accordance with the specification.

In this last connexion, it will be realized that in practice the joists may be spaced a fraction over the theoretical maximum in certain cases, in order to avoid an additional joist; and this might be reasonable if the spacing is exceeded by a small fraction of an inch over each space between joists. One must be on the safe side however, since the shortage in quantities of one joist could prove very embarrassing if the bill was subsequently used as a 'type' bill for an estate of (say) two hundred houses. The quantity surveyor would then have to explain away a shortage of two hundred joists!

The number of joists required to the floor shown is Fig. 77 would be calculated on waste as follows (400 mm centres assumed):

```
               7·50
2/·05           ·10
          ·40)7·40(19 + 1
```

giving 20 joists in all.

It will be appreciated that floors are measured net, i.e., deductions in the length of joists for hearths, etc., are made on waste, and the joists measured their actual lengths.

Joist descriptions

SM N4 groups floor and roof joists together and this classification would include both ground- and upper-floor joists. A typical item would be:

linear 125 × 50 *mm Softwood joists.*

Note that the single word 'joists' is used in the description. If the descriptions of upper-floor and roof joists are similarly worded, the effect will be to group all joists together, in accordance with the SMM classification.

A point worth mentioning in connexion with joists is that textbooks sometimes show the ends splayed back from the wall, as in Fig. 78.

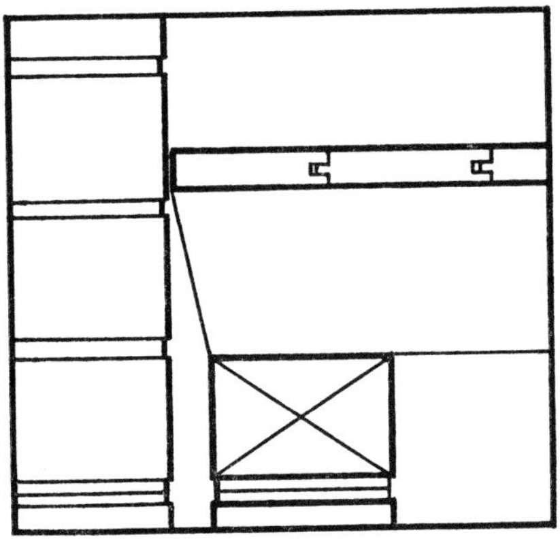

Fig. 78

272 Should this ever be done in practice however, the surveyor need not be perturbed. Splaying is deemed to be included under SM N6(a).

Floor boarding

The Standard Method divides this into two types (see SM P3):

BOARD FLOORING. This refers to boarding over 100 mm nominal width. Usually of softwood, and often square-edged, each board simply butt-jointed against the next; though heading-joints are splayed.

STRIP FLOORING. This term applies when the boarding is in nominal widths not exceeding 100 mm. Often of hardwood and usually tongued and grooved.

The method of measurement is the same in each case, the superficial area of the finished floor being taken, no allowance being made for the tongues in tongued and grooved flooring (see SM P2(b)). The following points regarding measurement are worth noting:

NET MEASUREMENT. As in the case of floor construction, floor coverings are usually measured net, i.e., any openings in the floor for hearths, etc., are adjusted when the flooring is measured, not in the 'fires' section, for example.

FLOORING IN OPENINGS. Where door openings occur, it will be necessary to run the floor-boarding through the opening, to maintain continuity, unless the flooring in the next room happens to be of a different kind. This will apply in the case of external door openings also.

Except in the case of partitions built up off the floor, no joists will exist underneath such door openings, and it will consequently be necessary to measure bearers to receive the flooring. These bearers are included in the description of the flooring under SM P3(d), which flooring must be kept separate as 'in openings'; but otherwise described as for the rest of the flooring.

In view of this last, it is usually more convenient to measure the flooring in openings in the 'floors' section of the work; not when the openings are adjusted, in the 'doors' section.

THRESHOLDS. These are measured when openings are adjusted in the 'doors' section; not with 'floors'.

NOSINGS AND MARGINS. Since flooring is measured net, any margins around, e.g., hearths, are measured with the floors. For nosings see 'upper floors' later in this chapter. For margins, see page 275.

SKIRTINGS. In the case of timber floors, skirtings are regarded not as floor finishings, but as wall finishings. They are measured in the 'internal finishings' section of the work not with 'floors'.

Descriptions of flooring

A typical description for softwood flooring follows, with measurements appropriate to Fig. 77:

7·50 5·25	25 *mm Softwood square-edged board flooring with splayed head joints, well cramped up; each board nailed to each joist with two flooring brads, and cleaning-off on completion.*
0·90 0·22	*Ditto in openings including bearers.* *(doorway*

SM P3(a) states that where boards or strips are required to be of a particular width, the width shall be stated. The specification notes must be examined on this point, but the following factors may be worth noting in connexion with the widths of flooring:

BOARD FLOORING. When in doubt, it is desirable to state a width, in order to avoid the possibility of over-wide boards being used, thus increasing the shrinkage factor, etc.

STRIP FLOORING. If no width is stated, the use of random widths is implied, with a consequent possible saving in cost which may be significant in the case of an expensive hardwood floor. The architect's instructions should therefore be obtained regarding permissive widths in the case of strip flooring, and the description worded accordingly.

A typical description for strip flooring would be as follows:

super	20 *mm Afrormosia tongued and grooved strip flooring in* 75 *mm (face) widths with splayed head joints, well cramped up; each strip secret-nailed to each joist with one oval wire nail, and cleaning-off on completion, and preparing for polisher.* *&* *Two coats 'Bournseal' on afrormosia strip flooring.*

274 *Note 1.* If flooring is of hardwood, the type of hardwood must be stated, e.g., afrormosia, oak, teak, mahogany, etc. It is usual to underline the name of the hardwood in the description, as shown, to emphasize its cost implication as opposed to softwood.

Note 2. When a board-width is given in the case of tongued and grooved work, it must be made clear whether the width includes the tongue. Hence the term '(face) width' above, meaning that the tongue is excluded—the normal manner of specifying.

Note 3. Flooring brads are normally used for square-edged boarding, but are unsuitable for tongued and grooved work if this is secret-nailed, as is usually the case. Instead, oval nails are used, one to each joist for secret-nailing.

Note 4. SM P1(c) requires timber kept clean for polishing to be so described. The actual polishing (or sealing) of hardwood floors must be anded-on as a separate item, since this is not joiners work. The type of floor should be stated in the polishing description, as this could affect the price of polishing.

Floor margins

These may occur around hearths, and are measured with the flooring, as previously noted. They are dealt with under SM P6, and the following points must be borne in mind:

EXTRA OVER FLOORING. Margins are ignored in the first instance; the flooring being measured over the margin. No subsequent deduction is made of the flooring for the margin, which is 'extra over'.

TONGUED EDGES. Margins will normally be tongued to the flooring. The size of the margin (which requires stating in the description) will be the extreme size, measured over the tongue.

GROOVES IN FLOORING. These must be given in the description of the margin (see SM P6). This is done even if the flooring is tongued and grooved, since the groove in T & G flooring may require to be re-run where the margin occurs.

CROSS-GRAIN GROOVES. Where the *ends* of the boards abut the margin, the grooves will require to be run across the grain, which increases the labour cost of the groove. Since both grooves and cross-grain grooves must be given in the description of the margin under P6, it will be necessary, where the margin abuts the ends of floor-boards, to keep this margin separate from that which runs parallel to the boards.

ENDS AND MITRES. Clause P1(k) applies here. In the case of (a) *softwood* margins not exceeding 0·004 square metre in sectional area and

(b) *hardwood* margins not exceeding 0·002 square metre, ends and mitres *shall be given in the description.*

In all other cases they must be enumerated separately.

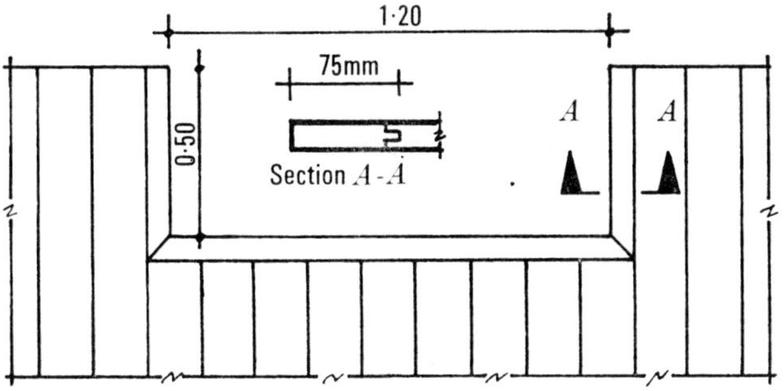

Fig. 79

We are now in a position to measure the margin to the hearth shown in Fig. 79, as follows:

1·20 0·50	*Deduct* 20 *mm Afrormosia flooring as before described.*	
		(*hearth* 0·50 0·075 0·575
2/0·58	*Add* *Extra over ditto for* 75 × 20 *mm afrormosia margin, tongued to edge of floor including groove, and including mitres and ends.*	
	2/·075	1·20 0·15 1·35
1·35	*Ditto, including groove cross grain.*	

Air-bricks

The whole of the timber ground floor having been measured, there remains the air vents to take. Vents to rooms, etc., will have been measured in the 'fires and vents' section; but those in connexion with under-floor ventilation are measured at this stage, in the 'floors' section of the work, as previously explained.

For the principles of measurement of vents generally see Chapter 21.

Floor vents will consist of an air-brick in the outer skin of the external wall, and a corresponding opening in the inner skin. No internal ventilator will be necessary, since the opening is below floor level, and not exposed to view. The cavity should theoretically be closed around the opening, to ensure a supply of fresh air from outside, rather than stale air from the cavity. A typical description is as follows:

number	225 × 75 *mm Square hole pattern terra-cotta air-brick and building in, including forming opening in* 270 *mm brick cavity wall, and lining with slates bedded and pointed in cement mortar* (1:3).

TIMBER UPPER FLOORS

In measuring timber upper floors, regard must be had for the necessity of trimming the floor construction around openings in the floor to accommodate stairwells, chimney-breasts, etc. If the floor-covering is taken-off in advance of the constructional work (joists, etc.) the surveyor will have familiarized himself with the positions and sizes of such openings by the time he is ready to consider the trimming problems involved; for it will be realized that the position of trimming timbers (or any other joists for that matter) will probably not be indicated on the floor plans.

For this reason it is customary to adopt the following order of taking-off in the case of upper floors:

(1) coverings,
(2) construction,

each section being denoted by a sub-heading in the taking-off; unless the amount of work involved is not sufficient to warrant such sub-headings.

Coverings

These will consist of either board or strip flooring, the problems of measurement and description involved being all as for timber ground floors, dealt with above, except for the presence of nosings.

Floor nosings

We have seen that ground floors may involve the measurement of margins around hearths, etc. These may occur in upper floors as well, and will be dealt with in the same manner. A *margin* is a member tongued on to the flooring where it abuts a vertical surface, e.g., the edge of a hearth.

A *nosing* on the other hand, is a similar member tongued on to the flooring where it stops around a void, e.g., at a stairwell. The only difference between a margin and a nosing is that the margin has a square edge and the nosing a rounded edge (or possibly a rounded and splayed, or moulded edge). The method of measurement is therefore exactly the same, and both are dealt with under SM P6.

The reader will recall from the section on 'floor margins' above that the following factors must be taken into account, and these factors apply also to nosings, viz.:

(1) nosings and margins are extra over flooring,
(2) a groove in the flooring must be included in the nosing description,
(3) nosings tongued to a cross-grain groove must be separated,
(4) Clause P1(k) governs the method of dealing with mitres and ends,
(5) the cross-sectional size of the nosing must include the tongue,
(6) the length of the nosing is its *extreme* length; not its mean length, or the girth of the stairwell.

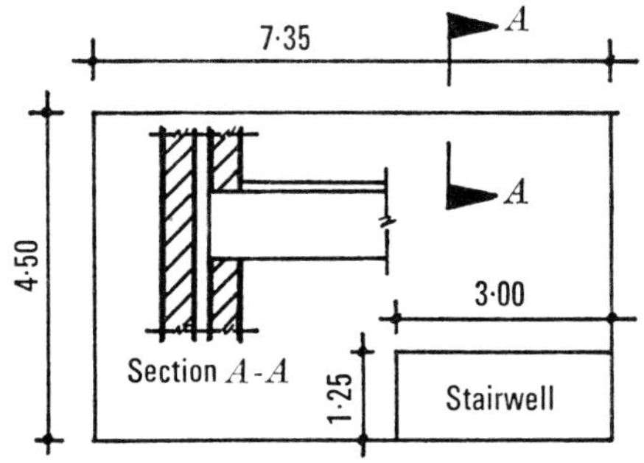

Fig. 80

Bearing these points in mind, we shall see that the adjustment of the floor-coverings for the stairwell shown in Fig. 80 for example, would appear as follows:

3·00 1·25	*Deduct* 25 *mm Softwood square-edged flooring as before described.*
	(stairwell

3·00
0·10
3·10

3·10 — *Add Extra over ditto for* 100 × 25 *mm softwood rounded nosing tongued to edge of floor including groove, and including mitres and ends.*

1·25
0·10
1·35

1·35 — *Ditto including groove cross grain.*

Since the nosing is taken as (a) softwood, and (b) not exceeding 0·004 square metre in sectional area, the mitres and ends must be included in the description under Clause P1(k). Note that a *hardwood* nosing of this size (i.e., over 0·002 square metre in sectional area) would require the mitres and ends to be measured separately as enumerated items under the above clause.

Upper-floor construction

Having completed the measurement of coverings (not forgetting any flooring in openings including bearers; as described on page 273), the floor construction can now be dealt with.

Upper floors may contain any or all of the following members:

(a) plates
(b) bridging joists
(c) trimmed joists
(d) trimming joists
(e) trimmers
(f) metal hangers or corbels
(g) strutting
(h) steel beams.

It will be appreciated that the ceiling on the underside of upper floors is measured in the 'finishings' section of the taking-off, not with 'floors'.

Taking-off items

For the principles of measurement of structural timbers generally, see

Chapter 22 (Carpentry Work), especially the section on *trimming* (pages 252–3).

Trimming timbers are not separated from bridging joists under the Standard Method. Items (b) to (e) above can therefore be regarded as a single item (joists) from the point of view of bill descriptions; the only difference in the taking-off descriptions as between bridging joists and trimming timbers being a difference in thickness only.

Bearing this in mind, together with the necessity of measuring certain labour items involved, we shall find that the actual bill items which require to be considered when measuring a simple upper floor are as follows:

(1) plates (N4(i))
(2) bedding of plates (G25(a))
(3) joists (N4(iii))
(4) extra labour trimming (N6(e)(vi))
(5) creosoting ends of timbers
(6) metal hangers or corbels
(7) cutting and pinning ends of timbers (G56(c))
(8) herringbone or solid strutting (N8).

In addition there may be steel beams to be dealt with.

Plates

Where these occur in upper floors, they are dealt with as for ground-floor plates (see page 269). In the case of upper floors, there are three main possibilities:

(1) PLATES RESTING ON BRICKWORK. Bedding will require to be measured separately as usual.

(2) PLATES WITH BRICKWORK ABOVE AND BELOW. The brick-work in this case will doubtless have been measured over the plate, in the 'brickwork' section of the taking-off. No bedding is measurable here (see G25(a)).

(3) PLATES RESTING ON IRON CORBELS. No bedding is measurable, but the iron corbels must be numbered.

The usual rules for measuring plates must be observed, i.e., allowance for laps must not be forgotten (150 mm every 6 metres in length) and the extreme girth taken if an angle should occur—unlikely perhaps in upper-floor construction. In fact, plates are frequently omitted altogether in the construction of present-day upper floors, the ends of the timbers simply resting on the brickwork, or the load-bearing block inner skins, or internal walls.

280 **Upper-floor joists**

The setting out of joists in upper-floor construction will be affected by the position of trimming members, which will require to be measured as joists. It might be as well, before proceeding further, to refresh the memory regarding definitions of members used in connexion with the trimming of openings in floors:

(a) COMMON JOIST. An ordinary joist which may or may not be interrupted by an opening.

(b) BRIDGING JOIST. A common joist running full span not interrupted by an opening.

(c) TRIMMED JOIST. A common joist which is interrupted by an opening, requiring it to be cut short and framed to a trimmer.

(d) TRIMMER. A joist forming the side of an opening, and supporting the ends of trimmed joists.

(e) TRIMMING JOIST. A bridging joist supporting the end of a trimmer (or trimmers), and consequently requiring to be strengthened, e.g., by being made thicker than the remaining bridging joists.

(f) CRADLING-PIECE (see Fig. 72). A short length of joist spanning from trimmer to wall, and supporting the edge of a hearth.

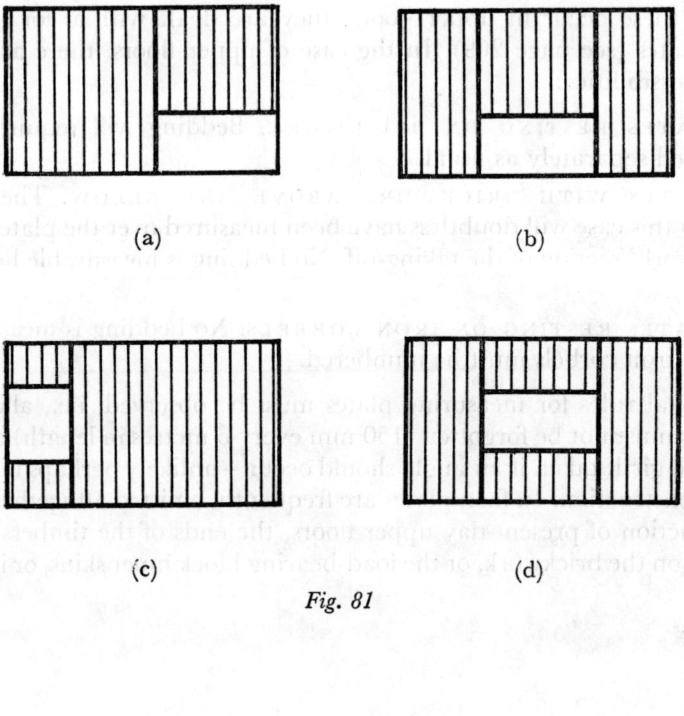

Fig. 81

Figure 81 shows four conditions under which trimming may occur:

(a) CORNER OPENINGS. One trimmer and one trimming joist are required. One end of the trimmer requires wall support.

(b) SIDE OPENINGS. Two trimming joists are required, in addition to the trimmer.

(c) END OPENINGS. Two trimmers are necessary, each requiring wall support one end.

(d) CENTRE OPENINGS. Two trimmers and two trimming joists are required.

Calculating the number of joists

It is possible that an upper floor may not require any trimming, in which case the number of joists is calculated in the manner described for ground-floor joists, i.e., using the formula:

$$N = \frac{d - 100}{S} + 1$$

where N = number of joists
d = internal dimension of room in millimetres
S = spacing of joists,

the internal dimension being that of the longest wall to give the joists the shortest span.

If the floor is trimmed however, the common joists will comprise (a) *bridging* joists, and (b) *trimmed* joists; each requiring separate measurement owing to their difference in length.

Again, the exact position of the *trimming* joists will be fixed by the dimensions of the opening, and the common joists will therefore require to be set out in relation to the trimming joists, in order to maintain maximum spacing.

Each set of *common* joists on either side of any *trimming* joist will therefore require to be dealt with separately in calculating the number required, one calculation for the *bridging* joists, and a separate calculation for the *trimmed* joists as follows:

(1) NUMBER OF BRIDGING JOISTS. The distance from the centre-line of the wall-joist to the centre-line of the trimming joist must be found. This distance may then be divided by the spacing to give the number of bridging joists. The trimming joist itself will require to be measured separately, since it will be thicker than the bridging joists. The required

number of bridging joists will therefore be *equal* to the number of spaces between them.

(2) NUMBER OF TRIMMED JOISTS. The above operation is repeated in respect of the trimmed joists. In the case of trimmed joists occurring *between* two trimming joists (as in Figs. 81(b) and (d)), the distance between the centre-lines of the two trimming joists will be the relevant dimension.

Note that in this last case, the number of trimmed joists required will be *one less* than the number of spaces between them.

The trimming joists and trimmers can now be measured, these normally being taken as 25 mm thicker than the common joists.

Openings for stairwells

If the opening is for a stairwell, we may assume that the nosing will project 40 mm from the face of the trimming member, to allow for the riser or apron-lining, as the case may be (measured with the staircase). The trimming member will be 75 mm thick, so that a further 37·5 mm will need to be added to obtain the distance from the front of the nosing to the centre-line of the trimming member, making 77·5 mm in all; say 80 for convenience (see Fig. 82).

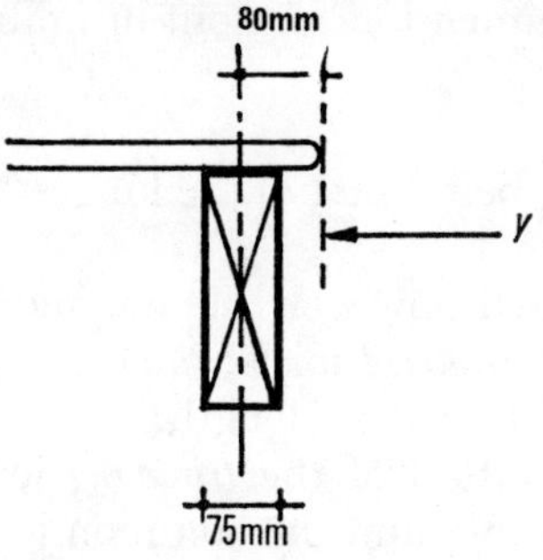

Fig. 82

If we now call the net length of the floor opening (as shown on the drawing) Y, the number of bridging joists in a floor containing one trimming joist can be obtained from the following formula:

$$B = \frac{d - Y - 130}{S},$$

where B = number of bridging joists,
d = internal dimension of room (millimetres),
Y = size of opening (millimetres),
S = spacing of joists (millimetres),

the parameter 130 being made up of 80 mm from nosing to centre of trimming joist, plus 50 mm from wall-face to centre of end joist.

When the opening requires two trimming joists, as in Fig. 81(b) and (d), the parameter will be doubled, 130 for each trimming joist, thus:

$$B = \frac{d - Y - 260}{S}.$$

Similarly, the number of trimmed joists is given by:

$$T = \frac{Y - 50 + 80}{S},$$

where one trimming joist is necessary (Fig. 81(a) and (c)). When two trimming joists occur (Fig. 81(b) and (d)), it will become:

$$T = \frac{Y - 2(50 + 80)}{S} - 1,$$

the number of joists being one less than the number of spaces between them in this instance.

Openings for fireplaces

It must be remembered that the *width* of hearths will probably be the distance between *cradling-pieces*, not trimming-joists. In this case the distance from edge of flooring to centre of trimming-joist will be determined by the distance from edge of hearth to side of chimney-breast.

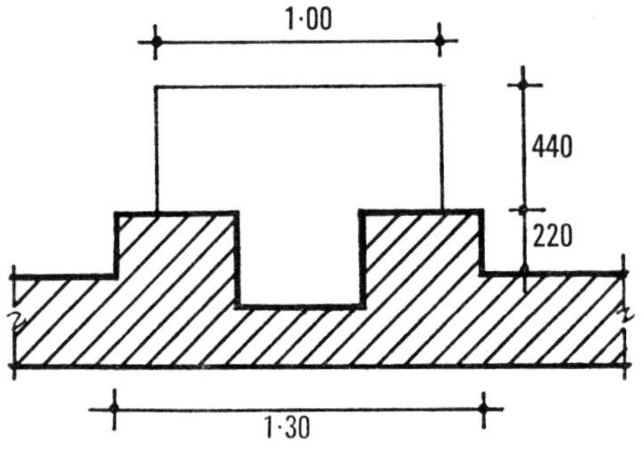

Fig. 83

Thus a hearth in the position of the opening shown in Fig. 81(b) would appear on the plan as shown (say) in Fig. 83. Here, the earth itself is only 1·00 wide, but the joists would require trimming round the 1·30 wide chimney-breast, and the distance between centre-lines of trimming-joists would need to be calculated on waste as follows:

breast		1·300
clearance	2/·025	·050
to joist centres	2/½/·075	·075
		1·425

This resultant dimension can now be divided by the joist spacing to give the number of *trimmed joists* (minus one from the number of spaces in this case). Similarly, when deducted from the room dimension, the result (after deduction of 50 mm at either end to arrive at centres of end joists) will give the number of *bridging joists*, when divided by the spacing.

Length of members

It will be recalled from Chapter 22 that timber is measured over the joints, the extreme length of each member being taken.

Trimmed joists are therefore measured to the face of trimmers (as shown in Fig. 84(a)), the trimmers themselves being measured over the trimming joists; a further 150 mm being allowed for the tusk in each case, (as in Fig. 84(b)).

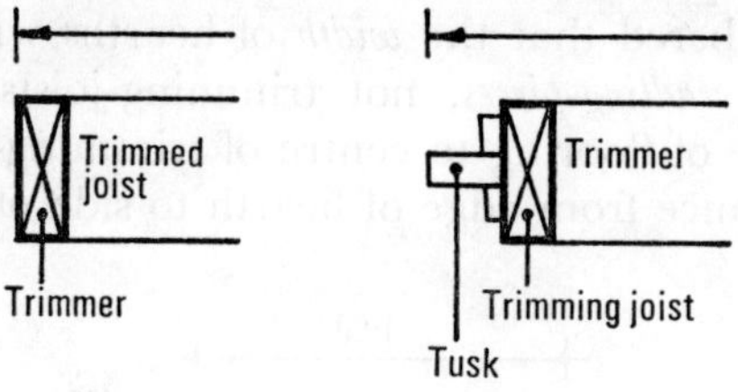

(a) (b)

Fig. 84

PROBLEM

Measure the joists for the trimmed floor shown in Fig. 80.

SOLUTION

Bridging joists $B = \dfrac{d - Y - 130}{S}$

$$d = 7{\cdot}35$$
$$Y = 3{\cdot}00$$
$$4{\cdot}35$$
$$- \quad {\cdot}13$$
$$S = {\cdot}40)\overline{4{\cdot}22}(11$$

Trimmed joists $T = \dfrac{Y - 50 + 80}{S}$

$Y = 3{\cdot}00$
$-\ {\cdot}05$
$2{\cdot}95$
$+\ {\cdot}08$
$S = {\cdot}40)\overline{3{\cdot}03}(8$

Lengths:

		4·50
bearing	2/·11	·22
bridging joists		4·72
		4·50
	1·25	
nosing	·04	1·29
		3·21
bearing		·11
trimmed joists		3·32

11/4·72 8/3·32	225 × 50 *mm Softwood joists*

	3·00
bearing	·11
nosing	·04
trimming joist	·075
tusk	·15
	3·375

4·72	225 × 75 *mm ditto* *(trimming joist*
3·38	*(trimmer*

Note especially that all members are described as 'joists' their particular functions (e.g., trimmers, etc.) being side-note material only.

Labour trimming

It will be recalled from Chapter 22 that a numbered item in respect of each opening to be trimmed must be given, in accordance with SM N6 (e)(vi), to pay for the extra cost of forming the special carpentry joints at the ends of the trimmers, trimmed joists, etc. Although the trimming members themselves will have been measured, they will have been included in with the descriptions of ordinary joists, as we have seen. No opportunity to allow for the extra labour involved in trimming the joists around the opening will be given to the estimator unless the extra labour item is measured.

The student may ask whether the complicated tusk tenon and dovetailed notch joints as illustrated in textbooks on construction are always carried out on the site in this way nowadays. It is true that trimmed joists are in practice sometimes merely notched to fillets spiked to the side of the trimmer instead, short trimmers often being fixed in the same manner. Whatever method of jointing is employed however, there is always an extra cost involved in this work, which must be covered by the labour item for trimming; which, in the case of the opening in the above floor (see Fig. 80) would be taken as:

1	*Extra labour trimming* 225 × 50 *mm joists around* *opening size* 3·0 × 1·25 *metres.*

Ends of joists

Having measured the joists themselves (together with labour trimming where appropriate) the ends of the joists must now be considered. If built into the brickwork, creosoting, or other wood preservative will probably be specified to the ends so built in. It is customary to number this item as:

number	*Twice creosote ends* *of timbers built into* *brickwork.*

all sizes being lumped in together for this purpose. Note in the case of the floor measured above, that even if a wall-plate was provided to take the ends of the common joists, one end of the trimmer would require to be built into the brickwork since no plate would exist in this position.

In situations where regulations require that timber shall not be built into brickwork (at fireplaces, for example), a metal hanger or corbel may be specified to take the end of a joist. These are numbered, e.g.,

number	75 × 15 *mm Galvanized mild* *steel corbel* 200 *mm* *long, and building in* *to brickwork.*

It will be observed from SM G56(b) that *building in* ends of timbers as the work proceeds is deemed to be included. *Cutting and pinning* is to be enumerated under G56(c).

The question is: are upper-floor joists cut and pinned, or built in as the work proceeds?

All upper-floor joists will be at the same level, and the ends can therefore rest on the same course of brickwork. It will therefore be convenient for the bricklayers to build up to underside of joist level, and to then allow the carpenters to position the joists; after which the bricklayers can then continue, filling in between the joists as necessary.

If this process is regarded as 'building in as the work proceeds', a note should be made in the dimensions after measurement of joists, e.g.,

[*no cut and pin:*
joists assumed
built in.

which indicates that a decision has been made, and the cutting and pinning not just forgotten about altogether by the taker-off.

Herringbone or solid strutting

This will probably not be shown on the plan, but must on no account be overlooked by the taker-off.

Strutting should be measured at 2·5 metre centres across the joists. Herringbone strutting, as normally specified, spans from wall to wall; the spaces between the end joists and the wall being closed with wedges, regarded as part of the strutting from the measurement point of view. The room dimension is therefore taken as the linear dimension under SM N8; the depth (not thickness) of the joists being given, together with the size of the strutting. Strutting to the floor shown in Fig. 80 would appear as follows:

7·35	50 × 50 *mm Softwood herringbone strutting to* 225 *mm joists.*

Note that strutting should not be wedged against the sides of openings. Thus, two rows of strutting would require to be taken to the construction shown in Fig. 81(d), if the room were (say) 4·50 metres wide, in order to avoid the opening in the middle; unless this was small enough to permit one row slightly off-centre of the room.

Solid strutting is measured in the same way, but may be accompanied by a long bolt, passing through the centres of the joists next to the

288 strutting, in order to tighten the floor rigidly. Such bolts are usually measured as tie-rods under SM N29, thus:

4·50	*12 mm Diameter mild steel tie-rod.*
2	*Extra for threaded end including nut and washer*
13	*Hole through 50 mm joist for 12 mm diameter bolt (measured separately)*

Note that the holes through the joists require enumerating under SM N26(a).

Steel beams

For measurement of steel beams generally; see page 190.

If a timber floor spans a greater distance than about 5 metres, it will probably be considered economical to use steel beams at 5-metre centres,

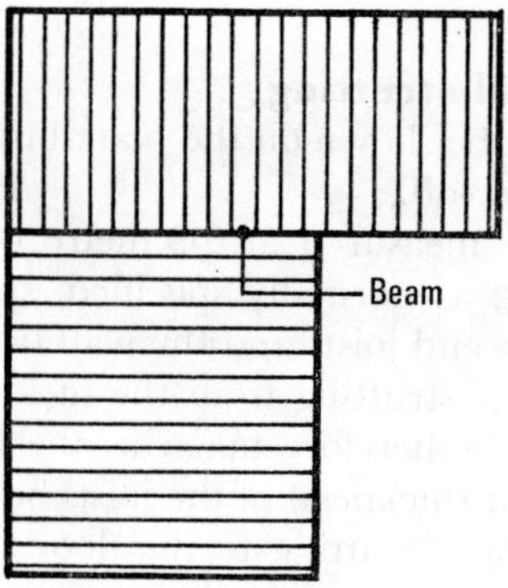

Fig. 85

and span the joists between these. Again, in the case of an L-shaped living-room, it may be desirable to use a steel beam as a trimming-joist, as shown in Fig. 85.

Beams below joists

The simplest form of construction involving the use of a steel beam in connexion with a timber floor is that shown in Fig. 86(a). The joists are lapped across the top flange of the beam, which is entirely below underside of joist level. The beam will require bracketing to receive the ceiling finish, as shown. This bracketing, like the ceiling finish, is *measured in the 'finishings' section*—not with 'floors'.

Only the beam itself (together with its painting, and cutting and pinning of ends to brickwork) is measurable in the 'floors' section. The additional length of timber required by the lap of the joists will, of course, need to be added to the length of each joist.

Beams partly below joists
In order to reduce the amount of beam projecting below the ceiling, the beam may be positioned as shown in Fig. 86(b). Here, the top flange is below top of joist level, to facilitate fixing of floor-boards, which require a bearer fixed across the top of the beam where each joist occurs. The joists rest on plates bolted to the web of the beam.

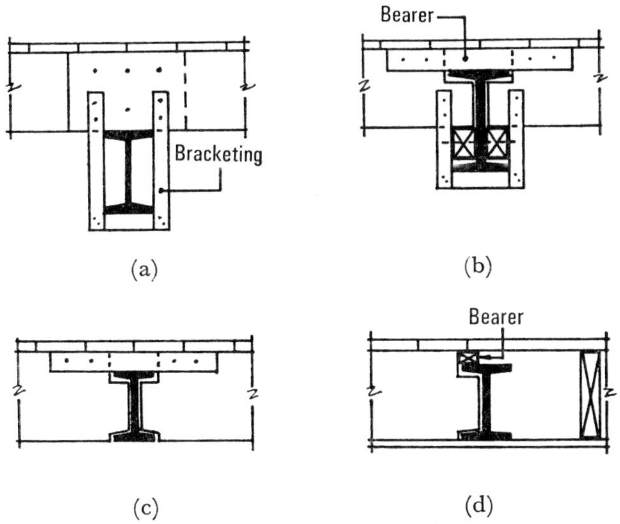

Fig. 86

Measurement of the beam in this case would involve the following items:

The following in unframed steelwork:

linear *Plain girder of joist section, over* 20 *but not exceeding* 50 *kilogrammes per linear metre*
× 30 *kg* =

(end of unframed steelwork)

number *Cut and pin or build in end of small steel section to brickwork.*

linear	60 × 60 *mm Softwood plate bolted on (bolts and holes measured separately)*
number	12 *mm diameter bolt* 175 *mm long, with head, nut, and washers.* (*SM N*31(*a*)
number	*Hole through* 60 *mm softwood for* 12 *mm diameter bolt (measured separately)* (*SM N*26(*a*)
number	12 *mm diameter hole through web of* 200 × 100 *mm steel joist.* (*SM Q*17
linear	50 × 35 *mm Sawn softwood bearers not exceeding* 1·25 *metres long.* (*SM N*4(*ii*) & *N*3(*d*)

Note that the bearers have been described as 'sawn', to ensure that they are billed under Carpentry instead of Joinery work which may also contain bearers. Note also that notching of the joists is not measurable (see SM N6).

No paint has been measured to the beam, but this would need taking if specified (for painting steelwork, see pages 191–2).

Beams in the floor thickness

This construction, illustrated in Fig. 86(c), will still require bearers to be measured, since the top flange of the beam will not be suitable as a fixing for the floor-boards, and will require to be kept below floor level, unless it supports a partition above, or is otherwise concealed from view.

Beams used as trimming joists

In the case of a beam used to trim joists in the manner shown in Fig. 85, the problem of securing the ends of the floor-boards running at right-angles to the beam arises.

Either short lengths of bearer may be used, or a continuous bearer, as shown in Fig. 86(d) can be run along the top flange of the beam, skew-nailed to the ends of the joists. Such details will probably not be shown on the drawings, but must nevertheless be accounted for by the quantity surveyor.

Roof Coverings

Roofs in domestic work may be divided into two main kinds, pitched roofs and flat roofs. In either case the general order of taking-off is as follows:

1. coverings
2. construction
3. eaves and rainwater goods.

The coverings are measured first so that when the time comes to deal with the construction necessary for their support, the surveyor will have familiarized himself with the general layout and constructional requirements. Then, having measured the main roof construction, the eaves construction (fascias and soffites, etc.) is dealt with, followed by the rainwater goods.

This chapter deals with pitched roof coverings only, flat roofs being considered separately in Chapter 27.

Types of covering

The traditional covering used for pitched domestic roofs in this country is either slates or tiles. Other coverings are available (e.g., copper, asbestos, aluminium, etc.), but are not frequently met with in small houses, and will not be dealt with at this stage.

The term 'slates' refers to materials cut from the quarry, and should not be confused with a concrete product known as 'interlocking slates', which are a new type of interlocking tile giving the appearance of a slated roof, and are best thought of as tiles rather than slates. Slates are not widely used as a roof covering today, and are unlikely to be met with in new work very frequently. They are measured in the same way as tiles, under Section M of the Standard Method of Measurement, and should present no special problem to the student who has mastered the measurement of tiled roofs.

This leaves only tiles to be considered, and these will be found to fall into two main categories:

(a) single lap tiles, and
(b) double lap tiles.

292 Although the measurement of these two categories is basically the same, they differ as to the treatment required at hips and valleys, etc., and this affects the items to be measured. It is well therefore to get the difference between these two main types of tile clear in one's mind to start with.

Single lap tiles

These are tiles which to a greater or lesser extent rely on interlocking with each other to keep out the weather. Because of their shape, and their interlocking qualities, it is sufficient to lap each course of tiles over the top of the one below. Hence the term single lap. Examples of single lap tiles are: pantiles, Italian tiles, Spanish tiles, Roman tiles, double Roman tiles, interlocking tiles and interlocking 'slates'.

Double lap tiles

These are plain tiles and have a flat (slightly cambered) surface. They are laid like slates, i.e., each course laps not only over the one below, but also over the next one below. Hence the term double lap. Plain (or double lap) tiles may be of either clay or concrete. Today they are often made of concrete, and so coloured as to be indistinguishable from the clay product, which is more expensive.

Subsidiary coverings

Underneath the tiles it will be necessary to have one or all of the following:

(a) underfelting (SM M15)
(b) roof boarding (SM N11)
(c) counter-battening (SM M14(b)),

and the drawings and specification notes must be examined to see whether or not all three of the above items are required. Quite often underfelting alone is specified, and boarding and counter-battening should not be taken unless specifically called for.

Method of measurement

Order of taking-off is as follows:

(1) main areas of roof coverings,
(2) linear items to boundaries of areas (i.e., eaves, verges, ridges, etc.).

In dealing with the areas of coverings it is usual to measure the tiling and assume the same area for underfelting, etc., simply anding-on the subsidiary coverings to the item for roof tiling. This is considered sufficiently accurate for all practical purposes.

The area required for each section of roof will be the length on plan by the slope height *twiced for both sides of the roof.* Care must be taken not to forget this last requirement (all too easily done if in a hurry) which may result in a very embarrassing interview with a contractor who has discovered that half the roof is missing!

Unless a dimensioned roof plan is available it will be necessary to calculate the length of roof from brickwork dimensions on the floor plans, and care should be taken to include a suitable allowance on waste for the overhang of the roof at the verges, or eaves at gable ends. The slope height of the roof coverings should be carefully scaled from the section.

Areas of hips and valleys

The measurement of pitched roof areas will be greatly facilitated if it is appreciated that *provided the pitch remains constant*, hips and valleys will cancel out, and therefore not affect the area, and should be ignored.

Thus, the gabled roof in Fig. 87(a) has the same area as the hipped

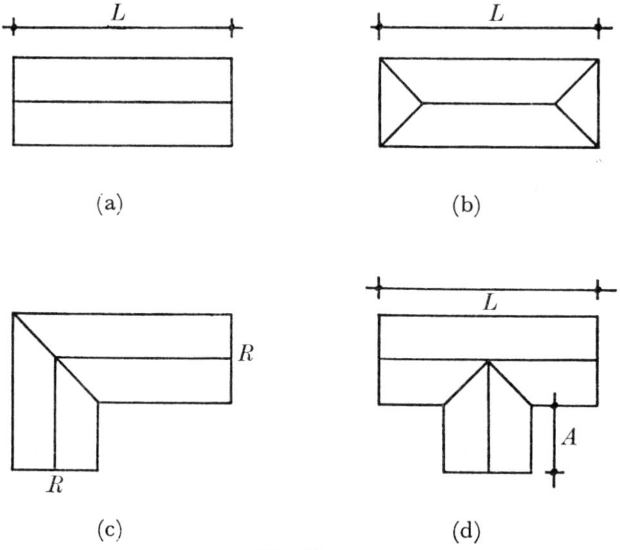

Fig. 87

roof in Fig. 87(b); the length L by the slope height (twiced) giving the correct result in each case. Again, in Fig. 87(c) the length of the ridge R–R by the slope height (twiced) will give the area; while in Fig. 87(d) the length L plus the additional length A by the slope height (twiced) will give the area in this case. Thus not only may hipped ends be measured as gable ends, but a roof broken by hips and valleys may be measured as a series of separate roofs each with gable ends.

It should be remembered that this will only apply if the pitch of the slope is constant; a quick way of checking this being to notice whether the angle *on plan* made between the line of adjacent hips (or valleys) is a right angle. Should the pitches vary, it will be necessary to measure the main areas net, and add the areas of the remaining triangles formed by hips and valleys. This will not often be necessary, since the pitch will usually be found to be constant in order to simplify the construction.

Openings in the roof for chimney-stacks, etc., should be ignored at this stage, and left for later adjustment (see Chapter 26).

Descriptions of roof coverings

It will be noticed that the Standard Method calls for the lap to be given (not the gauge) for slate or tile roofing (SM M2(b)). If the gauge is specified, this will require converting to lap for the purpose of the description, by means of the formula:

$$\text{lap} = L - 2G$$

where L = length of tile,
G = gauge.

The lap for head-nailed slates is sometimes given as $L - 2G - 25$ mm, the allowance of 25 mm being made for the distance from the nail-hole to the head of the slate. This last formula is only correct if 'lap' is defined as the distance from the tail of a slate to the nail-hole of the slate in the next course but one below. British Standard Code of Practice CP 142, however, defines 'lap' as 'the amount by which the tails of slates in one course overlap the heads of slates in the next course but one below'. If we accept this definition, then lap = $L - 2G$ as given above will be true not only for plain tiles, but also for both centre-nailed and head-nailed slates as well. It will not, of course, apply to single lap tiles, where the lap will simply equal the length minus the gauge.

Typical descriptions are as given on page 295.

Note 1. Battens are included with the tiles, but counter-battens must be measured separately under SM M15.

Note 2. Roof boarding must be classified under SM N11 as either (a) flat, (b) sloping not exceeding 50 degrees, or (c) vertical or sloping over 50 degrees.

super	270 × 160 *mm Concrete plain roofing tiles laid to a* 60 *mm lap and nailed every fourth course with galvanized nails to and including* 32 × 19 *mm softwood battens.*
	&
	32 × 19 *mm Softwood counter-battens at* 400 *mm centres.*
	&
	Layer of reinforced bituminous roofing felt lapped 150 *mm at joints and closely nailed.*
	&
	19 *mm Softwood boarding to sloping roofs not exceeding* 50 *degrees from horizontal.*

Double course at eaves

Having measured the whole of the areas of roof coverings, the surveyor must now turn his attention to the linear items. As each one is measured it should be ticked off on the plan to ensure none is missed, an item being measured in respect of each boundary of every roof area on plan. The first of these items will be the eaves, under SM M5. Certain types of single lap tile will require a plain tile undercourse at eaves, bedded and pointed in cement mortar. Plain tiles require a double course instead, thus:

linear	*Extra for double course at eaves.*
	&
	Softwood splayed tilting-fillet (2 *out of* 100 × 50 *mm*)

Note 1. The words 'extra for' indicate that the roof area has already been measured to edge of eaves.

Note 2. The tilting-fillet at eaves, which includes mitres (see SM N17(b)) is described as two fillets cut from the same scantling, this being common practice.

Verges

Eaves are measured from the plan, but care must be taken not to inadvertantly measure verges in this way, the slope height up the gable ends being the length required. Extra undercloak course is mentioned in SM M6, and this will be required with plain tiles, as well as bedding and pointing in cement mortar. The description will run as follows:

linear	*Extra for verge including undercloak and bedding and pointing in cement mortar* (1:3).

Ridge cappings

These are measured under SM M9, and it will be noticed that filled ends, mitred angles and mitred intersections are to be enumerated. These should be dealt with as follows:

FILLED ENDS. Taken at all gable ends, and at hipped ends *unless hip cappings are used*, in which case take '*mitred intersection of ridge and two hips*' instead.

MITRED ANGLES. Taken at changes in direction of the ridge unless this causes a hipped angle which is *covered by hip cappings*, in which case take '*mitred angle at intersection with hip capping*' instead.

MITRED INTERSECTIONS. The ridge intersection shown in Fig. 87(d) would be described as a '*three-way right-angled mitred intersection*'. If two ridges were to cross each other, this would be a '*four-way ditto*'.

A typical description for the ridge itself, followed by fair ends would be:

linear	250 *mm diameter half round ridge tiles to match general tiling, bedded and pointed in cement mortar* (1:3).
number	*Fair ends, filled in solid with tile slips in cement mortar* (1:3).

Measurement of hips

Care must be taken not to scale these off the plan, but to measure the *actual slope length*. This may be done by means of a simple projection on the roof plan by pencil lines as shown in Fig. 88.

(1) draw a perpendicular *AB* at one end of hip *AC*, and equal in length to the vertical height of the roof from eaves to ridge,

(2) connect *BC*.

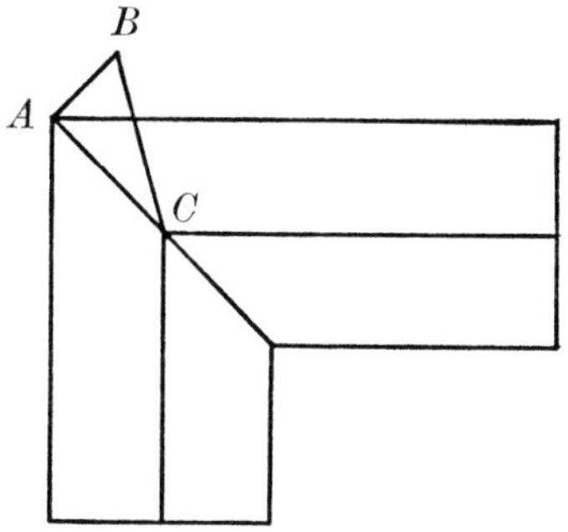

Fig. 88

The line *BC* is now equal in length to the true slope height of the hip *AC*, and can be scaled to give the required dimension.

Treatment of hips

Three main methods of dealing with hips may be encountered in specifications:

(1) MITRED HIPS (SM M8(b)). In this case the roof coverings are simply mitred at the angle of hip, metal soakers being necessary underneath to render the hip waterproof. This method is normally only used with slating, and is not often encountered.

(2) HIP TILES (SM M8(c)). These are special tiles used in conjunction with plain tiling, with which they are bonded-in. The two main types are (a) angular and round hip tiles, and (b) bonnet hip tiles. The former lie flat, while the latter project slightly, giving a pleasing emphasis to the hip.

(3) HIP COVERINGS (SM M9). These may be regarded as an extension of the ridge down the length of the hip, since they usually match the ridge capping. The tiles or slates are simply mitred underneath, as in (1) above, no soakers being necessary in this case however. Hip coverings are essential in the case of single lap tiles, which do not permit of hip tiles. They are often used with slates also, in preference to soakers.

Descriptions for hips

For plain tiling angular or round hip tiles are common. Typical items are as follows:

linear	*Extra over plain tiling for angular hip tiles to match general tiling, nailed at every course.*

linear | *Raking cutting on 25 mm roof boarding.*

&

Ditto on roofing felt.

Note 1. Unlike the general tiling, hip tiles require nailing at every course.
Note 2. Raking cutting is required to *both sides* of each hip, the length of which must therefore be twiced (see SM M15(b) and N15(a)).
Note 3. No raking cutting is measurable on the roof tiling, since cutting and bonding is deemed included under M8(c).

Bonnet hip tiles are measured in a similar manner, followed by the raking cutting items as above, typical descriptions for bonnet hip tiles being as follows:

linear — *Extra over plain tiling for bonnet hip tiles to match general tiling, nailed every course and bedded and pointed in cement mortar* (1:3).

number — *Fair ends filled in solid with cement mortar* (1:3).

Bonnet hip tiles should be bedded in cement, and are often pointed in addition, as described above. The ends of such hips are filled solid, and enumerated under SM M8(c).

If hip cappings are used instead of hip tiles, they are measured in the same way as ridge cappings under M9. In this case 'cutting to hips' must be measured on either side, in accordance with M8(a), and this item can be anded-on to the raking cutting on the felt, since the dimensions will be the same. It will also be necessary to measure a hip iron at the foot of each hip covered with hip cappings, under SM M11. The painting of hip irons is enumerated under SM W12(b), and should be anded-on to the hip iron, as follows:

number | *Galvanized hip irons and screwing to feet of rafters.*

&

Prepare and paint two coats oil colour on galvanized hip iron.

Treatment of valleys

As with hips, it must not be forgotten that the *actual slope length*, not the plan length, is the measurement required, and a similar pencil plot will be required to establish this. Where valleys are in conjunction with hips, their lengths will normally be the same, and the valley lengths can then be taken direct from the hip dimensions.

Disregarding laced and swept valleys, which are rare, two treatments may be met with:

(1) VALLEY TILES. These are special tiles used in conjunction with plain tiles, with which they are bonded-in. They may be regarded as *inverted hip tiles* and are measured in exactly the same way.

(2) VALLEY GUTTERS. These are usually necessary with single lap tiles, and slates—but not with plain tiles.

Valley tiles

These are measured under SM M7(b), followed by raking cutting items to the boarding and felt, all as for hip tiles, thus:

linear	*Extra over plain tiling for valley tiles to match general tiling, nailed every course.*
linear	*Raking cutting on 25 mm roof boarding.* & *Ditto on roofing felt.*

Once again, raking cutting is required on *both sides* of each valley for boarding and felt, but cutting and bonding to the tiling itself is deemed to be included under M7(b).

Valley gutters

These are not necessary with plain tiling, but may be encountered with single lap tiles. A valley-board is laid in the valley, and covered with metal (usually lead) which is dressed up under the tiles on either side. Typical items to be measured are as follows:

linear	175 × 25 *mm Softwood valley-board prepared for lead.*

300 *linear* — *1·70 mm Lead valley gutter 500 mm girth lapped 100 mm every 2 metres, dressed over tilting-fillets and fixed with copper nailing on both sides (no allowance for laps).*

linear — *Cutting on roof tiling to valleys, including bedding and pointing in cement mortar (1:3).*

&

Raking cutting on 25 mm roof boarding.

&

Ditto on roofing felt

Note 1. Splayed edges to the valley-board are deemed included under N15(b). 'Prepared for lead' indicates something more than a sawn finish.

Note 2. For lead gutter descriptions see SM M54 and M58. An allowance of 150 mm for end passings should be made on waste, making the gutter 300 mm longer than the valley-board.

Note 3. All items of cutting must be measured to *both sides* of each valley.

Sections through valleys

The above descriptions are fairly typical, but the construction of valley gutters may vary slightly, and it may sometimes be necessary to set up a section through the valley in order to visualize the construction specified, and to scale off the width of the lead, etc. In the event of this being necessary, the angle which the jack rafters make with each other will be required, as shown by the 'angle of valley' in Fig. 89. To arrive at this angle, proceed in the following way (Fig. 90 refers):

(a) At one end of the valley *AB* set up a perpendicular *AC* equal in length to the vertical height of the roof from eaves to ridge. Join *CB*.
(b) Draw *DE* so as to intersect *AB* at any point *O*. From *O* draw *OF* perpendicular to *CB*.
(c) Rotate *OF* to intersect *AB* at *P*.
(d) Join *PD* and *PE*. The angle *DPE* is the required angle of valley.

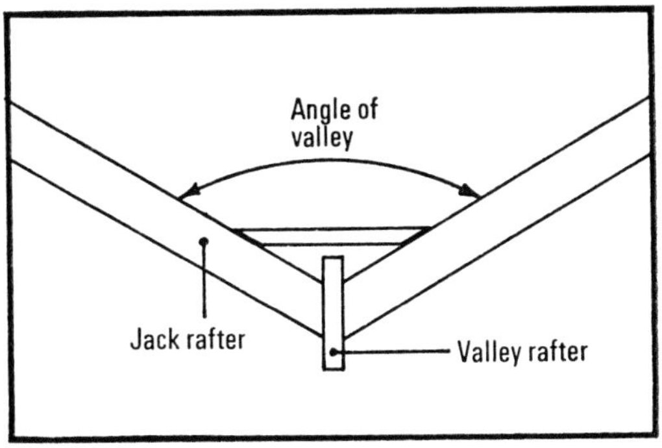

Fig. 89

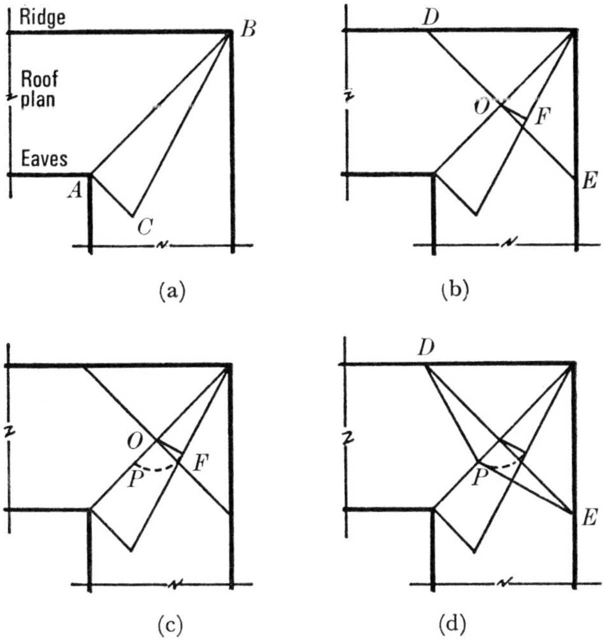

Fig. 90

Abutments

Should a pitched roof abut a brick wall at the side, a stepped flashing will be required. In the case of a lean-to roof abutting a wall at the back, an apron flashing will be required at the top edge. Such flashings are dealt with in the same way as those to chimney-stacks, described in the chapter dealing with stack adjustments.

Chapter Twenty-five

Roof Construction

A domestic pitched roof may contain any or all of the following members:

(a) rafters
(b) ceiling joists and collars
(c) plates
(d) ridges
(e) hips and valleys
(f) purlins
(g) struts and hangars
(h) sole-plates
(i) roof trusses and trussed rafters.

Other members are also possible, such as trussed purlins, dragon ties, etc., but these are less likely to be encountered in simple domestic work nowadays, and in any case should present no special difficulties. In addition to the above members, eaves construction will also have to be considered, together with rainwater goods.

In taking-off a pitched roof it is usual to start with the rafters and ceiling joists, and then follow with plates and ridges. Other structural members are then measured, taking the most important-looking ones first, odd struts, etc., being left until the end. Trusses may be left to deal with last of all, or they may be measured in advance, depending on personal choice. Most domestic roofs avoid the use of trusses, and rafters will normally be the first item measured.

Rafters

The length of rafter is scaled from the drawing, bearing in mind the necessity to measure each timber its overall length, as shown in Fig. 91. The length of a typical rafter when entered in the dimension column can usually be timesed by the total number of rafters.

Where hips and valleys occur, the jack rafters (i.e., the shortened rafters) are ignored, and the hips and valleys treated as gable ends. This is because the total length of jack rafters in a hip and valley is equal to the total length which would be required if the roof were gabled at this point, as may be seen from the alternative rafter plans shown in Fig. 92. Where two sections of roof meet by means of a hip and valley therefore, one

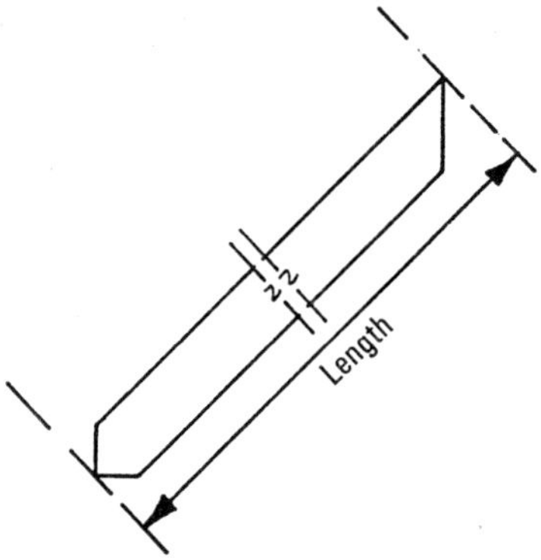

Fig. 91

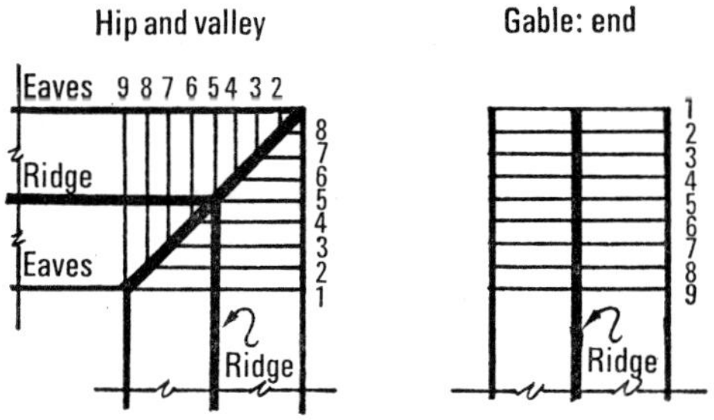

Fig. 92

section is measured as if it ran straight through to a gable end, and the other section measured to a point where the eaves meet. In this way the measurement of rafters is simplified.

A similar situation arises in the case of a hipped end; but in this instance the number of rafters required is *one greater* than that needed for a gable end. This may be seen from Fig. 93, the rafter shown by a broken line being the extra one needed. It should be noted that only one extra rafter is involved here, not a *pair* of rafters, and care should be taken to

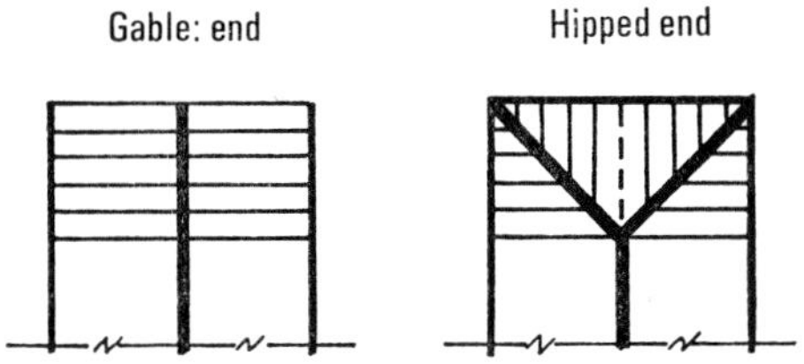

Fig. 93

deal with this extra rafter in such a way that it is not inadvertently twiced when twicing the remaining rafters.

Although jack rafters are ignored in the case of normal hips and valleys, and hipped ends are measured as gable ends in the manner described above, the junction between a main roof and subsidiary roof *with a lower ridge* may involve the main rafters being carried down to eaves level underneath the jack rafters of the subsidiary roof. In such a case the jack rafters must be separately measured and their lengths added to the rest of the rafters in the dimension column.

Calculating the number of rafters

We are now in a position to consider how best to calculate the number of rafters in a pitched roof. In the unlikely event of every rafter being shown on the drawing, they may be simply counted of course. Otherwise, the length of the roof must be divided by the spacing of the rafters in order to achieve the result. There are three possible configurations for each length of roof, and the procedure varies accordingly:

(1) HIPPED BOTH ENDS. Measure from extreme edge of roof coverings one end to extreme edge of coverings the other end. Divide by the spacing and *deduct one* from the result.

(2) GABLED BOTH ENDS. Measure from the centre-line of the first rafter one end to the centre-line of the last rafter the other end. Divide by the spacing and *add one* to the result.

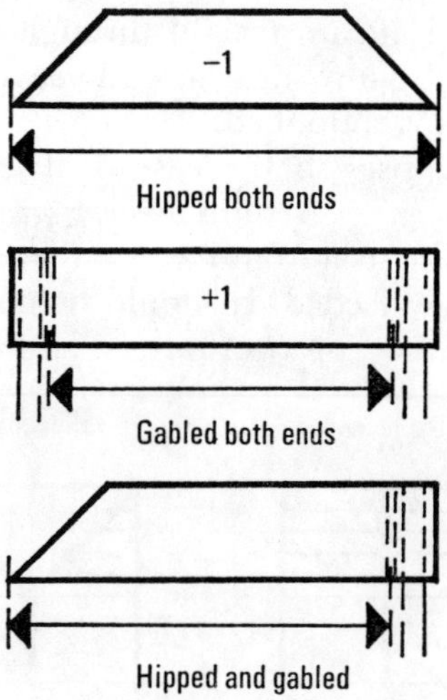

Fig. 94

(3) HIPPED ONE END AND GABLED OTHER END. Measure from extreme edge of roof coverings at the hipped end to the centre-line of the last rafter at the gable end. Divide by the spacing. This gives the correct result.

The above three cases are shown diagrammatically on elevation in Fig. 94.

Example 1

PROBLEM

Measure the rafters to the roof shown in Fig. 95, assuming 125 × 50 mm rafters 4·5 metres long at 350 mm centres.

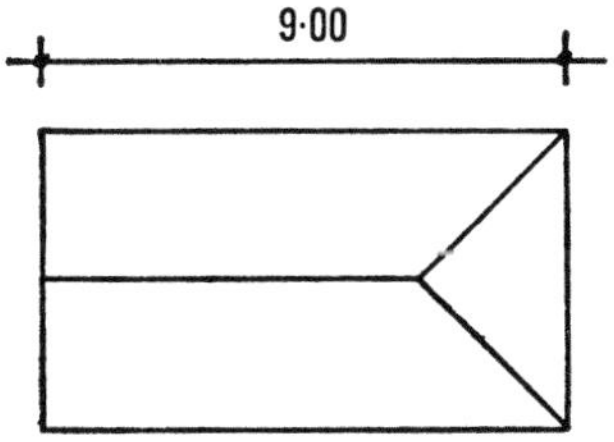

Roof plan

Fig. 95

SOLUTION

		9·00
verge	·15	
wall	·27	
to centre of rafter	·05	·47
		·35)8·53(25

2/25/4·50	125 × 50 *mm Softwood rafters.*
4·50	*(extra at hip*

Note 1. From the figured dimension is deducted (a) the verge, (b) the wall thickness, and (c) the distance from wall-face to the centre-line of the last rafter at the gable. The result is divided by the spacing.

Note 2. The additional rafter at the hipped end (as shown in Fig. 93) is entered separately, as this is not twiced.

Example 2

PROBLEM

Measure the rafters to the roof shown in Fig. 96 assuming 125 × 50 mm rafters 6·75 metres long at 400 mm centres.

306

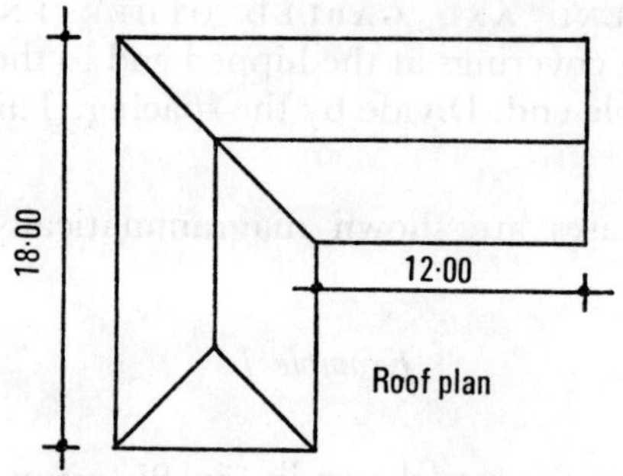

Fig. 96

SOLUTION

		12·00
verge	·15	
wall	·27	
to centre of rafter	·05	·47
		·40)11·53(29 + 1
		·40)18·00(45 − 1

2/44·30/6·75	125 × 50 *mm Softwood rafters in lengths exceeding 6 metres and not exceeding* 7·50 *metres.*	
6·75		[*extra at hipped end*

Note 1. The 12-metre length of roof is treated as being gabled at both ends; the 18-metre length as hipped at both ends.

Note 2. See SM N3(c) regarding structural timbers over 6 metres long. Rafters would be in one length, not scarfed.

Ceiling joists and collars

Ceiling joists are often spiked to the rafter feet, one joist to each pair of rafters. They are measured the extreme length, and described as (e.g.),

linear	100 × 50 *mm Softwood ceiling joists.*

Where hips and valleys occur, one end of a set of ceiling joists may be unsupported. In Fig. 96 for example, the ceiling joists to the 18-metre length of roof will be unsupported one end, where the roof is joined by the 12-metre length of roof. Unless a suitable load-bearing partition happens to be located underneath to support the ends of these joists they will need to be trimmed into the end joist of the adjoining roof. This joist will therefore become a trimmer, and require to be thickened to 75 mm. To pay for the trimming involved it is usual to measure an item analogous to N6(e)(vi), as follows:

linear	*Deduct* 100 × 50 *mm Softwood ceiling joists.* & *Add* 100 × 75 *mm ditto.*
number	*Extra labour trimming* 100 × 50 *mm ceiling joists at* 400 *mm centres for a length of* 5·50 *metres.*

Roof collars, where these occur, are measured in the same way as ceiling joists.

Plates

Roof plates are dealt with in the same way as floor plates, and the student will recall that certain rules must be observed in the measurement of all plates, i.e., allowance for laps must not be forgotten (150 mm every 6 metres in length) and the extreme girth taken at angles. The lengths of plate are collected on waste, care being taken to measure a plate at hipped ends but not to inadvertently take the plate round any gable ends shown on the roof plan.

In the case of plates to pitched roofs under discussion, the brickwork at eaves will normally have been measured over the plate, to top of eaves filling. In this case therefore, no bedding of the plate is measurable (see SM G52(a)).

Ridges

Lengths of ridge over 6 metres long are scarfed, and this involves an allowance on waste under SM N3(b)(ii) *and* the enumeration of the scarf joint itself, under N6(a). Thus a 10-metre continuous length of ridge would be measured as follows:

		10.00
scarf	2/·18	·36
		10·36

308	10·36	180 × 30 *mm Softwood ridge, including cutting and fitting ends of rafters both sides.*
	1	*Extra over ditto for scarf joint, including wedges.*

Note that ridges are measured linear under N5, and that the description must include cutting and fitting ends of rafters.

Where the ridge bears on brickwork at gable ends, an item must be taken for cutting and pinning under G56(c), as follows:

number	*Ends of timbers cut and pinned to brickwork.*

Hips and valleys

Like ridges, the cutting and fitting to ends of rafters must be included in the description, under SM N5.

As with hip and valley tiles in the 'coverings' section of the taking-off, it is the *true slope length* of timber hips and valleys which is required, and as these will be for all practical purposes the same length as the hip coverings, the taker-off can at this point turn back the dimensions and simply copy down the appropriate figures.

It should be noted that timber hips and valleys are required to be in one length, for structural reasons. Should the length exceed 6 metres, therefore, no scarfing should be measured; instead, the length of the members must be described in stages of 1·50 metres in accordance with SM N3(c). A typical hip description would run:

linear	250 × 32 *mm Softwood hips including cutting and fitting ends of rafters both sides.*

Valleys are similarly described.

Purlins, and struts, etc.

Having measured the rafters, ceiling joists, ridges, hips and valleys, the taker-off will now examine the roof for evidence of any other roof timbers, and measure these in turn, classifying them as far as possible in accordance with SM N4. All such timbers must be measured their extreme length, and halved joints allowed for under N3(b)(1) in the case of any bearing timbers over 6 metres in length. No labours will normally be measurable, since splaying, notching, halving, etc., are deemed to be included under N6(a).

Where purlins occur, it must not be forgotten that they will normally return round any hipped ends. Where they bear on the brickwork at verges, an item of 'ends of timbers cut and pinned to brickwork' must be taken in accordance with SM G56(c). A purlin over 6 metres in length may possibly be scarf-jointed, but it is often possible to use a simple halved joint at the point where the purlin is supported by a strut. The surveyor should use his discretion here, having regard to the number and position of any struts supporting a long length of purlin.

ROOF TRUSSES

The type of timber roof truss encountered in modern domestic work will usually be of the 'batten' variety. Pioneered in England by the Timber Development Association, this type of truss is often referred to as a TDA truss. It consists of relatively thin members bolted together, the joints being strengthened by means of timber connectors.

The principal rafter of a normal roof truss supports the purlins of the roof, which in turn support the rafters. In some cases, however, the purlins are supported below the level of the principal, which lies in the same plane as the common rafters. The truss is then technically known as a 'trussed rafter', rather than a roof truss. It is not usually regarded as necessary to distinguish between the two types when taking-off, however, both being described as roof trusses.

Measuring procedure for trusses

A roof truss will usually be found to contain the following parts; which are taken-off in the order given:

(a) timber members (principals, tiles, struts, etc.)
(b) timber packing-pieces and cover-pieces
(c) timber cleats
(d) bolts, with nuts and washers
(e) metal timber connectors.

The usual procedure is to take off a typical truss all complete, and then to times the dimensions in red ink for the total number of such trusses. By using a different colour for the timesing of trusses it is easy to spot whether the truss timesing has been missed for a particular item; which event will otherwise be obscured by the timesing of items within the individual truss.

Start by measuring the main members first (e.g., principals, ties) and

310 then follow with subsidiary members (struts, hangars, etc.). Principals and ties will usually be found to be double members, requiring timesing. Typical descriptions are as follows:

linear	*80 × 40 mm Softwood roof truss members.*	*(tie beam*
linear	*50 × 30 mm ditto*	*(principals*
linear	*80 × 40 mm ditto not exceeding 1·25 metres long.*	*(struts*

Note 1. See SM N4(x) for roof truss members, which are classified together as such. The functional descriptions should appear as side-notes only.

Note 2. Watch out for any members not exceeding 1·25 metres long, which must be separated out under N3(d).

Packing-pieces, etc.

Packing-pieces, cover-pieces and cleats now follow. Packing and cover-pieces are similar, and are usually put together in the same description. They are numbered, as being analogous to the items in SM N9, which makes provision for the cleats. Typical descriptions are:

number	*75 × 25 mm Softwood cover or packing-piece 0·35 metres long.*
number	*100 × 50 mm Softwood cleat 0·25 metres long, splayed one end.*

Bolts and connectors

Having accounted for all the cover-pieces, packing-pieces and cleats, the bolts are dealt with. SM N31 makes provision for these, and it will be necessary to group them according to their diameters and lengths. The length of a bolt is the length of its shank, as shown by *L* in Fig. 97. To

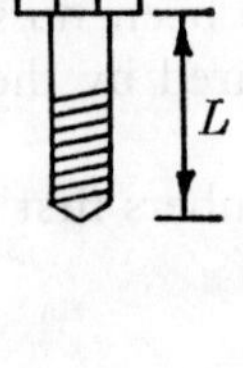

Fig. 97

establish this length it will be necessary to add together on waste the thicknesses of the members being bolted, and then add allowances for the nuts and washers, plus a clearance below the nut. The following allowances are standard: 311

nut = *diameter of bolt*
each washer = 3 *mm*
clearance = 3 *mm*

The length of a 12-mm diameter bolt required to connect two 50-mm members together would therefore be calculated as follows:

timber	2/·05	·10
	nut	·012
washers	2/·003	·006
	clearance	·003
		·121

This would be rounded-off and described as a 130-mm bolt. It should be noted that two washers will always be required, one under the head and one under the nut, square plate washers being the type invariably used with batten roof trusses.

Holes for the bolts are measurable under N26(a), which makes mention of pellating. This refers to filling in a counter-sunk hole over the top of a bolt, and will not be required in the case of roof trusses. Note that the holes cannot be anded-on to the bolts, since more than one hole will be required for each bolt; though in practice they may be drilled together. Typical descriptions are as follows:

number	12 *mm diameter bolt* 200 *mm long with head, nut and* 50 × 50 × 3 *mm plate washers.*
number	*Hole through* 50 *mm softwood for* 12 *mm diameter bolt.*
number	*Ditto through* 30 *mm softwood for ditto.*

Between each face of timber in each bolted connexion will occur a timber connector. These are now counted up and enumerated under SM N31(b) thus:

number	50 *mm diameter approved double-sided toothed plate connector.*

312 **Hoisting**

When the truss is complete, and all the parts are accounted for, there remains the item of hoisting to be taken, before finally timesing for the number of trusses. This is required by SM N7, and the description must include the size of the truss and its height of hoisting above ground level. Some trusses, especially of the trussed-rafter type, are brought to the site and hoisted into position in two halves; and where the design appears to indicate this procedure, it is usual to state this in the description, as follows:

number	*Hoist timber batten roof truss* 10 *metres span and* 3 *metres high overall (in two sections) and fix* 6 *metres above ground level.*

THERMAL INSULATION

Having measured the main roof construction for roof timbering, there remains the possibility of insulating materials which may be specified to the roof space. These are usually measured before proceeding with the eaves and rainwater goods. Insulating quilt laid on top of the rafters will have been anded-on to the general roof coverings, but the disadvantage of insulating over the rafters in this way is that the roof-space then requires to be allowed for in the volume of space to be heated. It is therefore frequent practice to insulate over (or rather between) the ceiling-joists instead; either by means of loose exfoliated vermiculite poured between the joists, or fibreglass or other quilting cut and fitted between them. Simply laying quilting *over* the joists obscures the electrical wiring in the roof, and is thus usually avoided.

The practice is to measure the overall area, including the joists. SM N27 states that insulating quilts shall be measured 'the area covered', but this is taken to mean that no allowance for laps is to be made, rather than that joists should be deducted. It is usual to allow sufficient width of strip to give a turn-up against the joists on either side, so that the total area less the joists would not be correct in any case, with quilting fixed in this way.

Two typical descriptions follow, one for exfoliated vermiculite, and an alternative one for fibreglass quilting:

super	*Layer of exfoliated vermiculite* 75 *mm thick, spread and levelled in roof space between ceiling joists at* 400 *mm centres (measured over joists).*

super	*50 mm thick fibreglass quilting cut into strips 400 mm wide and layed between ceiling joists at 350 mm centres in roof space (measured over joists).*

EAVES CONSTRUCTION

In measuring the main roof construction, any openings in the roof for chimney-stacks, skylights, etc., will have been ignored and left for later adjustment. Likewise the work to eaves, including gutters and down-pipes, which must now be dealt with. Together with the eaves construction will be taken any constructional work which may be necessary to verges, by way of barge-boards and the like.

Eaves construction will normally consist of a fascia and soffit, together with bearers to support the soffit. If there is no soffit, and the rafter feet are exposed, a superficial item of 'extra for wrought face of timber' must be measured to the surfaces of the exposed ends under SM N6(b); unless these exposed ends are not exceeding 600 mm in length, in which case they are enumerated under N6(e)(iii). The area of paint to the rafter feet and adjoining woodwork is then measured in accordance with W3(a)(vi). Usually the eaves is boxed-in by means of a fascia and soffit however, and these are measured in turn, followed by the painting.

Eaves fascias

These should be measured the extreme girth. This may not be quite the same as the girth on the external face, since twice the thickness of the timber must be allowed for each external angle, when these exist, in order to comply with the general principle of measuring each individual piece of timber its extreme length as fixed. If no dimensional roof plan is available, it may be convenient to take the mean girth of external walls and adjust this by adding 8 times the distance from the wall centre-line to the outside face of the fascia-board. To this must be added twice the fascia thickness for each internal angle of fascia. This method will be found suitable if the fascia goes all round the building, but care must be taken not to inadvertently measure a fascia to any gable ends; and if the building is gabled rather than hipped, the fascia is probably best measured piecemeal, each section being dealt with separately in establishing its length.

Having carefully collected the lengths of fascia on waste, the total can be transferred to the dimension column and described in accordance with SM P11:

linear 150 × 25 *mm Softwood grooved fascia including mitres and ends.*

Note 1. Mitres and ends are to be given in accordance with SM P1(k), and must be measured separately to fascias exceeding 0·004 metre in sectional area.

Note 2. Returned ends are in any case measured separately under P11. Where these exist they will generally take the form of spandril-shaped boxings at gable ends to cover the entire eaves construction, and are best measured after the soffit. Returned mitred ends to the fascia may be specified in addition to boxed ends, in order to conceal the fascia end grain. They are often omitted in practice however.

Note 3. Beware of the temptation to and-on the paint to the fascia. The girth for paint must include the soffit, and painting is therefore best left until the entire eaves construction has been measured.

Note 4. The groove in the above description is on the assumption that the soffit is shown tongued to the fascia.

Eaves soffits

These are measured in accordance with SM P9. Superficial soffits (over 300 mm wide) are measured on their centre-line, but linear soffits the extreme girth, as for fascias. SM P8(b) requires that the method of jointing shall be given, and this clause must be read in conjunction with P1(e) which assumes that all joinery members over 200 mm wide are jointed, unless specified as being in one width. The type of jointing normally required in members over 200 mm wide is *cross-tonguing*, and this should always be assumed, failing directions to the contrary. A general rule with joinery members therefore is:

always describe as cross-tongued if over 200 *mm in width.*

This rule would also apply to fascias, if deep enough, for these are also measured under Section P (Joinery), and not under Section N (Carpentry).

Both fascias and soffits (and barge-boards) will require to be wrought, unlike other roof members. There is no need to state this in the description however, since all items measured under Section P are deemed to be wrought in any case (see Clause P1(b)). Typical items in respect of 200 × 25 mm soffit (one edge rebated to form a tongue to be received by the groove in the fascia) are as follows:

linear	*200 × 25 mm Softwood cross-tongued rebated eaves soffit.*
linear	*Raking cutting on 25 mm boarding.*
linear	*50 × 25 mm Softwood bearers for soffit boarding.*

Note 1. The raking cutting measured here is diagonally across the corners at mitred angles to the soffit. Angles and ends are not measured as such under P9(a), but the raking cutting involved in the case of such angles is measured linear under P10(a).

Note 2. Eaves bearers will probably be spiked to the rafter-feet. The total length for one rafter can be calculated on waste and set down in the dimension column and then timesed by the number of rafters involved. All such bearers are measured linear under P9(b) however short, and are not described as in short lengths.

End treatment to eaves

There are three main ways of treating the eaves at a gable end, (a) boxed ends in timber, (b) brick or tile corbelling, (c) stopping against a barge-board. These are illustrated in Fig. 98. The drawing must be examined

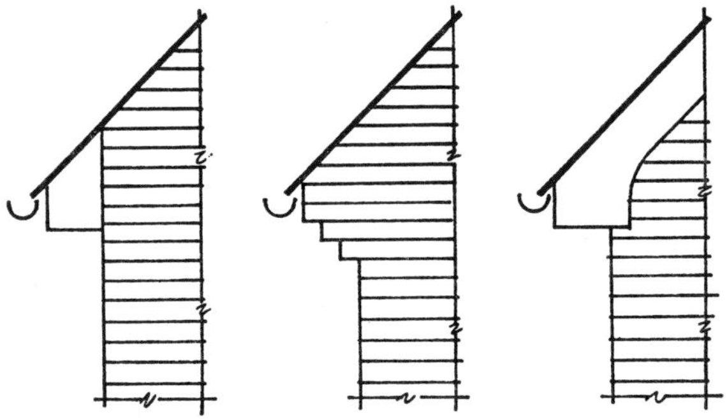

(a) (b) (c)

Fig. 98

316 to see which treatment applies. In the case of boxed ends a typical description would be:

number	*25 mm Softwood spandril-shaped boxed ends size 230 × 400 mm overall to eaves.*

In the case of brick corbelling, a description similar to the following would be appropriate:

number	*Extra labour and material forming one-brick thick corbel 330-mm projection and 6 courses high in facings to eaves at gable end, including pointing to match.*

Where barge-boards are used, their shaped ends will cover the eaves construction, dealt with after completion of eaves measurement.

Painting to eaves

When eaves construction has been measured, there remains the external painting to take.

Without going into the measurement of paintwork too deeply at this stage, it will be seen that SM W3(a) provides for paint to be measured superficially if over 300-mm girth, and linear (in 100-mm stages) if not exceeding 300-mm girth. It should be especially noted that the girth of the *paint* is the relevant factor, not the girth of the member to be painted. This, the *total* girth of paint, must be found by studying the section through the eaves, and this will usually be established by adding together on waste the exposed parts to be painted, i.e.,

(a) face of fascia (behind gutter)
(b) lower edge of fascia
(c) back of fascia below soffit level
(d) exposed face of soffit
(e) girth of any cover-fillets, etc.

Having thus established the total girth for paint, it is measured super or linear accordingly.

In either case the length will have to be established, and the usual practice here is to take the total fascia length from the dimensions for the actual fascia. Paint is a relatively cheap item, billed in metres, and meticulous arithmetical gymnastics to arrive at a notional mean length are often not worthwhile. On the other hand, calculation of the girth (width) must always be done with extreme care, as this can affect the description and method of measurement. Typical measurement of eaves totalling 30 metres in length would be as follows:

fascia	·15
	·03
	·03
	·20
	·41

30·00 0·41	*Knot, prime, stop and paint three coats oil colour on general surfaces of woodwork.* *(Externally)*
2/2/0·23 0·40	*boxed ends*

Note 1. The term knot, prime, stop . . . is the usual description of the preparation of wood surfaces for the application of paint. The abbreviation KPS followed by the number of coats written as a digit inside a small circle is a shorthand form sometimes used by takers-off when writing descriptions of paint.

Note 2. Clause W1 (a) requires external work to be separated from internal painting. Hence the word *externally* (in brackets, underlined) should always be written after the description of any painting to eaves or verge construction, or any other external work.

Verge construction

Where verges occur, the tiles may simply be bedded on a plan tile undercloak on top of the gable wall, set projecting 100 mm or so, and pointed up. In this case the last rafter will occur inside the building, and no verge construction will be necessary. Alternatively, however, the ridge and plates may be extended past the gable end (together with purlins perhaps) and finished with a barge-board. Soffit boarding may or may not be provided in connexion with an overhanging verge of this kind.

No difficulty should be found in measuring verge construction; to which the principles of eaves construction will apply. Clause P9 covers verges as well as eaves, barge-boards being measured similarly to fascias under P11, which mentions shaped ends, a familiar feature in the case of barge-boards. Finials are numbered and described, and paintwork to

318 verges is measured in the same manner (and using similar descriptions) as paintwork to eaves.

RAINWATER GOODS

Rainwater goods in simple domestic construction will consist of gutters and down-pipes, together with their fittings. Unless self-coloured, painting must also be measured. Since there will not normally be any superficial items of paintwork, the painting can usually be anded-on to the items of gutters and pipes as measuring proceeds.

Gutterwork is covered by Clauses S2 to S6 of the SMM, and S8(a) deals with rainwater pipes, fittings being enumerated under S11(a). The student should read through these clauses prior to attempting measurement. The usual order of taking-off is as follows:

(a) gutters and painting
(b) gutter fittings
(c) down-pipes and painting
(d) pipe fittings.

Gutters and pipes are measured the mean girth on the centre-line, and it may therefore be convenient to re-use the fascia girth for the gutters, suitably adjusted on waste to bring it out to the centre-line of the eaves guttering on the section. Typical items for gutterwork are as follows:

linear	*100 mm Diameter cast iron half round eaves gutter with socketted joints bolted together in red lead and fixing with standard pattern galvanized steel brackets screwed to fascia.*
	&
	Prepare and paint three coats oil colour on 100 mm diameter half round eaves gutter, inside and out.
number	*Extra over 100 mm diameter eaves gutter for square angles.*

number	*Ditto for stop ends.*
number	*Ditto for* 75 *mm diameter nozzle outlets.*

Note 1. See SM W9(b) for painting on eaves gutters.

Note 2. Angles are referred to in the SMM as 'elbows' (Clause S5). The BSS term 'square angles' is used here instead, but this is a matter of choice. A nozzle outlet is taken at the head of each down-pipe, where it joins the gutter.

Note 3. At one time the practice was to and-on a balloon grating to each nozzle. It is doubtful whether these are very often supplied nowadays however.

It should also be noted that any isolated lengths of gutter not exceeding 2 metres must be enumerated in accordance with SM S3(b).

Down-pipes and fittings

The total length of down-pipes must be carefully measured, the additional amount due to swan-neck bends needed at eaves level being estimated and included in the length, since the swan-necks themselves are measured 'extra over' the pipe. Typical descriptions are as follows:

linear	75 *mm Diameter medium grade cast iron rainwater pipe with socketted joints and projecting ears cast on, and plugging to brickwork with galvanized nails, including making good facings.* & *Prepare and paint three coats oil colour on large pipe.* *(Externally)*
number	*Extra for swan-neck* 200 *mm projection.* & *Extra for rainwater shoe.*

Note 1. For painting on pipes see SM W10. This clause divides pipes into small, large, and extra large, and this terminology must be used in the description.

320 *Note 2.* Where the lower end of rainwater pipes are connected to back-inlet gulleys, an item of 'cement joint of 75-mm pipe to back inlet of gulley' must be taken, in lieu of the rainwater shoe.

Plastic rainwater goods

Cast iron as a material for gutters and down-pipes is rapidly giving way to polyvinyl chloride, and if this material is specified no painting is required to be measured, since it is self-coloured. Fittings for PVC gutters and pipes are measured in the same way as those for cast iron, using similar descriptions. Typical descriptions for the gutters and pipes themselves are as follows:

linear	100 *mm Diameter self-coloured heavy grade rigid PVC eaves gutter with neoprene joints, on and including moulded PVC brackets screwed to fascia at* 1 *metre centres.*
linear	75 *mm Diameter self-coloured heavy grade rigid PVC rainwater pipe with socketted joints, and fixing with moulded PVC pipe clips plugged and screwed to brickwork, including making good facings.*

It is desirable to give the spacing of brackets in the case of plastic gutters, since this is more critical than with cast iron, in order to avoid sagging.

Chapter Twenty-six

Stack Adjustments

During the measurement of a pitched roof for coverings and construction, any openings in the roof for chimney-stacks will have been ignored, to allow the taking-off to proceed unimpeded. When general measurement has been completed, however, the surveyor must check the drawings for the presence of chimney-stacks projecting through the roof, and make the necessary adjustments. For this purpose, a heading of 'Adjustments for Chimney-Stacks' is written in the dimensions, and each stack dealt with in turn.

It will be recalled that the actual stacks themselves were measured for brickwork and facings earlier in the taking-off, and the flue-linings and chimney-pots taken in the 'fires and vents' section. All that remains to be taken in the roofs section therefore, is the adjustment of roof coverings and roof timbering around the base of the stack. This will, of course, involve the provision of metal flashings, etc., at this point; the general order of taking-off being as follows:

(1) Coverings:
  (a) deductions for opening
  (b) linear items to tiling
  (c) provision of back gutter
  (d) leadwork
(2) Construction:
  (e) deduction of roof timbers
  (f) provision of trimming timbers.

Deductions for opening

SM M1(e) states that no deduction of any kind of roofing shall be made for voids not exceeding 0·50 square metre. 'Voids' are defined by SM A2(b) as being 'openings or wants which are wholly within the boundaries of measured areas'. Openings which are *at* the boundaries of measured areas (not wholly *within*) must be deducted irrespective of size. Thus the opening for the chimney-stack shown on the roof plan in Fig. 99(a)

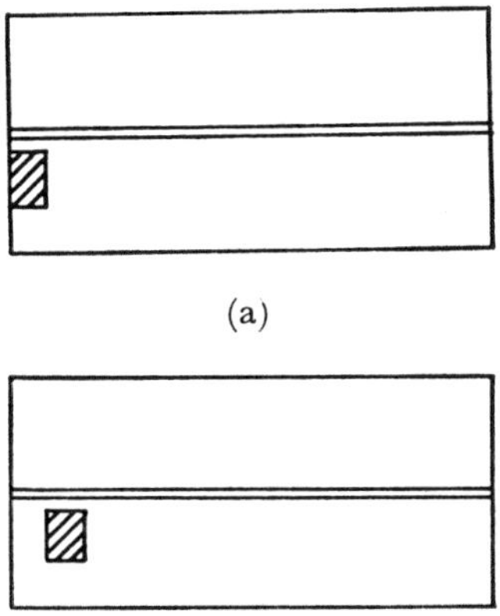

(a)

(b)

Fig. 99

322 is at the boundary of the measured roof area, and must therefore be deducted irrespective of size. The opening in Fig. 99(b), however, is wholly within (i.e., detached from) the boundary and so becomes a 'void' within the meaning of the clause, and not deductable unless over 0·50 square metre in area.

The rule for voids applies to 'any kind of roofing', which is generally held to include underfelting and counter-battening as well. It also applies to the roof boarding under N10(b).

The area for the deduction will not, of course, be the plan area, but the width multiplied by the *length up the slope*. In order to determine this it may be desirable for the surveyor to set up a section through the stack himself (if one is not available), in pencil, to $\frac{1}{20}$ scale. If the position of the trimmers and back gutter are also marked in, this will facilitate measuring generally. Figure 100 shows such a sketch, from which the appropriate deduction of roof coverings may be readily scaled off

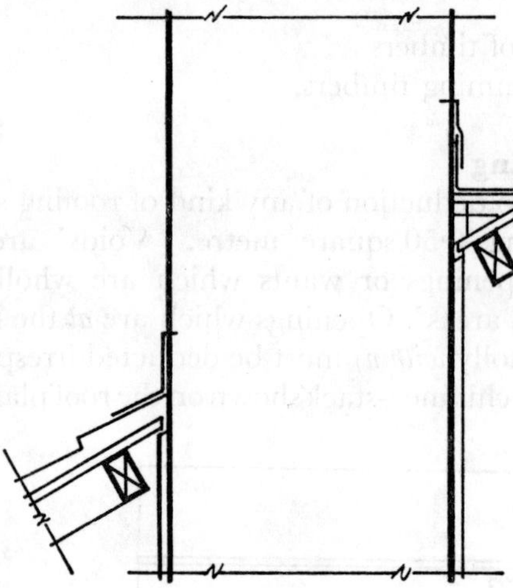

Fig. 100

As we have seen, to be *not* deductable, the opening must possess two qualities. It must be (a) not exceeding 0·50 square metre and (b) detached from the boundary of the roof. If it possesses both these qualities the deduction is ignored, and measurement of the next sections (the back gutter and leadwork) can proceed accordingly. Otherwise the deductions are made, as shown at the top of page 323. Note that in the case of deduction items, only brief descriptions are normally required, sufficient to identify the item previously measured, followed by 'a.b.' (as before). Examination questions sometimes require that full descriptions be given

super	*Deduct*	*Roofing tiles a.b.*
		&
	Deduct	*Counter-battens a.b.*
		&
	Deduct	*Roofing felt a.b.*
		&
	Deduct	25 *mm Boarding to roofs a.b.*

in all cases however, and in examinations the student must read the question-paper carefully and act accordingly.

Having deducted the roof coverings, where appropriate, the linear items on tiling must now be considered.

Linear items on tiling

The three items to be considered are (a) top edges, (b) abutments, and (c) eaves.

TOP EDGES. This refers to the additional work in forming the top edge of the tiling to the front of the stack. Dealt with under SM M4(a) it is only measurable to openings over 0·50 square metre. The inference here is that it is only measurable to openings which have been deducted. Openings not exceeding 0·50 square metre which are not classifiable as 'voids', and have consequently been deducted, are therefore generally considered to be eligible for measurement of top edges in compensation for the deduction.

ABUTMENTS. This is where the tiling abuts the brickwork at the sides of the stack. The provisions of M4(a) apply here also, with the same implications as for top edges.

EAVES. At the back of the stack an eaves course will be necessary, with a tilting-fillet at the rear of the back-gutter (see Fig. 100). Eaves courses are not considered to be measurable where openings have not been deducted; this being the implication of M4(a) (second sentence), which refers eaves to 'large openings' only.

Assuming the opening has been deducted under the rules, the linear items for an opening 0·50 metre wide and 1·25 metres up the slope would be:

324

0·50	*Extra for top edge of tiling.* & *Extra for double course at eaves.* & *Softwood eaves fillet a.b.*
2/1·25	*Square cutting on tiling at abutments.*

Back gutter

This is measured for the width of the stack under SM N14. Bearers are included under N14(b), but gusset ends (triangular filling-pieces at either end) must be enumerated under N14(c), thus:

0·50	175 × 25 *mm Softwood gutter board prepared for lead, including bearers.*
2	*Gusset ends.*

Note 1. Splayed edges to the gutter-board are deemed included under N15(b).

Note 2. Gusset ends are 'written short' in the bill; hence the brief description.

Leadwork

The metalwork around a chimney-stack is usually carried out in sheet lead, as being a material combining flexibility with durability. The work comprises the following:

(a) apron flashing to front of stack
(b) gutter covering and cover flashing at rear of stack
(c) stepped flashings and soakers to both sides of stack.

It should perhaps be mentioned here that a back gutter, with covering and flashing, would not of course, be necessary if the stack were projecting through the ridge, with the roof sloping down from it on either side; neither would an eaves course be appropriate. Instead, a deduction of ridge cappings would be necessary in such a case. Again, if the stack were on the edge of the roof, as in Fig. 99(a) stepped flashings would obviously be necessary to one side only. For the purpose of discussion we are assuming that the stack is positioned as in Fig. 99(b), with the back gutter as shown in Fig. 100.

Apron flashing to front of stack

Aprons are measured linear under SM M55(a), stating the girth. Allowing for 150 mm vertically, and another 150 mm down the slope (over the tiles) a girth of 300 mm will be appropriate. It is the custom to lap the apron 100 mm round the sides of the stack, underneath the stepped-flashing. This lap is a 'passing at angle', and must be added to the length of the apron in accordance with the last sentence of M54(d). For an 0·5-metre wide stack therefore, the measurements would be as follows:

		0·50
passings	2/·10	·20
		0·70

0·70	1·70 *mm Lead apron* 300 *mm girth including wedging.*
0·50	*Rake out joint of brickwork for flashing and point in cement mortar* (1:3).

Note 1. Wedging must be included in the description in accordance with M55(b).

Note 2. See G55(a) for raking out joints and pointing, which must always be measured in such cases; but not so as to include the passings, which will be accounted for when the stepped flashing is measured.

Gutter covering and flashing

The lead lining to the back gutter ranks as a 'flat gutter lining' under Clause M57, rather than a sloping or secret gutter under M58, which refers to gutters going down the roof slope. It is therefore measured super, the girth being carefully measured (lead is an expensive material) from the section. The cover flashing with its pointing is then dealt with under M55(a), as for aprons, as follows:

upstand	·15	*length*		·50
gutter	·18	*ends*	2/·20	·40
up slope	·30			·90
	·63			

326	0·63 0·90	1·70 *mm Lead gutter*	
		passings 2/·10	·50 ·20 ·70
	0·70	1·70 *mm Lead cover flashing* 150 *mm girth including wedging*	
	0·50	*Rake out joints of brickwork for flashing all a.b.*	

Stepped flashings

At the sides of the stack stepped flashings will be necessary owing to the roof being sloping while the brick joints of the stack remain horizontal. Two methods are in general use:

(1) STEPPED COVER FLASHINGS. Here the flashings, which are stepped at the top edge, simply have the lower edge dressed over the tiles. In this case a girth of 300 mm will be necessary.

(2) STEPPED FLASHINGS (FOR USE WITH SOAKERS). In this case the flashing is dressed *under* the tiles, so that the lower edge is not visible. A total girth of 200 mm is sufficient; but soakers are necessary in addition, and require to be measured separately.

In measuring either type of stepped flashing, the slope length, where the roof coverings abut the side of the stack, is scaled, and passings added at top and bottom, to arrive at the overall length of the stepped flashing, which is measured linear under M55(a), but kept separate as a *stepped* flashing (or stepped cover flashing). The following descriptions refer to a stepped flashing (for use with soakers):

	passings 2/·10	2·00 ·20 2·20
2/2·20	1·70 *mm Stepped flashing* 200 *mm girth, including wedging and tacking with lead tacks.*	
2/2·00	*Rake out joint of brickwork for stepped flashing and point in cement mortar* (1:3).	

Note 1. Do not forget to twice the items for both sides of the stack (where appropriate).

Note 2. Tacks are included in the description of the flashing in accordance with M54(c). They would not be required to the short cover and apron flashings previously measured. Tacks are frequently omitted altogether in practice.

Note 3. Pointing to stepped flashings must be so described (see SM G55(a)), since it is the sloping dimension which is measured, not the actual sum of the horizontal grooves.

Stepped cover flashings are similarly measured, but when used with tiles which are not flat (e.g., pantiles, Roman tiles, etc.) mention of this fact must be made in the description, in accordance with M55(d); which incidentally applies to aprons also. It does not apply to ordinary stepped flashings (for use with soakers) however, since these are not dressed over the tiles. Stepped flashings must always be used with single lap tiles, since soakers are not suitable for use with these. A typical description would be:

linear	1·70 *mm Stepped cover flashing* 300 *mm girth, including wedging at top and dressing over double Roman tiles.*

Raking out joints and pointing would follow as usual.

Soakers

These are individual pieces of lead laid underneath each tile, and turned up 50 mm against the stepped flashing. They are only used with double lap tiles, and slates. Single lap tiles demand cover flashings instead.

Soakers are enumerated under M59, stating the size. It will therefore be necessary to establish both the size and the number of soakers. This is done (on waste) as follows:

SIZE OF SOAKERS. For width allow 150 mm under the tiles plus 50 mm turn-up = 200 mm wide. The length of a soaker is given by the formula:

$$\text{length} = \text{lap} + \text{gauge} + 25$$

Thus with plain tiles laid to a 100-mm gauge and a 60-mm lap, the soaker length would be 100 + 60 + 25 = 185 mm long.

NUMBER OF SOAKERS. Since one soaker is laid underneath each tile, the total number will be given by the formula:

$$\text{number} = \frac{\text{distance}}{\text{gauge}},$$

328 the distance being the distance up the slope for which soakers are required. This will be the same as the dimension for the pointing to the flashing. If this dimension is (say) 2 metres, then the number of soakers required would be 2 metres divided by the gauge:

$$2{\cdot}000 \div 0{\cdot}100 = 20 \text{ soakers required.}$$

Typical measurement would be as follows:

lap	·06	·100)2·000(20
gauge	·10	
	·025	
	·185	

2/20	*Supply only* 1·70 *mm lead soakers size* 200 × 200 *mm for fixing by tiler.*
	&
	Fix only soakers, provided by plumber.

Note 1. Soakers must be twiced for both sides in the case of the chimney-stack under discussion.

Note 2. Two separate items are always necessary (a) supplying the soakers under M59, and (b) fixing in accordance with M12. This is because two separate trades are involved, though both items come under the same SMM work section.

ADJUSTMENT OF CONSTRUCTION

Having completed the adjustment of coverings, the roof construction may now be dealt with. A heading of 'construction' should be inserted in the dimensions at this point, and the drawings examined to see what adjustments to the roof timbering are necessary for the chimney-stack in question. The case shown in Fig. 101 would involve:

(a) deduction of common rafters
(b) substitution of trimming rafters
(c) measurement of trimmers
(d) extra labour trimming.

It should be noted that ceiling joists, collars, etc., may also require trimming, and if so must be similarly dealt with afterwards. No deduc-

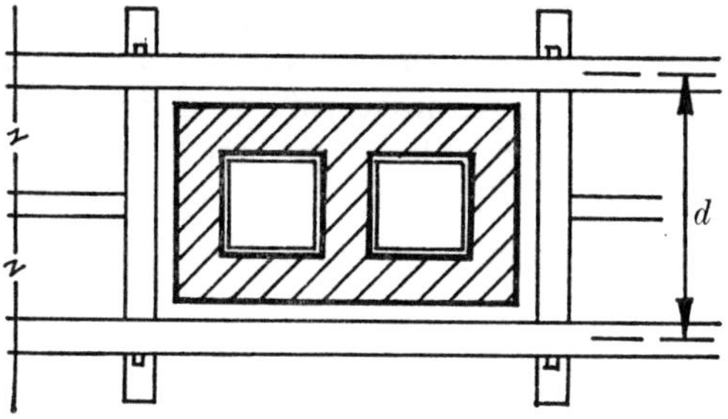

Fig. 101

tion of the ceiling itself will be necessary however, since this will be measured net in the internal finishings section of the work.

Deduction of common rafters

The number of common rafters displaced by a chimney-stack within the boundaries of a roof is given by the following formula:

$$n = \frac{d}{s} - 1$$

where n = number of rafters displaced
d = distance from centre to centre of trimmers
s = spacing of rafters.

It will be apparent from Fig. 101 that the distance d will be the stack-width plus clearance plus two halves of the trimming-rafter thickness.

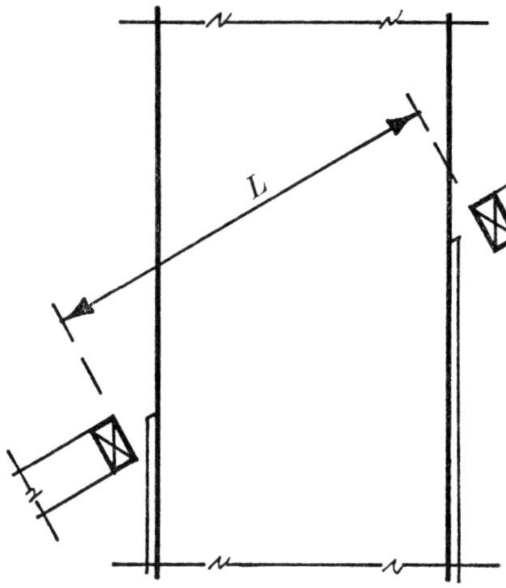

Fig. 102

330 The length of the displaced rafter(s) must now be established, and this is scaled from the section. Since the lengths of common rafter remaining require to be the lengths (of trimmed rafters) as fixed *measured over joints*, the length for the deduction will be the net length, as shown at *L* in Fig. 102. Assuming this length to be 2 metres, and the stack 0·5 metres wide, the measurements would be as follows (400 mm rafter spacing assumed):

	stack		·50
	clearance	2/·05	·10
	to c/l rafters	2/½/·075	·075
			·40)·675(2 − 1
2·00	*Deduct*	130 × 50 *mm Softwood*	
	rafters.		

Note. When dividing by the spacing the answer is always taken to the nearest integer above in case of a remainder, since the spacing is always a specified maximum.

Substitution of trimming rafters

It will have been noticed that ·075 is taken on waste above as being half the trimming rafter thickness, and these 75-mm wide trimming rafters must now be substituted for the 50-mm common rafters they displace, thus:

2/4·50	*Deduct* 125 × 50 *mm Softwood* *rafters*
	&
	Add 125 × 75 *mm ditto*
	[*trimming rafters*

Note 1. The rafter length is taken direct from the roof construction dimensions, previously measured.

Note 2. Trimming timbers are not separately described as such, the functional description being relegated to a side-note.

Measurement of trimmers

These are measured in the same way as floor trimmers, an allowance of 50 mm being made for the tusks:

	stack		·50
	clearance	2/·05	·10
	rafters	2/·075	·15
	tusks	2/·150	·30
			1·05

2/1·05	125 × 75 *mm Softwood rafters.* (*trimmers*

Labour trimming

This is measurable as in the case of floor trimming, under SM N6(e)(vi):

1	*Extra labour trimming* 125 × 50 *mm rafters around opening size* 2·00 × 0·50 *metres.*

Other roof adjustments

In addition to stack adjustments, other adjustments to domestic roofs may also be necessary. Lantern-lights, dormers, skylights and trap doors are possibilities which may arise. Lantern-lights and dormers are not considered to come within the curriculum of elementary studies however; skylights and trap doors, requiring as they do a knowledge of joinery and internal finishing work, do not come within the scope of this particular volume.

Chapter Twenty-seven

Flat Roofs

Like pitched roofs, the measurement of flats is divided into (a) coverings, (b) construction and (c) eaves and rainwater goods. In domestic work the construction is generally confined to either timber joist or in-situ reinforced concrete construction, and in the United Kingdom flat roofs are usually reserved for secondary roofs over garages, carports, outbuildings, canopies, etc., dealt with after the main roof has been measured complete. The coverings to such flat roofs could in theory be any one of the following:

(a) lead
(b) zinc
(c) copper
(d) aluminium
(e) asphalt
(f) felt.

The type of covering may to some extent govern the construction; but we can simplify the general picture at this stage by eliminating (a) lead and (b) zinc as being obsolescent, and unlikely to arise except perhaps for very small canopies, and also (c) copper and (d) aluminium as being only used to a limited extent in domestic work, and in any case not coming within the scope of an elementary curriculum.

Asphalt and felt are the two forms of covering commonly used in domestic flat roof construction. Of the two, asphalt is possibly the more durable, but the post-war development of reliable bitumen-felt roofing techniques and materials has made this form popular for subsidiary domestic flats, owing to its relative cheapness.

BITUMEN-FELT ROOFING

Temporary buildings, or garden sheds, etc., detached from the main building may be roofed with single-layer felt. Flat roofs attached to the main building, however, call for multi-layer bituminous-felt roofing, and in this case each layer (three layers is usually specified) is bonded by means of bituminous compound applied hot. The top layer of felt (known as the capsheet) may be plain, or alternatively 'mineral surfaced', i.e., covered with granules at the factory in order to lengthen its life. If a plan capsheet is used, it may be surfaced in-situ with a layer of chippings bedded in hot bitumen, to protect the roof and add to its durability. Alternatively, if foot traffic is to be allowed for, the capsheet may be covered with tiles, macadam, or similar surface coverings.

Method of measurement

Whichever specification is called for, the work is measured in accordance with SM M39–45, where it will be seen that coverings over 300 mm wide are measured super (M40(a)); while flashings, aprons, turn-downs, etc., are measured linear (M42). Flashings may be of metal, but frequently skirtings, etc., are formed in the felt itself. From the point of view of measurement of coverings a felted flat roof is likely to consist of the following main parts:

(a) areas of roofing
(b) verges
(c) eaves
(d) abutments,

and this order is a convenient one to use for taking-off purposes.

Areas of roofing

These should be measured the extreme roof dimensions in accordance with M1 (e), including the thickness of any drip battens at eaves and verges, since the multi-layer covering will extend over these. In practice bitumen-felt roofing is often laid by a specialist sub-contractor, and the descriptions may include the catalogue reference of an approved specialist. Alternatively, the sub-contractor may be nominated, and made the subject of a P.C. item, or P.C. measured items. Failing this however, and having regard to likely examination requirements, full descriptions in accordance with the SMM must be given.

To avoid repetition of specification notes in the descriptions it is usual to start with a heading, under which the items are measured, the heading being subsequently transferred to the bill:

The following in bitumen-felt roofing of bitumen asbestos underlays finished with mineral surfaced capsheet, each layer lapped 75 mm at joints and bonded with bitumen bonding compound applied hot:

super | *Three layer flat coverings to falls, laid on softwood boarding (measured separately).*

&

25 mm Softwood tongued and grooved flat roof boarding on and including shallow firrings to falls.

It is customary to and-on the roof-boarding in this way. It will be abstracted-out from under the heading and billed with Carpentry Work (see SM N10–15). Note that N11(b) provides for shallow firrings (i.e., those not exceeding 50 mm average depth) to be included in the description, but those over 50 mm average depth to be measured separately. BS CP 144.101 gives a minimum fall of 25 mm in 1·5 metres for felt roofing. Allowing 12 mm minimum thickness at the lower end, the average thickness of the firrings for a 3-metre width of roofing falling sideways would be calculated on waste as follows:

3·00 =	50 *mm fall.*	
min.		·012
max.	·050	
min.	·012	·062
		2)·074(·037 *av.*

Should the average exceed 50 mm, the firrings must be measured separately and described as:

linear	50 *mm Softwood firrings to falls, average* 75 *mm deep.*

Roof screeds

Should the roof construction be of concrete instead of timber, a screed will require to be anded-on to the coverings, in lieu of the boarding. In this case the description of the coverings will need amending to read '. . . laid on screed (measured separately)', for M39(b) requires the nature of the base to be stated.

The screed itself is measured under SM U28(b), which refers to 'floated beds' for bitumen-felt roofing. The thickness must be stated under U27(b), and this will be an average thickness where the roof is to falls, as will normally be the case. This average thickness is obtained in a manner similar to the average thickness of the firrings given above, except that a minimum thickness of 20 mm would be more appropriate in the case of a screed; but the drawings and specification notes must be examined on this point.

The screed (or floated bed as it is more properly called, according to the SMM) is described as follows:

50 *mm (average) cement and sand* (1:4) *floated bed, top surface to falls.*

Verges

A typical verge detail is shown in Fig. 103. A splayed fillet is used to form a check, in order to stop rain driving over the edge of the verge. A drip batten is shown nailed to the fascia, but this is sometimes omitted. The

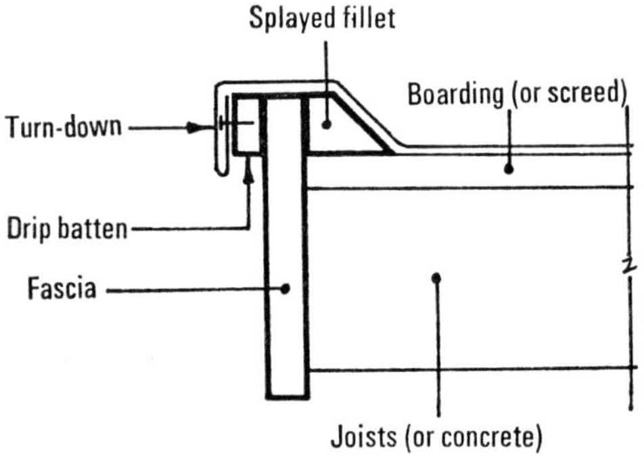

Fig. 103

capsheet is carried over the splayed fillet and top of the fascia and turned down to form an apron. The apron is folded inwards (or 'welted') and nailed to the batten. This is referred to as a 'turn-down at verge' in SM M42(a), and is measured linear, *angles and ends deemed included.* Typical descriptions would be as follows:

<u>*linear*</u>	*Extra for turn-down of capsheet* 75 *mm girth to verge, including dressing three-layer felt over check-fillet at top, and welted drip nailed to softwood at bottom edge.*
	&
	75 × 50 *mm Softwood splayed check-fillet, including mitres and ends.*
	&
	50 × 25 *mm Softwood drip batten including ditto.*

The check-fillet and drip batten are usually anded-on here, and the fascia itself left for measuring in the 'eaves and rainwater goods' section. Note that the detail would probably be the same if the roof were concrete instead of timber, and verge descriptions would also be the same in this case.

Eaves

Again a turn-down will be required, as shown in Fig. 104, which gives a typical eaves detail, descriptions for which would be:

linear	*Extra for turn-down of capsheet* 75 *mm girth to eaves, including welted drip nailed to softwood at bottom edge.*
	&
	50 × 25 *mm Softwood drip batten including mitres and ends.*

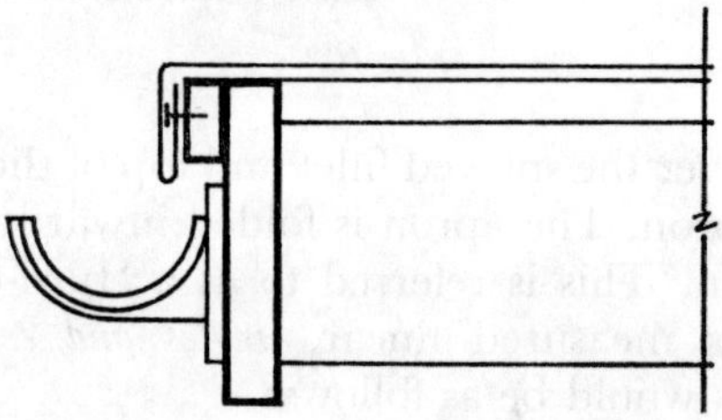

Fig. 104

Abutments

Where the flat roof abuts a vertical brick wall, a skirting will be necessary. This involves a turn-up tucked into a groove at the top edge. Angles and ends are again deemed included under M42:

linear	*Three-layer skirting* 150 *mm high, including wedging into groove at top edge and bonding into roof coverings at junction with flat.*
	&
	Rake out and enlarge joint of brickwork for turn-in of three-layer felt skirting and point in cement mortar (1:3).

Note. Three-layer felt would require more than just raking-out of the joint; it would require to be enlarged, more in the form of a chase, as for asphalt.

Sometimes a separate metal flashing is specified, to cover the top of a skirting or upstand at abutments. This would be measured linear under M55, e.g.,

linear	1·22 *mm super-purity aluminium flashing* 150 *mm girth, and lapped* 100 *mm at joints, bent and wedged into groove at top edge (no allowance for laps).*

Rake out joint and pointing would be taken as usual.

Triangular fillets of timber or cement may be encountered at the junction between flats and abutting walls, and these must be measured separately and described; cement angle-fillets are dealt with under G51. It should be noted that any holes through roof coverings for pipes, etc., are not measured at this stage, but when the pipes themselves are measured, i.e., with rainwater goods, or plumbing, as appropriate.

Having completed measurement of bitumen-felt roof coverings, the heading is closed thus:

(End of bitumen-felt roofing)

and the taker-off can then proceed with measurement of the construction.

ASPHALT ROOFING

As in the case of bitumen-felt roofing, the order of taking-off for asphalt covering to flats will be:

(a) areas of roofing
(b) verges
(c) eaves
(d) abutments.

Two main types of mastic asphalt are available for roofing, (a) natural rock aggregate and (b) limestone aggregate. The type selected should be stated in accordance with SM L1(b)(i) (kind and quality of asphalt).

It should be noted that under L1(a) asphalt work is to be given under an appropriate heading. We saw that such a procedure is desirable in the

338 case of bitumen-felt. It is mandatory in the case of asphalt; a suitable heading for roofing work being as follows:

The following to comply with
the British Standard for
Mastic Asphalt for Roofing
(natural rock aggregate):

after which the areas of roofing can be measured, taking extreme roof dimensions, as for bitumen-felt. An underlay is normally provided to isolate the asphalt from the structure, and this is to be included in the asphalt description under SM L1(c). The screed is anded-on, as for felt roofing, a minimum fall of 75 mm in 3 metres being necessary in the case of asphalt. Typical descriptions are as follows:

super	*20 mm Asphalt covering to flat in two coats laid to falls on screed (measured separately) on and including felt underlay.*
	&
	45 mm (average) Cement and sand (1:4) floated bed, top surface to falls.

Note 1. Both the thickness and the number of coats are required to be stated in the description under L1(b) as is the nature of the base. Where foot traffic is to be catered for, thicker asphalt may be required, two coats being still sufficient unless the roof is designed as a reservoir to hold water. In any case the specification notes must be read carefully and followed.

Note 2. Surface treatment, if with chippings, gravel, etc., is to be included in the description under L1(b). If no treatment is mentioned, the estimator will assume a lightly sanded self-finish, which may be regarded as the normal requirement.

In the event of an asphalt-covered timber flat being encountered, the roof-boarding and firrings would require to be dealt with in lieu of a screed at this stage. Procedure would be similar to the felted roof previously described.

Verges

An apron of some kind will be required here, and the detail must be

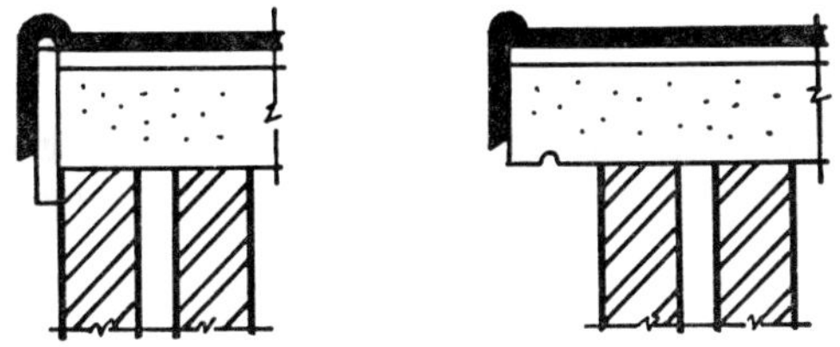

(a) (b)

Fig. 105

carefully studied. There may or may not be a timber fascia, and the check-roll may be a timber one, or formed solid in the asphalt itself. Two possible cases are shown in Fig. 105. The detail in Fig. 105(a) would be dealt with as follows:

linear	*Asphalt apron* 150 *mm high in* 2 *coats, including extra labour and material dressing over check-roll at junction with flat, and forming drip at bottom edge; on and including strip of expanded metal lathing stapled to softwood fascia.*
	&
	38 × 25 *mm Softwood rounded check-roll.*
number	*Extra for angles to last apron.*

Note 1. See SM L5(a) for asphalt aprons. Vertical asphalt will not adhere to timber without the provision of EML reinforcement, which must be included in the description of the asphalt under L1(c).

Note 2. Angles are enumerated, but ends are deemed included under L5(a).

Dealing now with the verge shown in Fig. 105(b), the following description would be suitable:

linear	*Asphalt apron* 150 *mm high in* 2 *coats on concrete, including extra labour and material forming solid asphalt check-roll* 30 *mm diameter at junction with flat, and forming drip at bottom edge.*

340 Any angles would be measured as before, but ends deemed included. In general, vertical asphalt on concrete requires the concrete to be hacked to form a key, and this is a measurable item under SM F62. In the above case, however, it is likely that removal of sawn shuttering would leave a sufficiently rough surface to provide an adequate key for a narrow apron of this kind.

In framing descriptions for asphalt aprons and the like, it may be helpful to note that the general order of describing such work is:

(1) describe the main item
(2) describe its top edge
(3) describe its bottom edge
(4) finish with any reinforcement, etc.

Eaves

The usual practice here is to dress a strip of metal under the asphalt and down over a drip batten into the gutter, as in Fig. 106. Measured under M55, the description would be as follows:

<u>*linear*</u>	1·70 *mm Lead apron* 300 *mm girth lapped* 100 *mm at joints, one edge double-welted and dressed under asphalt roofing, other edge dressed over batten in eaves gutter (no allowance for laps).*
	&
	50 × 25 *mm Softwood drip batten, including mitres and ends.*
	&
	Fair splayed edge and arris to asphalt covering.

Both the drip batten and the labour to edge of asphalt are anded-on, the asphalt labour being measurable under SM L4.

Abutments

Three items will usually require to be measured where an asphalt flat abuts brickwork:

(1) asphalt skirting with angle-fillet and turn-in (L5(a))

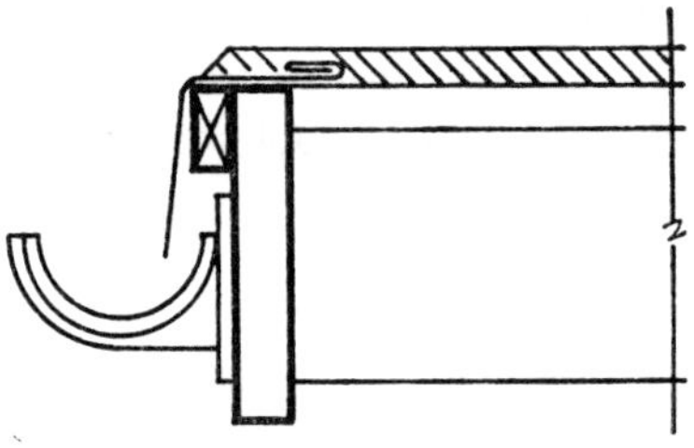

Fig. 106

(2) rake out joints of brickwork as key (G48(a))
(3) enlarged joint and pointing (G55(b)).

In addition, a metal flashing may be specified to the top of the skirting, and if so this will require to be measured also. Items for the skirting, etc., would be as follows:

linear	*Asphalt skirting* 150 *mm high in* 2 *coats on brickwork, with turn-in at top, and angle-fillet at bottom at junction with flat.*
	&
	Rake out joints of brickwork as key for vertical asphalt.
	super × 0·15 =
	&
	Rake out and enlarge joint of brickwork for turn-in of asphalt skirting, and point in cement mortar (1:3).

Any angles to the skirting would then follow; ends are deemed to be included.

It might be as well to note here that 150 mm should be regarded as a *minimum* height for skirtings, and where they run in the direction of the fall of the roof, an *average* height must be calculated on waste, taking 150 mm at the upper end. The description would then be 'Asphalt skirting average 200 mm high . . . etc.', or as appropriate. These remarks apply also to bitumen-felt skirtings.

342 Having closed the heading by inserting 'end of asphalt roofing' (in brackets, underlined) in the dimensions, the surveyor is now ready to proceed with the measurement of flat roof construction.

FLAT ROOF CONSTRUCTION

Flat roofs occurring in connexion with domestic work are likely to be constructed of either timber or in-situ reinforced concrete. Timber roof construction is likely to present little difficulty to the student who has mastered timber floor and pitched roof construction. Members will consist of plates and joists measured in the usual way; the boarding or other coverings having already been taken in the 'coverings' section. Eaves and verge construction will also be straightforward, comprising measurement of fascias, etc., and the paintwork. Rainwater goods will then follow all as for pitched roofs. At one time lead and other flat roof construction involved parapet gutters, with cesspools and rainwater heads, etc. This form of construction may be regarded as obsolescent today in domestic work, and will not be dealt with here.

Reinforced concrete flats

Order of taking-off for in-situ suspended slabs will be as follows:

(1) concrete
(2) formwork
(3) reinforcement
(4) beam adjustments

Beams supporting slabs will not be a very frequent occurrence in simple domestic construction, but they may occur over wide openings, e.g., the doors of double-garages. Beams of this kind, which form in-situ lintols over openings will be adjusted in the 'openings' section of the taking-off, rather than with roofs. Nevertheless, beams projecting from the underside of slabs merit discussion in connexion with the measurement of in-situ slabs generally, and will thus be dealt with in this chapter, in so far as elementary slabs may be affected.

Concrete

The measurement and description of elementary reinforced concrete ground slabs has already been considered in sub-structures, and suspended roof slabs will follow much the same general principle. They are measured superficially under SM F7. Any beams projecting from the underside (or kerbs, etc., projecting upwards) are ignored in the first instance; the slab being measured straight across, stating its thickness.

In the event of a variation in the thickness of the slab, due to its top surface being to falls (or 'cross-falls', i.e., falling in more than one direction) the *average* thickness must be calculated on waste, and stated in the description. Most likely, however, falls will be achieved by means of a separate screed, or 'floated bed', which will have already been accounted for in the 'coverings' section of the work. A typical description for a roof slab would be as follows:

<u>*super*</u>	150 *mm thick reinforced concrete suspended roof.*

Note 1. It will be recalled that SM F2 requires plain and reinforced concrete to be separated. Reinforced slabs must therefore be so described.

Note 2. A tamped finish to receive a screed will be assumed by the estimator as normal in the case of suspended slabs, unless you state otherwise.

Slabs must be measured from edge to edge, and the drawing carefully studied to determine the exact extent of the slab, suitable allowances being made in respect of its bearing on the brickwork; 110 mm being allowed in respect of any edge supported by a chase in an abutting brick wall. Here it should be noted that chases in brickwork (in new work) for edges of roofs are not measurable, but deemed included under SM G11. It is also worth remembering that a suspended slab is sometimes designed as a continuous beam spanning between two parallel walls, and may not require any support along the other two sides.

Formwork

This must be measured to 'the actual surfaces of the finished structure which require to be supported during the deposition of the concrete', to quote SM F20(a). In the case of a roof slab, this will usually involve formwork:

(a) to the soffit of the slab, and
(b) to the edges of the slab.

Formwork to soffits is measured super under F21(d). Only the net area of the concrete to be supported is measured. The area of concrete resting on brickwork at the edges of the roof will be supported by the brickwork, and will not therefore require any formwork. For this reason, the formwork to soffits cannot normally be anded-on to the concrete, since the areas will differ.

It will be seen from F20(a)(i) that voids in formwork not exceeding one square metre are not deducted. A void is by definition wholly within

344 the area of the formwork (A2(b)). Therefore deductions in respect of piers attached to the supporting walls should always be made, irrespective of size. Notches, however, are deemed to be included under SM F26. Is a deduction for a small pier a 'notch' in the formwork, and therefore not deductable? If it is so considered, a side-note to the effect should be made in the dimensions.

Any projecting beams are ignored at this stage, formwork being measured straight across.

Example

PROBLEM

Measure the formwork to the soffit of the roof slab shown in Fig. 107.

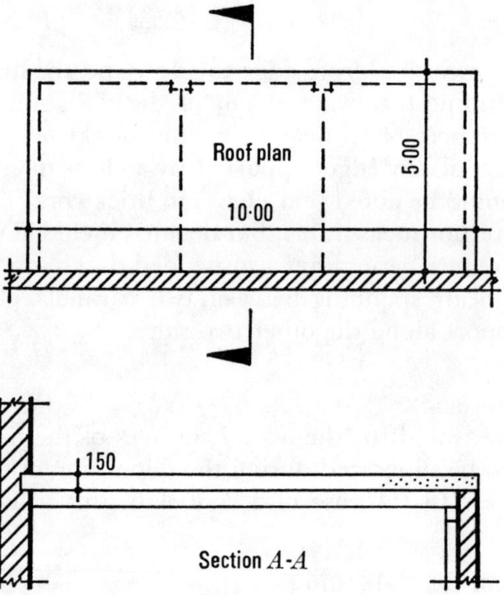

Fig. 107

SOLUTION

		5·00
	bearing	·22
		4·78
		10·00
bearings	2/·22	·44
		10·44

10·44 4·78	*Sawn formwork to horizontal soffits.*
2/0·33 0·11	*Deduct* *Ditto* (*piers*

Note 1. Formwork is described as 'wrought' (F20(f)) instead of sawn, if it is to be left exposed. Internal plastering has been assumed here.

Note 2. Formwork to soffits over 3·50 metres high must be kept separate in stages of 1·50 metres under F21(b).

Formwork to edges

The vertical edges of the slab, where unsupported by brickwork, etc., must now be measured for formwork. These are measured under F22(a), linear stating the width. Thus the edges of the slab in Fig. 107 would be measured for formwork as follows:

10·00 2/ 5·00	*Sawn formwork* 150 *mm wide to vertical edge of roof slab.*

Note. 'Sawn' formwork is again taken here, on the assumption that the concrete is to be covered by a timber fascia, and thus not left exposed to view.

Formwork to projecting eaves

Take the case shown in Fig. 108. Here the eaves (or verges) of the slab

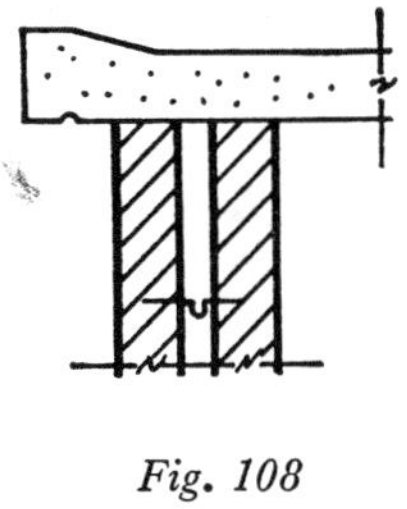

Fig. 108

project over the edge of the walls. Formwork to soffits will have been measured to the inside face of the wall, leaving the vertical edge and the narrow eaves soffit still remaining to be taken.

Eaves of this kind (together with verges) are measured under SM F22(b), linear stating the girth of the edge and soffit, thus:

linear	*Wrought formwork to eaves of concrete roof comprising*

(description continued overleaf)

346 *vertical edge* 150 *mm high and*
soffit 150 *mm projection, including*
fillet planted on to form
throat.

There is also the problem here of formwork necessary to bridge the cavity, in cases like the one shown. The usual solution is to do this with slates, which are subsequently left in position, the item being measured linear:

<u>*linear*</u> *Course of slates* 150 *mm wide*
bedded horizontally on
top of hollow walls as
formwork to bridge
cavity.

Kerbs and upstands

These are measured linear under SM F12. The slab will have been measured through, and only that concrete projecting above the general surface will require to be accounted for, together with its formwork. Formwork to sides of kerbs and upstands is measured linear under F22(a)(iii), but sometimes the shape of the item in question justifies including the formwork (if any) in the concrete description—a useful trick to remember when measuring projections on concrete of a rather odd shape. Thus the upstand in Fig. 108 might be measured as follows:

<u>*linear*</u> *Extra labour and material*
forming upstand 230 *mm wide*
and 50 *mm high in concrete*
(1:2:4) *to verge of*
suspended roof slab, with
splay 150 *mm wide on inner*
edge, including any
necessary formwork.

Reinforcement

Measurement and description of bar and fabric reinforcement to roof slabs follows the general principles of reinforcement discussed in the section of this book dealing with substructures. It will be recalled that bar reinforcement is measured linear, the extra girth due to hooked ends, etc., being added on waste; weighing of the bars being carried out at abstracting stage. In general, each diameter must be kept separate (for weighting) and the location of the bars given in accordance with the classifications in F17(b).

Fabric reinforcement is measured under F18; super in normal instances, but linear in the case of any tension strips under F18(c). Notching fabric round obstructions is enumerated under F19(b), irrespective of size.

A suitable description for bar reinforcement to a concrete flat would be:

linear	12 *mm diameter mild steel bars in roof slabs.*

Note. Avoid stating the spacing of the bars, as this is irrelevant. Any links or stirrups must be separately described under F17(g).

Fabric reinforcement would be dealt with as follows:

super	*Fabric reinforcement reference B*503, *lapped* 150 *mm at joints, and laying in roof slab.*
number	*Notch ditto around obstructions.*
	(piers

Beam adjustments

Having measured the slab for concrete, formwork and reinforcement, the surveyor must now inspect for the presence of any beams projecting below its surface, and make the necessary adjustments. Should any occur in domestic construction, they will almost certainly be simple steel beams, encased in concrete.

The measurement of steel beams encased in concrete was dealt with in the chapter on brickwork, earlier in this book. The following example should suffice to show how such beams are adjusted should they occur in connexion with a concrete flat roof.

Example

PROBLEM

Take off the beam adjustments to the roof shown in Fig. 107. Assume 200 × 100 × 30 kg joists projecting below the slab.

SOLUTION

	5·00
	·11
	4·89
bearing	·11
	5·00

348	2/5·00	*Plain girder of joist section, over* 20 *but not exceeding* 50 *kilogrammes per linear metre.* × 30 *kg* =	
	2/2/1	220 × 220 × 150 *mm Precast concrete* (1:2:4) *padstone and building in.* & *Cut and pin or build in end of small steel section to brickwork.*	
		beam ·20 *cover* 2/·05 ·10 ·30 *slab* ·15 *depth* ·15	*beam* ·10 *cover* 2/·05 ·10 *width* ·20
	2/5·00 ·20 ·15	*Concrete* (1:2:4) *in horizontal beam casing not exceeding* 0·05 *square metre in sectional area.*	
			sides 2/·15 ·30 *soffit* ·20 ·50
	2/5·00 ·50	*Sawn formwork to sides and soffits of horizontal beam casings.*	
	2/4·78 ·20	*Deduct* *Sawn formwork to horizontal soffits a.b.* (*slab soffit*	

Eaves and rainwater goods

As with timber flat construction, eaves and rainwater goods to concrete flats will present little difficulty to the student who has mastered eaves construction to pitched roofs. Fascias will require to be described as 'plugged' if they are fixed direct to the concrete, as well as any other timber members, fillets etc. Sometimes a fixing fillet may be cast into the concrete, in which case it should be described as (e.g.) '50 × 38 mm softwood twice splayed fixing fillet cast into concrete'. Since eaves and verge construction to flat roofs may well consist of little more than a fascia round the edge, the paintwork can usually be anded-on to the item concerned.

After completing the measurement of any fascias, etc., to eaves and verges, together with paintwork, rainwater goods are then dealt with in the usual way.

Appendix One

Constructional Elements Checklist

It is considered important that the student, at an early stage in his studies, should be able to form a concept of the boundaries of the subject which he is studying, since this helps him to get his problems into perspective, and to relate them to the subject as a whole.

What follows may be regarded as an attempt to define the boundaries of knowledge in the field of dimension-preparation required by the student of quantity surveying. Like all such boundaries, this one is constantly expanding, as new materials and constructional forms come into use within the industry. It is emphasized therefore that the resultant table makes no claim to be completely exhaustive, but is intended rather as a guide to indicate the scope of work entailed in the present-day quantity surveying curriculum, and to serve as a checklist to help students and lecturers to ensure adequate coverage of the subject at each level.

This checklist is based on the syllabus of the RICS examinations current in 1970. This syllabus is a broadly worded document, and the checklist is the interpretation of the author, based on his experience as a lecturer and examiner. The allocation of the taking-off elements as between Intermediate, Final Part 1, and Final Part 2 material therefore, should not be interpreted as official RICS examination policy, but merely as a guide to the student showing the approximate scope of work in each section of the syllabus.

The constructional elements are, of course, *taking-off* elements, not to be confused with the structural elements into which an elemental bill is subdivided. They are annotated as follows:

I = Intermediate Syllabus
F1 = Final Part 1 Syllabus
F2 = Final Part 2 Syllabus.

CONSTRUCTIONAL ELEMENTS CHECKLIST

GROUP 1. SUBSTRUCTURES

(a)	Strip foundations	I
(b)	Stepped foundations	I
(c)	Deep strip foundations	I
(d)	Raft foundations	I
(e)	Short bored piling	F1
(f)	Basements	F1
(g)	Stancheon-base foundations	F1
(h)	Composite foundations[1]	F1
(i)	Sloping site foundations[2]	F1
(j)	Underpinning	F2
(k)	Sheet piling	F2
(l)	Timber piling	F2
(m)	Precast concrete piling	F2
(n)	In-situ concrete piling	F2

GROUP 2. BRICKWORK AND FACINGS

(a)	Pitched-roof construction with stacks	I
(b)	Flat-roof construction with parapets	I
(c)	Hollow-wall construction with projections	I
(d)	Plinths and string courses	I
(e)	Fence walls with piers and copings	I
(f)	Battering walls and circular work	I
(g)	Boiler flues and chimney shafts	F1
(h)	Ornamental brickwork	F2
(i)	Half timbering	F2

[1] Any combination of the preceding types.
[2] Involving problems of reduced-level excavation.

GROUP 3. PARTITIONS

(a)	Blockwork	I
(b)	Timber stud	I
(c)	Timber trussed	F1
(d)	Slab partitioning	F1
(e)	Glass-block work	F2
(f)	W.C. cubicles	F2

GROUP 4. FIRES AND VENTS

(a)	Ground-floor slabbed fireplaces	I
(b)	Upper-floor ditto	I
(c)	Domestic air vents	I
(d)	In-situ fire surrounds	F1
(e)	Panel fires and stoves	F1
(f)	Corner fireplaces	F1
(g)	Ventilating turrets	F2

GROUP 5. FLOORS

(a)	Timber ground floors	I
(b)	Timber upper floors	I
(c)	Timber framed floors	F1
(d)	Concrete ground floors	I
(e)	Filler-joist floors	F1
(f)	Hollow-block suspended construction	F1
(g)	Contractor-designed construction	F1
(h)	Reinforced concrete suspended floors	F1

GROUP 6. ROOFS

(a)	Lean-to	I
(b)	Simple pitched roofs	I
(c)	Purlin roofs	I
(d)	Roofs with tapering gutters	I
(e)	Roofs with varying pitches	F1
(f)	Thatching	F2
(g)	Lead flats (obsolescent)	I
(h)	Copper flats	F1
(i)	Zinc flats (obsolescent)	I
(j)	Aluminium flats	F1

352 (k)	Asphalted concrete flats	I
(l)	Felted concrete flats	I
(m)	Asbestos sheet roofing	I
(n)	Asbestos and other roof decking	F1
(o)	Shell concrete barrel vaulting	F2
(p)	Paraboidal construction	F2
(q)	Space frames	F2

GROUP 7. ROOF WORK

(a)	Stack adjustments	I
(b)	Skylights	I
(c)	Dormers	F1
(d)	Lanterns	F1
(e)	Eyebrows	F2
(f)	Dome lights	F1
(g)	Patent roof glazing	F1
(h)	Traditional timber trusses	F1
(i)	Timber batten trusses	I
(j)	Turrets, spires and domes	F2

GROUP 8. STEELWORK

(a)	Plain girders	I
(b)	Unframed construction	F1
(c)	Plain stancheons	F1
(d)	Compound stancheons and grillages	F1
(e)	Multi-storey frames, with stancheon- and beam-casings, etc.	F1
(f)	Welded construction	F1
(g)	Castellated girders and portal frames	F2
(h)	Trusses	F1
(i)	Staircases	F2

GROUP 9. STONEWORK

(a)	Rubble walling	I
(b)	Simple stone dressings	F1
(c)	Dressed walls and piers	F1
(d)	Stone facework	F1
(e)	Dressed stone facades	F1
(f)	Tracery windows and oriels	F2
(g)	Columns, balustrading and ornamental work	F2

GROUP 10. REINFORCED CONCRETE CONSTRUCTION

(a)	Basements and ducts	F1
(b)	Columns beams and slabs	F1
(c)	Retaining walls	F1
(d)	In-situ and precast staircases	F1
(e)	Open-riser staircases	F1
(f)	Trusses	F2
(g)	Prestressed concrete beams and slabs	F2
(h)	Chimney shafts	F1
(i)	Swimming baths	F2
(j)	Use of bar-bending schedules	F1

GROUP 11. CLADDING

(a)	Timber cladding	I
(b)	Timber curtain walling	F1
(c)	Metal curtain walling	F2
(d)	Asbestos cladding	F1
(e)	System building	F2

GROUP 12. INTERNAL FINISHINGS

(a)	Simple plaster finishings with skirtings beams and cornices	I
(b)	Wall tiling and papering	I
(c)	Floor finishings in tiles and granolithic	I
(d)	Terrazzo and mosaic work	F1
(e)	Ornamental plastering	F2
(f)	Suspended ceilings	F2
(g)	Fibrous plaster coves and cornices	F1
(h)	Wall panelling	F1
(i)	Ornamental joinery	F2
(j)	Use of schedules of finishings	F1

GROUP 13. WINDOWS

(a)	Timber casements	I
(b)	Timber sliding sashes	I
(c)	Windows with mullions and transomes	I
(d)	Centre-hung lights	I
(e)	Bulleye sashes	F1

(f)	Windows with shaped heads	I
(g)	Metal casements	I
(h)	Composite metal windows; metal trim; opening gear	F1
(i)	Oriel windows and bays	F1
(j)	Double glazing units	F2
(k)	Leaded lights and shutters	F2
(l)	Use of window schedules	I

GROUP 14. DOORS

(a)	Panelled doors	I
(b)	Flush doors	I
(c)	Glazed doors	I
(d)	Ledged and braced doors	I
(e)	Casement doors with sidelights	I
(f)	Entrance doors with steps	I
(g)	Double doors and double-margin doors	I
(h)	Glass doors	F2
(i)	Swing doors	F1
(j)	Sliding and other garage doors	F1
(k)	Revolving doors	F2
(l)	Shaped and curved doors	F1
(m)	Use of door schedules	I

GROUP 15. OPENINGS

(a)	Rough and fair arches	I
(b)	Rebated reveals	I
(c)	Hollow-wall adjustments	I
(d)	Finishings adjustments	I
(e)	Boot lintols	I
(f)	Canopies and porticos	F1

GROUP 16. JOINERY FITTINGS AND STAIRS

(a)	Cupboards and shelving	I
(b)	Counter fittings and bookcases	F1
(c)	Standard joinery units	I
(d)	Draining-boards and working-tops	I
(e)	Glazed screens	F1
(f)	Shop fittings	F2

(g)	Shopfronts	F2
(h)	Laboratory benches	F2
(i)	Gymnasium fittings	F2
(j)	Fire-fighting equipment and keyboxes	F2
(k)	Coat-rails and door-stops	I
(l)	Closed-string staircases	I
(m)	Open-string staircases	I
(n)	Panelled balustrading	F1
(o)	Geometric and spiral staircases	F2
(p)	Bars and pulpits	F1

GROUP 17. PLUMBING

(a)	Domestic schemes in iron, lead and copper	F1
(b)	Showers, urinals, ranges of fittings	F1
(c)	Plastic tubing and fittings	F1
(d)	Hospital and laboratory work	F2
(e)	Water mains and dry-risers	F1
(f)	Plumbing to multi-storey blocks	F1
(g)	Restaurant kitchen fittings	F2

GROUP 18. DRAINAGE

(a)	Domestic schemes with depths given	F1
(b)	Schemes with levels given	F1
(c)	Iron drainage and suspended systems	F1
(d)	Land drains and road drainage	F1
(e)	Large schemes with tumbling-bays, etc.	F1
(f)	Sewers and sewage disposal plant	F2
(g)	Pitch-fibre drainage	F1
(h)	Use of drainage schedules	F1

GROUP 19. SERVICES

(a)	Heating and hot water installations	F1
(b)	Ventilation and plenum systems	F2
(c)	Gas supplies	F2
(d)	Electrical work	F2
(e)	Lifts and escalators	F2
(f)	Lightning conductors and television aerials	F2
(g)	Telephone installations and bell systems	F2

(h) Ticket machines and turnstiles F2
(i) Amplifier and public address systems F2
(j) Firefighting and sprinkler installations F2

GROUP 20. EXTERNAL WORKS

(a) Estate roads in concrete and tarmacadam F1
(b) Paths and paved areas F1
(c) Grassed areas; spoil heaps; earth moving F1
(d) Use of planimeter F1
(e) Clothes posts; playgrounds; sand-pits F1
(f) Fencing and gates; site clearance F1

GROUP 21. ALTERATION WORK

(a) Simple work to existing structures F1
(b) Adaptation schemes F2
(c) Demolition work F2
(d) Measurement of shoring F2

CIVIL ENGINEERING WORK

Note: In the RICS examination structure all civil engineering work is confined to the Final Part 2 syllabus.

GROUP 22. ENGINEERING SITE WORKS

(a) Site investigation
(b) Site clearance
(c) De-watering schemes
(d) Geotechnic work
(e) Boreholes and wells

GROUP 23. ANCILLIARY ENGINEERING BUILDINGS

Pump-houses and sub-stations, ejector-chambers, warehouses, etc., in brickwork, concrete, stonework and steelwork, including foundations.

GROUP 24. REINFORCED CONCRETE ENGINEERING STRUCTURES

(a) Water towers
(b) Bunkers and silos

(c) Footbridges
(d) Highway bridges and culverts
(e) Flyovers
(f) Portal frames
(g) Prestressed concrete work
(h) Chimney shafts
(i) Hyperboidal cooling towers
(j) Dams
(k) Reactor shields
(l) Reservoirs

GROUP 25. STEEL ENGINEERING STRUCTURES

(a) Portal frames and cribs
(b) Bridges
(c) Gantries
(d) Pylons and Skylons
(e) Water tanks

GROUP 26. MARINE WORK

(a) Jetties and wharves
(b) Caissons and cofferdams
(c) Timber piling and fendering
(d) Concrete piling and box-piling
(e) Mooring dolphins
(f) Beacons and lighthouses
(g) Sea defence walls and groins
(h) Dredging

GROUP 27. RAILWAY TRACKWORK

(a) Bolted trackwork, with timber, metal and concrete sleepers, including crossings, turnouts, etc.
(b) Welded trackwork
(c) Crane tracks

GROUP 28. EARTHWORKS

(a) Cuttings and embankments for roads and railways
(b) Tunnels, shafts, headings and adits
(c) Grids and contouring

(d) Use of planimeter for civil engineering work
(e) Schedules for mass-haul diagrams

GROUP 29. PIPELINES, ETC.

(a) Pipelines in steel and concrete
(b) Main sewers in brickwork
(c) Penstocks, tidal flaps, etc.
(d) Settling tanks and sewage plant
(e) Filter beds

Appendix Two

Mathematical Formulæ

Where the measurement of areas and volumes, etc., requires the use of mathematical formulae it will be necessary for the taker-off to express his dimensions in a form which ensures that the result of the normal squaring process properly reflects the use of the correct formula. To achieve this, the selected formula must be translated into a dimensional form suitable for displaying in the timesing and dimension columns in the usual way. Simple cases, such as the area of a triangle, will present no problem. Certain other cases, such as the volume of a prismoid may require the conventional formula to be expanded prior to conversion into dimensions. The following notes deal with formulae which may be required in practice, and give examples of conventional translation into dimensional form.

AREA OF TRIANGLE

Formula: $\dfrac{b \times h}{2}$

Example: $b = 3{\cdot}00$
$h = 2{\cdot}25$

Dimensions: $\frac{1}{2}/3{\cdot}00$
$2{\cdot}25$

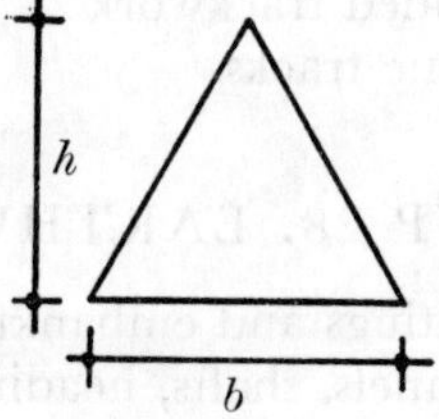

Appendix Fig. 1

AREA OF CIRCLE

Formula: πr^2

Example: $r = 1{\cdot}50$

Dimensions: $3\frac{1}{7}/1{\cdot}50$
$\underline{1{\cdot}50}$

AREA OF SECTOR

Formula: $\pi r^2 \times \dfrac{\theta}{360}$

Example: $r = 2{\cdot}50$
$\theta = 78°$

Dimensions: $\frac{78}{360}/3\frac{1}{7}/2{\cdot}50$
$\underline{2{\cdot}50}$

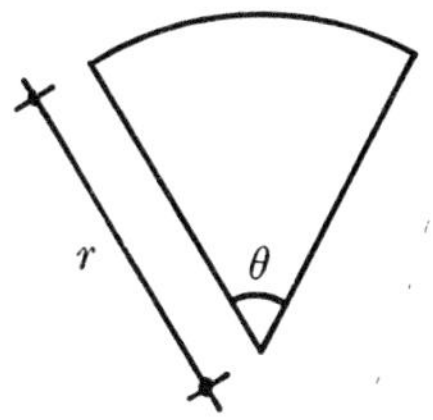

Appendix Fig. 2

SURFACE AREA OF SPHERE

Formula: πD^2

Example: $D = 3{\cdot}00$

Dimensions: $3\frac{1}{7}/3{\cdot}00$
$\underline{3{\cdot}00}$

Note. Surface area of hemisphere would be as above but timesed by $\frac{1}{2}$.

SURFACE AREA OF ZONE OF SPHERE [1]

Formula: $\pi D \times H$ (D = diameter of whole of sphere)

Example: $D = 7{\cdot}25$
$H = 1{\cdot}00$

Dimensions: $3\frac{1}{7}/7{\cdot}25$
$\underline{1{\cdot}00}$

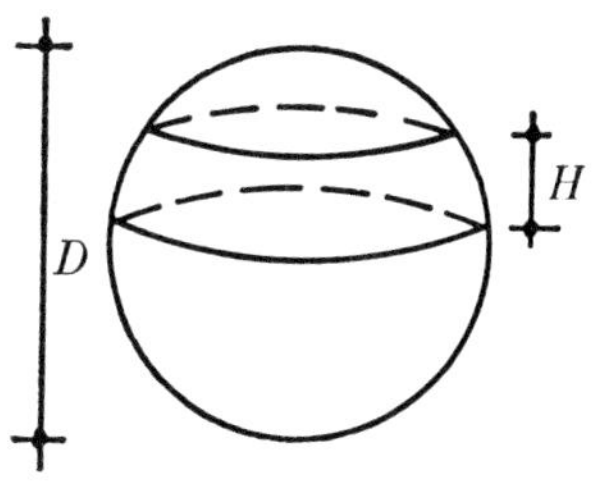

Appendix Fig. 3

[1] Excluding areas of base and top of zone.

SURFACE AREA OF SEGMENT OF SPHERE [1]

Formula: $\pi D \times H$ (as for zone of sphere)

Example: $D = 7{\cdot}25$
$H = 1{\cdot}00$

Dimensions: $3\frac{1}{7}/7{\cdot}25$
$\underline{1{\cdot}00}$

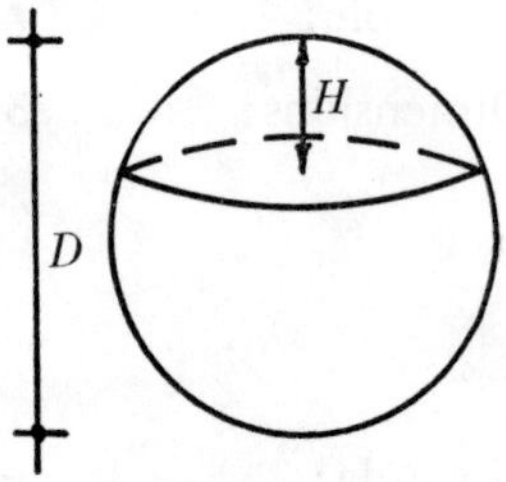

Appendix Fig. 4

SURFACE AREA OF CYLINDER [1]

Formula: $\pi D \times H$

Example: $D = 1{\cdot}50$
$H = 3{\cdot}75$

Dimensions: $3\frac{1}{7}/1{\cdot}50$
$\underline{3{\cdot}75}$

SURFACE AREA OF CONE [1]

Formula: $\dfrac{\pi D \times S}{2}$ (S = slope height)

Example: $D = 3{\cdot}75$
$S = 5{\cdot}50$

Dimensions: $\frac{1}{2}/3\frac{1}{7}/3{\cdot}75$
$\underline{5{\cdot}50}$

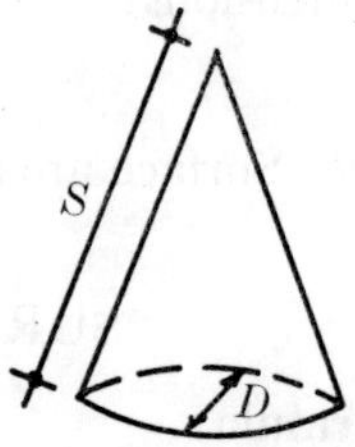

Appendix Fig. 5

SURFACE AREA OF CONIC FRUSTRUM [1]

Formula: $\dfrac{\pi D_1 + \pi D_2}{2} \times S$ (S = slope height)

Example: $D_1 = 1{\cdot}25$
$D_2 = 1{\cdot}75$
$S = 3{\cdot}50$

[1] Excluding areas of bases.

Formula expansion: $\left(\frac{\pi D_1 S}{2}\right) + \left(\frac{\pi D_2 S}{2}\right)$

Dimensions:

$\frac{1}{2}/3\frac{1}{7}/1{\cdot}75$
$\underline{3{\cdot}50}$
$\frac{1}{2}/3\frac{1}{7}/1{\cdot}25$
$\underline{3{\cdot}50}$

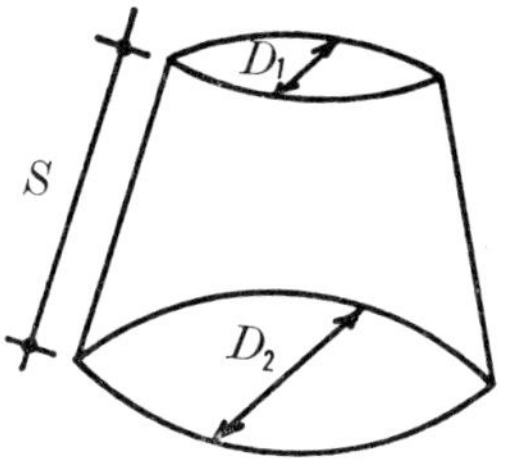

Appendix Fig. 6

AREA OF SEGMENT

Formula: $\frac{R^3}{2S} + \frac{2}{3}RS$

Note. This formula gives results to within the region of 1% accuracy.

Example: $S = 25{\cdot}00$
$R = 3{\cdot}00$

Formula expansion: $\left(\frac{1}{2} \times \frac{R}{S} \times R \times R\right) + (\frac{2}{3}RS)$

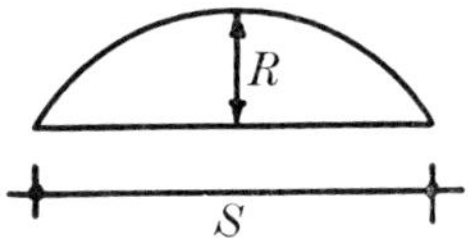

Appendix Fig. 7

Dimensions:

$\frac{1}{2}/\frac{1}{8}/3{\cdot}00$
$\underline{3{\cdot}00}$
$\frac{2}{3}/3{\cdot}00$
$\underline{25{\cdot}00}$

Note. Alternatively the superimposed rectangle may be timsed by a factor (K) which varies depending on the relationship of the span (S) to the rise (R) as shown in the following table.

R/S	$\frac{1}{8}$	$\frac{1}{4}$	$\frac{3}{8}$
K	$\frac{2}{3}$	$\frac{11}{16}$	$\frac{3}{4}$

When the area involved is small (e.g., arch deductions) a value for K, selected according to the shape of the segment, may be regarded as sufficiently accurate.

Example: $S = 2{\cdot}50$
$R = 0{\cdot}75$

Dimensions:

$\frac{11}{16}/2{\cdot}50$
$\underline{0{\cdot}75}$

LENGTH OF ARC OF SEGMENT

Formula: $\frac{8c - a}{3}$ (c = chord of arc; a = chord of half arc).

Example: $c = 6{\cdot}00$
$a = 3{\cdot}50$

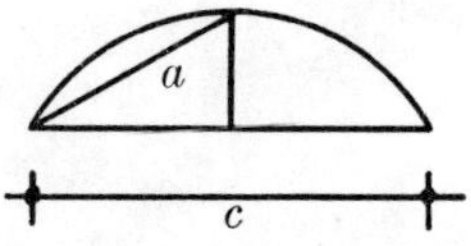

Appendix Fig. 8

Formula expansion: $(\frac{1}{3} \times 8c) - (\frac{1}{3}a)$

Dimensions: $\frac{1}{3}$/8/6·00 *Add*

$\frac{1}{3}$/3·50 *Deduct*

Note. This formula is accurate to about $1\frac{1}{2}\%$ maximum error, depending on the shape of the segment.

AREA OF ELLIPSE

Formula: $\frac{\pi AB}{4}$ (A = major axis; B = minor axis.)

Example: $A = 1{\cdot}75$
$B = 0{\cdot}75$

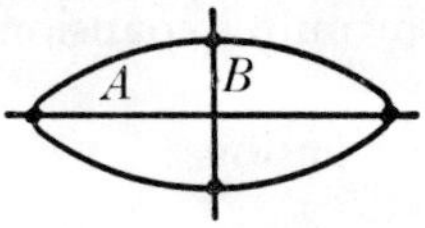

Appendix Fig. 9

Dimensions: $\frac{1}{4}$/$3\frac{1}{7}$/1·75
0·75

Note. Deduction of brickwork for an elliptical arch would be as above but timesed by $\frac{1}{2}$.

PERIMETER OF ELLIPSE

Formula: $$\frac{\text{perimeter}}{4} = \left(\frac{b^3 - a^3}{b^2 - a^2}\right)\left(1 + \frac{6a}{100b}\right)$$

where $a = \frac{1}{2}$ major axis
$b = \frac{1}{2}$ minor axis

Note. This formula is accurate to within about 1%. For small ellipses met with in taking-off (e.g., arches) it is customary to measure the perimeter by stepping round with dividers on the drawing.

AREA OF PLANE SURFACE WITH IRREGULAR SIDE

Preparation: Divide into an *even* number of strips, of equal length L by means of an *odd* number of ordinates. Number the intermediate ordinates.

Formula: (Simpson's Rule) $\frac{L}{3}(A + K + 2E + 4S)$

where A = length of first ordinate
K = length of last ordinate
E = sum of even intermediate ordinates
S = sum of odd intermediate ordinates
L = distance between ordinates.

Example: $A = 1{\cdot}25$
$K = 1{\cdot}00$
$L = 0{\cdot}75$

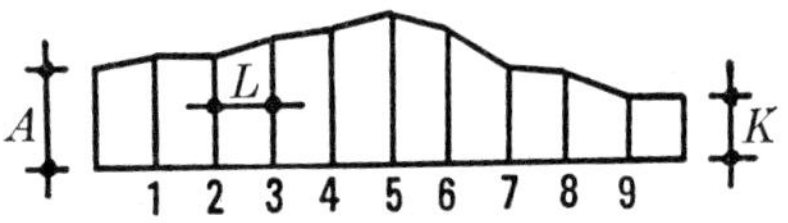

Appendix Fig. 10

Waste calculation: (length of intermediate ordinates assumed)

		A	1·25
		K	1·00
(2)	1·50		
(4)	1·60		
(6)	1·55		
(8)	1·15		
	2/5·80		11·60
(1)	1·35		
(3)	1·55		
(5)	1·70		
(7)	1·25		
(9)	1·00		
	4/6·85		27·40
			41·25

Dimensions: $\frac{1}{3}$/0·75
41·25

IRREGULAR PLANE SURFACE (*See Fig. 11, overleaf*)

Preparation: Draw parallel tangents *AB* and *CD*. Draw *odd* number of parallel ordinates (dividing *XY* into an even number of equal parts). Number the ordinates.

Formula: $\frac{L}{3}(2E + 4S)$ (variation of Simpson's rule with A and K both zero, therefore ignored).

where E = sum of even ordinates
S = sum of odd ordinates

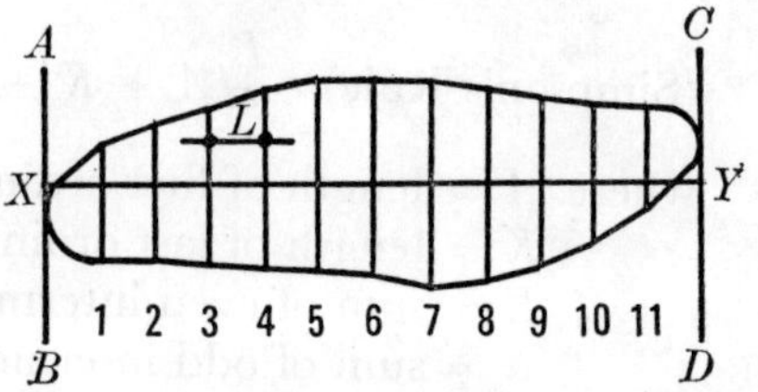

Appendix Fig. 11

Example:	$L = 1{\cdot}50$		
Waste calculation:	(length of ordinates assumed)		
	(2)	3·35	
	(4)	3·80	
	(6)	3·90	
	(8)	3·95	
	(10)	3·35	
		2/18·35	36·70
	(1)	3·25	
	(3)	3·75	
	(5)	3·95	
	(7)	4·00	
	(9)	3·70	
	(11)	3·25	
		4/21·90	87·60
			124·30

Dimensions: $\frac{1}{3}$/ 1·50
124·30

Note. Alternative methods of measuring the area of an irregular plane surface are (1) by triangulation, with 'give and take' lines at the boundaries, or (2) by use of a planimeter.

AREA OF HYPERBOLIC PARABOLOID

Formula: $L^2 + \dfrac{H^2}{3}$ (the Bellis formula for hyperbolic roofs).

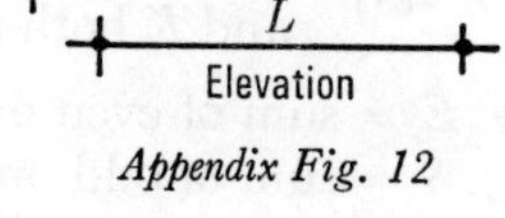

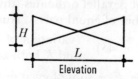

Appendix Fig. 12

Example: $L = 7{\cdot}50$
$H = 1{\cdot}50$

Dimensions:
$7{\cdot}50$
$\underline{7{\cdot}50}$
$\tfrac{1}{3}/1{\cdot}50$
$\underline{1{\cdot}50}$

Note. The Bellis formula gives results to an accuracy of $+1\%$ maximum error where H does not exceed $\tfrac{1}{2}L$.

SURFACE AREA OF HIPPED DOME

Formula: $2a^2$

Example: $a = 7{\cdot}50$

Dimensions:
$2/7{\cdot}50$
$\underline{7{\cdot}50}$

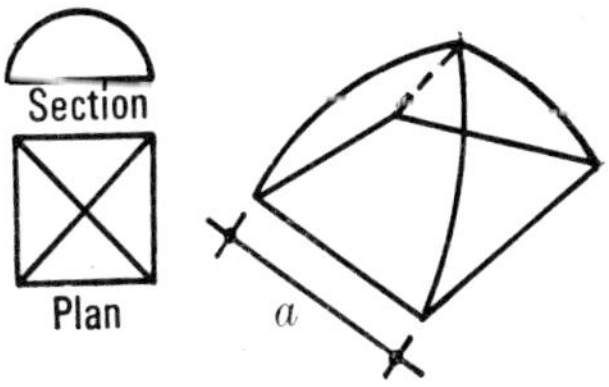

Appendix Fig. 13

Note. This formula assumes a square plan with a rise equal to half the least span. If the base is any regular polygon, the surface area of the dome may be taken as *twice the area on plan.*

SURFACE AREA OF SEMI-CIRCULAR BARREL-VAULT INTERSECTION

Preparation: Measure the semi-circular barrels up to points *ABCD*. The surface area *ABCD* is then calculated as follows.

Formula: $S^2 \quad (\pi - 2)$

Example: $S = 10{\cdot}00$

Waste calculation:
$3\tfrac{1}{7}$
$\underline{-2}$
$\underline{1\tfrac{1}{7}}$

Dimensions:
$1\tfrac{1}{7}/10{\cdot}00$
$\underline{10{\cdot}00}$

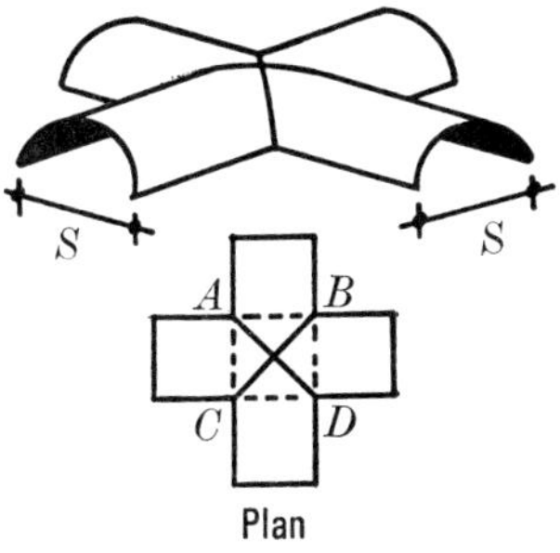

Appendix Fig. 14

VOLUME OF SPHERE

Formula: $\dfrac{\pi D^3}{6}$

Example: $D = 4{\cdot}50$

Dimensions:

$\frac{1}{6}/3\frac{1}{7}/4{\cdot}50$
4·50
4·50

Note. Volume of hemisphere would be as above but timesed by $\frac{1}{2}$.

VOLUME OF ZONE OF SPHERE

Formula: $\dfrac{\pi H}{6}(3R^2 + 3R_1{}^2 + H^2)$

Example:

$R = 1{\cdot}75$
$R_1 = 2{\cdot}15$
$H = 1{\cdot}25$

Formula expansion: $(\frac{1}{6}\pi H 3R^2) + (\frac{1}{6}\pi H 3R_1{}^2) + (\frac{1}{6}\pi H^3)$

Dimensions:

$\frac{1}{6}/3\frac{1}{7}/3/1{\cdot}25$
1·75
1·75
$\frac{1}{6}/3\frac{1}{7}/3/1{\cdot}25$
2·15
2·15
$\frac{1}{6}/3\frac{1}{7}/1{\cdot}25$
1·25
1·25

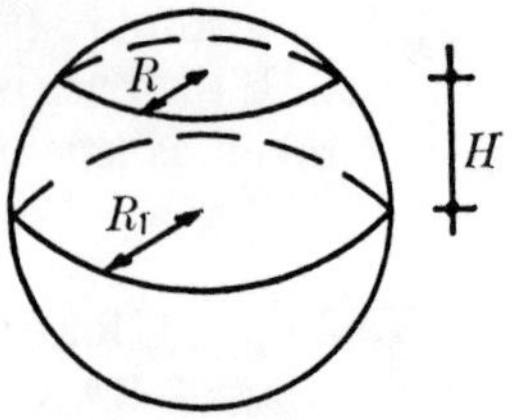

Appendix Fig. 15

VOLUME OF SEGMENT OF SPHERE

Formula: $\dfrac{\pi H}{6}(3R^2 + H^2)$

Example:

$R = 1{\cdot}50$
$H = 0{\cdot}75$

Formula expansion: $(\frac{1}{6}\pi H 3R^2) + (\frac{1}{6}\pi H^3)$

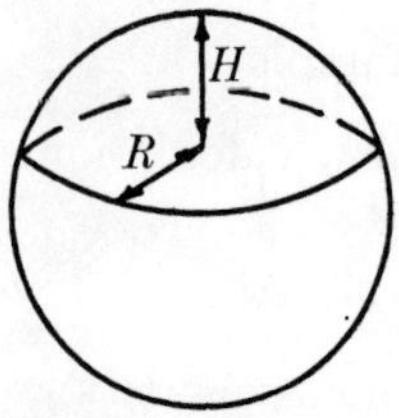

Appendix Fig. 16

Dimensions: $\frac{1}{6}/3\frac{1}{7}/3/0{\cdot}75$
$1{\cdot}50$
$\underline{1{\cdot}50}$
$\frac{1}{6}/3\frac{1}{7}/0{\cdot}75$
$0{\cdot}75$
$\underline{0{\cdot}75}$

VOLUME OF CYLINDER

Formula: $\pi R^2 \times H$

Example: $R = 1{\cdot}35$
$H = 2{\cdot}00$

Dimensions: $3\frac{1}{7}/1{\cdot}35$
$1{\cdot}35$
$\underline{2{\cdot}00}$

VOLUME OF CONE

Formula: $\dfrac{\pi R^2 \times H}{3}$

Example: $R = 1{\cdot}35$
$H = 2{\cdot}15$

Dimensions: $\frac{1}{3}/3\frac{1}{7}/1{\cdot}35$
$1{\cdot}35$
$\underline{2{\cdot}15}$

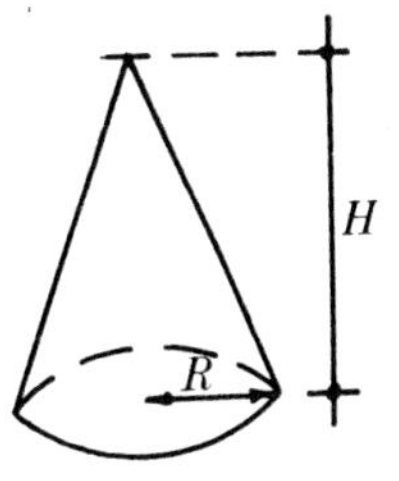

Appendix Fig. 17

VOLUME OF PYRAMID

Formula: $\dfrac{\text{area of base} \times \text{vertical height}}{3}$

Example: $a = 3{\cdot}85$
$b = 2{\cdot}75$
$H = 5{\cdot}30$

Dimensions: $\frac{1}{3}/3{\cdot}85$
$2{\cdot}75$
$\underline{5{\cdot}30}$

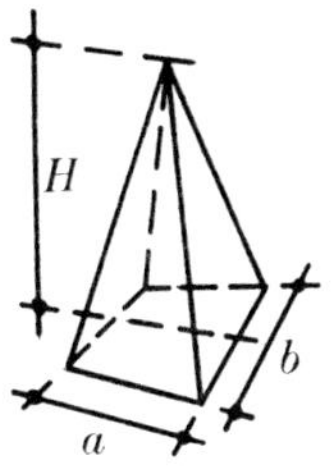

Appendix Fig. 18

Note. The above formula applies to a pyramid with any rectilinear shape of base.

VOLUME OF CONIC FRUSTRUM

Formula: (Prismoidal formula) $\dfrac{H(\mathrm{A} + a + 4m)}{6}$

Example:
A = area of base 1·85 radius
a = area of top 1·25 radius
m = area of midsection 1·50 radius
H = 2·50

Formula expansion: $(\frac{1}{6}HA) + (\frac{1}{6}Ha) + (\frac{1}{6}H4m)$

Dimensions:

$\frac{1}{6}/3\frac{1}{7}/1{\cdot}85$
1·85
2·50

$\frac{1}{6}/3\frac{1}{7}/1{\cdot}25$
1·25
2·50

$\frac{1}{6}/4/3\frac{1}{7}/1{\cdot}50$
1·50
2·50

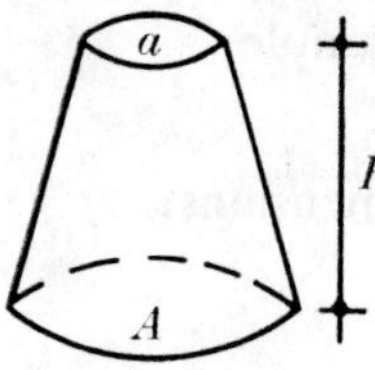

Appendix Fig. 19

Note. The above prismoidal formula applies also to the volume of the frustrum of a pyramid or prismoid.

VOLUME OF PRISMOID

Formula: (Prismoidal formula) $\dfrac{L(\mathrm{A} + a + 4m)}{6}$

Example:
A = area of one end (triangle 3·00 base and 1·75 high)
a = area of other end (triangle 1·75 base and 1·00 high)
m = area of midsection (triangle 2·50 base and 1·25 high)
L = 30·00

Formula expansion: $(\frac{1}{6}LA) + (\frac{1}{6}La) + (\frac{1}{6}L4m)$

Dimensions:

$\frac{1}{6}/\frac{1}{2}/$ 3·00
1·75
30·00

$\frac{1}{6}/\frac{1}{2}/$ 1·75
1·00
30·00

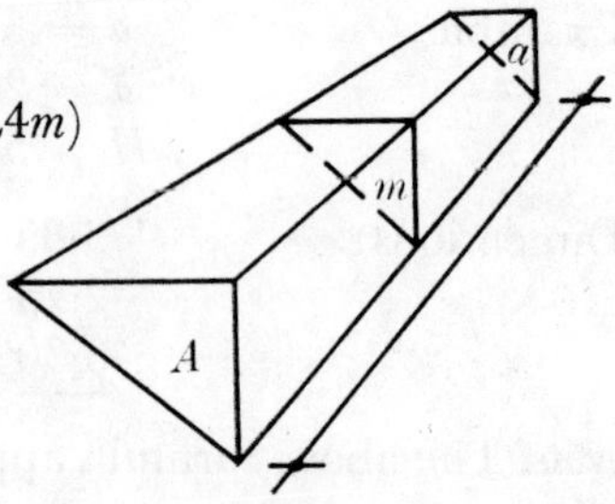

Appendix Fig. 20

$\frac{1}{6}/4/\frac{1}{2}$ 2·50
1·25
30·00

Note. The volume of earth in respect of cuttings and embankments for roads, etc., may form a series of prismoids of equal length between cross sections. In this case the application of Simpson's rule will give the total volume:

$$\frac{L}{3}(A + K + 2E + 4S)$$

where A = area of first section
K = area of last section
E = sum of areas of even sections
S = sum of area of odd sections
L = distance between sections.

This formula assumes an odd number of cross-sections, numbered sequentially from the first intermediate cross-section, as shown in the diagram, which represents a longitudinal section.

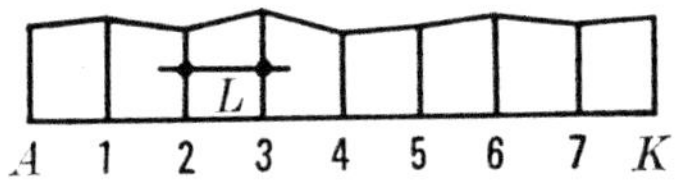

Appendix Fig. 21

LENGTH OF HELIX

Formula: Length of wire per unit of pitch for helical reinforcement = $\sqrt{(\pi D)^2 + P^2}$

where D = diameter of helix
P = pitch.

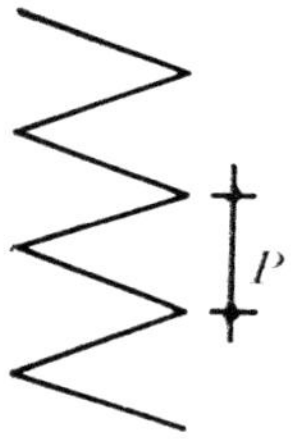

Appendix Fig. 22

Index to SMM Clause Number References

General Index